# Emerging Pollutants in the Environment: Occurrence, Fate, Risk Assessment and Degradation Methods

# Emerging Pollutants in the Environment: Occurrence, Fate, Risk Assessment and Degradation Methods

Guest Editors

**Qingwei Bu**
**Yuning Ma**

MDPI

Basel • Beijing • Wuhan • Barcelona • Belgrade • Novi Sad • Cluj • Manchester

*Guest Editors*

Qingwei Bu
School of Chemical &
Environmental Engineering
China University of Mining &
Technology
Beijing
China

Yuning Ma
College of Environmental &
Resource Sciences
Zhejiang University
Hangzhou
China

*Editorial Office*
MDPI AG
Grosspeteranlage 5
4052 Basel, Switzerland

This is a reprint of the Special Issue, published open access by the journal *Toxics* (ISSN 2305-6304), freely accessible at: https://www.mdpi.com/journal/toxics/special_issues/909Q92MI9T.

For citation purposes, cite each article independently as indicated on the article page online and as indicated below:

Lastname, A.A.; Lastname, B.B. Article Title. *Journal Name* **Year**, *Volume Number*, Page Range.

**ISBN 978-3-7258-4883-6 (Hbk)**
**ISBN 978-3-7258-4884-3 (PDF)**
**https://doi.org/10.3390/books978-3-7258-4884-3**

# Contents

# About the Editors

**Qingwei Bu**

Qingwei Bu is an environmental chemist whose work involves applying a range of tools to understand the risk of emerging contaminants and their future effects. He holds a Ph.D. from the Chinese Academy of Sciences (2012) and a professorship at the China University of Mining & Technology-Beijing (since 2022). He is an Associate Editor for *Emerging Contaminants* (since 2024). He has published over 80 papers in internationally renowned journals, such as *Environmental Science & Technology*, *Environment International*, and *Journal of Hazardous Materials*.

**Yuning Ma**

Yuning Ma is an environmental chemist with over a decade of experience in both academia and industry who specializes in understanding the environmental fate of emerging persistent organic pollutants. He holds a Ph.D. in environmental sciences (2013). He has published over 40 SCI papers in internationally renowned journals, such as *Environment International*, *Environmental Science and Technology*, and *Nano Research*.

# Preface

Humanity's reliance on synthetic chemicals has transgressed planetary boundaries, unleashing a complex array of unregulated emerging contaminants into global ecosystems. Chemical pollution is therefore not merely a localized issue but a systemic challenge demanding coordinated global action. In this context, the concept of emerging pollutants and contaminants has arisen. Emerging pollutants are defined in the literature as unregulated chemicals raising ecological or human health concerns, including persistent organic pollutants, pharmaceuticals, personal care products, industrial additives, microplastics, etc. These pollutants persist in the air, water, soil, and food chains, posing subtle threats to ecological integrity and public health. This reprint synthesizes critical research on their environmental occurrence, distribution, pathways, and risks, while also detailing pioneering mitigation strategies. It addresses the urgent need for science-informed policy and innovation.

**Qingwei Bu and Yuning Ma**
*Guest Editors*

toxics

MDPI

*Editorial*

# Emerging Pollutants in the Environment: Occurrence, Fate, Risk Assessment and Degradation Methods

**Qingwei Bu [1,*] and Yuning Ma [2]**

1 School of Chemical & Environmental Engineering, China University of Mining & Technology-Beijing, Beijing 100083, China
2 College of Environmental and Resource Sciences, Zhejiang University, Hangzhou 310058, China; julius.yuningma@gmail.com
* Correspondence: qingwei.bu@cumtb.edu.cn

## 1. Introduction

The use of chemical substances has brought great benefits to human well-being; however, the dark side of the coin is that they may pose risks to human health and the environment. As a consequence of the uncontrolled or unregulated use of many anthropogenic chemicals, humanity has exceeded the safe operating space (defined as the planetary boundary) for these chemicals [1,2]. In this context, emerging pollutants (also commonly known as emerging contaminants) are defined as chemicals that are not currently regulated but raise ecological or human health concerns. These pollutants include, but are not limited to, endocrine-disrupting compounds, pharmaceutical and personal care products, disinfection by-products, microplastics, and persistent organic chemicals, etc.—along with their degradation products—detected in air, water, soil, and food sources [3].

In recent decades, extensive efforts have been exserted to understand the occurrence, fate, and toxicity of these chemicals, as well as to develop new technologies for mitigating the associated risks. This Special Issue presents the most recent advancements in emerging pollutant research, including studies on their occurrence, distribution, fate, and the associated environmental risks, along with technological developments aimed at reducing and controlling their environmental presence.

## 2. An Overview of Published Articles

There are seven articles focused the occurrence and associated ecological/health risks of typical emerging pollutants, e.g., antibiotics, pesticides, and endocrine-disrupting chemicals.

Industrial activities heavily rely on organic solvents, which release volatile organic compounds (VOCs) linked to environmental pollution and occupational health hazards, particularly in manufacturing sectors like electronics and chemicals. Guo et al. (contribution 2) investigated the temporal trends and industry-specific differences in VOC usage and the associated health risks in Bao'an District, Shenzhen, China, from 2018 to 2023. Analyzing 1335 solvent samples and 1554 air samples, the research found that VOC usage declined during the COVID-19 pandemic but rebounded thereafter. Alkanes and aromatic hydrocarbons (e.g., toluene, n-hexane, and xylene) were the most prevalent VOCs, with detection rates highlighting toluene (22.5%) and n-hexane (22.0%) as the dominant components. Air monitoring identified trichloroethylene and xylene as high-risk compounds, exceeding acceptable health thresholds. Post-2020 data revealed a trend toward reduced solvent diversity, potentially lowering mixed-exposure risks but raising concerns about


https://doi.org/10.3390/toxics13070521

new substitutes. The study underscores the need for the targeted monitoring of high-risk VOCs and improved workplace safety measures to mitigate occupational hazards in industrial regions.

E-waste dismantling is a significant source of bisphenol chemicals (BPs) due to the release of additives from plastics and epoxy resins during processing. Zhao et al. (contribution 3) investigated the spatial distribution and health risks of BPs in surface soil from e-waste dismantling facilities and surrounding areas in South China. Using non-targeted screening and targeted analysis, 14 BPs were identified, including bisphenol A (BPA), tetrabromobisphenol A (TBBPA), and novel structural analogs. Total BP concentrations in e-waste soil (median: 6970 ng/g) far exceeded those in surrounding areas (median: 197 ng/g), with BPA, TBBPA, and bisphenol F being dominant. Spatial analysis revealed declining TBBPA and its debromination product concentrations with increasing distance from e-waste sites, indicating facility emissions as a primary source. Risk assessment via soil ingestion showed daily intakes for workers and residents were below current tolerable thresholds; the exception being BPA in workers, which surpassed stricter recent guidelines. The findings highlight e-waste activities as critical BP emission sources, necessitating ongoing monitoring to address potential health risks from emerging analogs and cumulative exposures. In another study, by Chen et al. (contribution 6), the authors investigated the contamination of BPs in aquatic products from South China, focusing on human exposure risks. The researchers analyzed 245 samples (of fish, crustaceans, and bivalves) from Shenzhen markets using liquid chromatography–mass spectrometry. All BPs except bisphenol AF were detected, with bisphenol S showing the highest detection rate. Enzymatic hydrolysis revealed that 49–96% of the BPs existed in bound forms, significantly increasing post-treatment concentrations. Bisphenol F and bisphenol S dominated the contamination profiles, likely due to their being increasingly used as BPA substitutes. Correlation analysis suggested shared pollution sources for certain BPs, and health risk assessments indicated low exposure risks, though females exhibited slightly higher risks than males. The study underscores the necessity of enzymatic hydrolysis for accurate BP detection and highlights the need for stricter regulations to mitigate contamination, despite the health threat currently being low.

Yang et al. (contribution 4) addressed the urgent need to regulate emerging pollutants in wastewater treatment plant effluents in China, focusing on pharmaceuticals and endocrine-disrupting chemicals. These pollutants, detected in aquatic ecosystems globally, pose ecological and human health risks due to their inadequate removal by conventional treatment processes. Analyzing data from 2012 to 2022, the research team identified 140 emerging pollutants in Chinese wastewater treatment plant effluents, with concentrations ranging from undetected levels to 706 μg/L. High-risk regions included Gansu, Hebei, Shandong, Guangdong, and Hong Kong. Using risk assessment methods, eighteen high-risk emerging pollutants were prioritized, but only carbamazepine, ibuprofen, and BPA met the conditions needed to derive long-term water quality criteria via species sensitivity distribution—their long-term water quality criteria values were 96.4, 1010, and 288 ng/L, respectively. Notably, carbamazepine and BPA concentrations frequently exceeded these thresholds, highlighting critical regulatory gaps. The study underscores the necessity for targeted monitoring and science-based discharge limits to mitigate ecological and health impacts, providing a foundational framework for the management of emerging pollutants in China's urban wastewater systems.

Dimethylcyclosiloxanes, widely used in silicone polymers for their thermal stability and flexibility, are linked to endocrine disruption, reproductive toxicity, and environmental persistence. Xing et al. (contribution 5) investigated the environmental and health implications of dimethylcyclosiloxanes in mobile smart terminal devices, focusing on their concentrations, distribution patterns, and emissions. Analyzing silicone rubber, adhesives, and plastics from devices like headphones and smartphones, the researchers found D5–D9 to be prevalent in silicone rubber (detection rates: 91–95.5%) and adhesives (50–100%), with total concentrations reaching 802.2 mg/kg in silicone rubber. Meanwhile, plastics exhibited higher detection rates of the lower-weight D3 and D4 (61.1%). Environmental emissions from silicone rubber in China were estimated to exceed 5000 tons annually, raising concerns about long-term ecological and human exposure. The study highlights the dominance of the understudied higher-weight D7–D9 in silicone materials and calls for toxicological assessments of these compounds. It underscores the need for optimized manufacturing processes to reduce residual dimethylcyclosiloxanes and improved regulatory frameworks to mitigate risks from widespread device disposal and environmental release.

With extensive agricultural activity around Xingkai Lake—a critical freshwater ecosystem—pesticide residues pose potential threats to water quality and aquatic life. Wang et al. (contribution 9) investigated the distribution and ecological risks of 57 pesticides in farmland soil and surrounding water bodies in the Xingkai Lake area, Heilongjiang Province, China. The researchers analyzed soil and water samples across three periods (sowing, vegetative, and maturity stages). Their key findings revealed 43 pesticides and 3 degradation products in the soil, with atrazine and acetochlor dominant in dry fields, while oxadiazon, mefenacet, and chlorpyrifos prevailed in paddy fields. The analyzed water samples showed peak contamination during the vegetative period, with atrazine, simetryn, and buprofezin as primary pollutants in drainage and lake water. Correlation analysis ($r > 0.8$) indicated shared contamination sources between the drainage systems and the lake, and ecological risk assessments highlighted significant risks from atrazine, chlorpyrifos, and prometryn, with potential affected species fractions exceeding 5%. The study underscores the impact of agricultural pesticides on freshwater ecosystems, emphasizing the need for targeted management to mitigate long-term ecological harm in ecologically sensitive regions like Xingkai Lake.

Organophosphorus flame retardants (OPFRs) are widely detected in indoor dust and air due to their additive nature in consumer products like furniture and electronics, leading to prolonged human exposure through ingestion, inhalation, and dermal contact. Song et al. (contribution 10) reviewed the distribution and health impacts of OPFRs in indoor environments, focusing on their role as replacements for restricted polybrominated diphenyl ethers. Key findings highlight higher concentrations of OPFRs in dust compared to air, with tris(2-butoxyethyl) phosphate, tris(1-chloro-2-propyl) phosphate, and tris(1,3-dichloro-2-propyl) phosphate being predominant. Regional variations in OPFR profiles reflect differences in usage patterns and regulations, such as increased tris(1-chloro-2-propyl) phosphate levels in Europe following tris(2-chloroethyl) phosphate restrictions, and health risk assessments indicate toddler exposure via dust ingestion in Japan reaching concerning levels, though overall combined exposure typically remains below reference thresholds. The study underscores the need for integrated evaluations of OPFR toxicity and exposure pathways, proposing pollutant equivalency factors to prioritize risk management, and provides a foundational framework for understanding OPFR-related health risks and informing regulatory strategies.

The remaining three articles focused on developing treatment technologies for the removal of emerging pollutants.

Silva et al. (contribution 1) addressed the environmental challenge posed by tenofovir disoproxil fumarate (TDF), a widely used antiretroviral drug for HIV treatment, whose stable metabolite, tenofovir, persists in aquatic ecosystems and poses risks to aquatic organisms. The authors evaluated the biodegradation potential of a cyanobacteria–bacterial consortium (*Microcystis novacekii* and *Pseudomonas pseudoalcaligenes*) for TDF removal. The process occurred in two phases—abiotic and enzymatic de-esterification of TDF into tenofovir monoester (TMF) within 72 h, followed by the intracellular removal of TMF over 16 days. The consortium achieved a 88.7–94.1% removal efficiency across TDF concentrations (12.5–50 mg/L), with optimal performance at 25 mg/L. Notably, tenofovir itself was not detected, but residual TMF—a partially active antiviral intermediate—remained, highlighting incomplete degradation. The findings underscore the potential of microbial consortia for sustainable pharmaceutical wastewater treatment while emphasizing the need for further research to address persistent metabolites. This work contributes to strategies for mitigating pharmaceutical pollution in aquatic environments, particularly for high-persistence drugs like TDF.

Conventional methods like advanced oxidation processes are energy-intensive and costly, prompting the exploration of adsorption using carbon aerogels (CAs) as a sustainable alternative. Lu et al. (contribution 7) addressed the environmental challenge of removing 1,4-dioxane, a carcinogenic and highly water-miscible pollutant, from contaminated water. The research synthesized CAs through controlled pyrolysis conditions (e.g., temperature and heating rate) to optimize their porous structure, achieving a specific surface area of 673.89 $m^2/g$ with enhanced mesoporosity. The optimized CAs demonstrated exceptional adsorption performance, removing over 95% of 1,4-dioxane, following quasi-second-order kinetics and Langmuir isotherm models, indicating monolayer adsorption. The maximum capacity reached 67.28 mg/g at 318 K, which was attributed to the material's mesoporous network and microporous synergy. Notably, competitive adsorption tests with trichloroethylene showed no significant inhibition, and regeneration experiments confirmed stable performance over five cycles. These findings highlight CAs as a cost-effective, reusable adsorbent for water purification, offering a practical solution for persistent organic pollutants like 1,4-dioxane while minimizing secondary environmental impacts.

Traditional methods like membranes and adsorbents face limitations in flexibility and degradability. Yu et al. (contribution 8) addressed the challenge of VOC emissions during soil remediation, which pose significant health and environmental risks. The research introduces an aqueous foam stabilized by $SiO_2$-$TiO_2$ nanoparticles to enhance VOC suppression. By modifying silica nanoparticles with hydrophobic groups and integrating $TiO_2$ for photocatalysis, the foam exhibits improved stability and functionality, and the experimental results show that the modified nanoparticles increase foam liquid half-life by 4.08 h and volume half-life by 4.44 h, compared to nanoparticle-free foam. Under UV irradiation, the foam maintained a 90% suppression rate for dichloroethane, n-hexane, and toluene for nearly 12 h, outperforming other variants. Characterization confirmed enhanced dispersibility and oxygen vacancies in the composite nanoparticles, boosting photocatalytic efficiency. This innovation offers a dual-function solution—blocking VOC emissions while degrading contaminants—providing an eco-friendly, adaptable alternative for soil remediation with reduced secondary pollution risks.

## 3. Future Development Prospects

The collection of articles in this Special Issue provides valuable contributions to our understanding of the occurrence and risk profiles of emerging pollutants; however, several critical research gaps warrant attention. Notably, there is a paucity of systematic investigations into ecotoxicological impacts, epidemiological correlations, exposomic interactions, and metabolic pathways, which are all crucial dimensions of comprehensive risk assessment. These aspects present significant challenges for environmental researchers given the diverse physicochemical properties and biological activities exhibited by the expanding spectrum of emerging pollutants. Particularly noteworthy are compounds with distinct mechanisms of action, such as endocrine disruptors, which require specialized evaluation frameworks. Furthermore, while three papers addressed contaminant mitigation technologies, there remains a notable absence of research on advanced oxidation processes in this collection. This omission is particularly conspicuous as advanced oxidation processes represent state-of-the-art advancements in contaminant degradation and are poised to remain a focal point of environmental engineering research given their effectiveness in addressing persistent organic pollutants.

**Data Availability Statement:** Not applicable.

**Conflicts of Interest:** The author declares no conflicts of interest.

**List of Contributions:**

1. Silva, S.R.; Souza-Silva, G.; Moreira, C.P.d.S.; Vasconcelos, O.M.d.S.R.; Silveira, M.R.; Barbosa, F.A.R.; Magalhães, S.M.S.; Mol, M.P.G. Biodegradation of the Antiretroviral Tenofovir Disoproxil by a Cyanobacteria/Bacterial Culture. *Toxics* **2024**, *12*, 729. https://doi.org/10.3390/toxics12100729.
2. Guo, Y.; Zhu, L.; Zhang, L.; Tang, X.; Li, X.; Ge, Y.; Li, F.; Yang, J.; Lu, S.; Chen, J.; et al. Temporal Variation and Industry-Specific Differences of the Use of Volatile Organic Compounds from 2018 to 2023 and Their Health Risks in a Typical Industrially Concentrated Area in South China. *Toxics* **2024**, *12*, 634. https://doi.org/10.3390/toxics12090634.
3. Zhao, L.; Zhou, F.; Wang, S.; Yang, Y.; Chen, H.; Ma, X.; Liu, X. Bisphenol Chemicals in Surface Soil from E-Waste Dismantling Facilities and the Surrounding Areas: Spatial Distribution and Health Risk. *Toxics* **2024**, *12*, 379. https://doi.org/10.3390/toxics12060379.
4. Yang, W.; Bu, Q.; Shi, Q.; Zhao, R.; Huang, H.; Yang, L.; Tang, J.; Ma, Y. Emerging Contaminants in the Effluent of Wastewater Should Be Regulated: Which and to What Extent? *Toxics* **2024**, *12*, 309. https://doi.org/10.3390/toxics12050309.
5. Xing, Y.; Ge, Y.; Lu, S.; Yang, T.; Peng, X. Dimethylcyclosiloxanes in Mobile Smart Terminal Devices: Concentrations, Distributions, Profiles, and Environmental Emissions. *Toxics* **2024**, *12*, 287. https://doi.org/10.3390/toxics12040287.
6. Chen, Y.; Chen, X.; Lin, W.; Chen, J.; Zhu, Y.; Guo, Z. Bisphenols in Aquatic Products from South China: Implications for Human Exposure. *Toxics* **2024**, *12*, 154. https://doi.org/10.3390/toxics12020154.
7. Lu, T.; Huang, H.; Lv, G.; Li, F.; Song, R.-j.; Cai, Y. Adsorption Behavior and Kinetics of 1,4-Dioxane by Carbon Aerogel. *Toxics* **2024**, *12*, 145. https://doi.org/10.3390/toxics12020145.
8. Yu, J.; Xuan, Y. $SiO_2$–$TiO_2$ Nanoparticle Aqueous Foam for Volatile Organic Compounds' Suppression. *Toxics* **2024**, *12*, 99. https://doi.org/10.3390/toxics12020099.
9. Wang, W.; Wang, D.; Liu, Q.; Lin, L.; Xie, Y.; Du, C. Distribution Characteristics and Risk Assessment of 57 Pesticides in Farmland Soil and the Surrounding Water. *Toxics* **2024**, *12*, 85. https://doi.org/10.3390/toxics12010085.
10. Song, X.; Zhu, S.; Hu, L.; Chen, X.; Zhang, J.; Liu, Y.; Bu, Q.; Ma, Y. A Review of the Distribution and Health Effect of Organophosphorus Flame Retardants in Indoor Environments. *Toxics* **2024**, *12*, 195. https://doi.org/10.3390/toxics12030195.

## References

1. Persson, L.; Carney Almroth, B.M.; Collins, C.D.; Cornell, S.; de Wit, C.A.; Diamond, M.L.; Fantke, P.; Hassellöv, M.; MacLeod, M.; Ryberg, M.W.; et al. Outside the safe operating space of the planetary boundary for novel entities. *Environ. Sci. Technol.* **2022**, *56*, 1510–1521. [CrossRef] [PubMed]
2. Steffen, W.; Richardson, K.; Rockström, J.; Cornell, S.E.; Fetzer, I.; Bennett, E.M.; Biggs, R.; Carpenter, S.R.; de Vries, W.; de Wit, C.A.; et al. Planetary boundaries: Guiding human development on a changing planet. *Science* **2015**, *347*, 1259855. [CrossRef] [PubMed]
3. Wang, F.; Xiang, L.; Sze-Yin Leung, K.; Elsner, M.; Zhang, Y.; Guo, Y.; Pan, B.; Sun, H.; An, T.; Ying, G.; et al. Emerging contaminants: A one health perspective. *Innovation* **2024**, *5*, 100612. [CrossRef] [PubMed]

toxics MDPI

*Article*

# Biodegradation of the Antiretroviral Tenofovir Disoproxil by a Cyanobacteria/Bacterial Culture

**Sandra Regina Silva [1], Gabriel Souza-Silva [1], Carolina Paula de Souza Moreira [2], Olívia Maria de Sousa Ribeiro Vasconcelos [2], Micheline Rosa Silveira [1], Francisco Antonio Rodrigues Barbosa [1], Sergia Maria Starling Magalhães [1] and Marcos Paulo Gomes Mol [2,*]**

[1] Faculdade de Farmácia, Universidade Federal de Minas Gerais, Belo Horizonte 30510010, Brazil; srssandrars@gmail.com (S.R.S.); silva_gs@yahoo.com (G.S.-S.); michelinerosa@gmail.com (M.R.S.); barbosa.ufmg@gmail.com (F.A.R.B.); sergia.starling@gmail.com (S.M.S.M.)

[2] Fundação Ezequiel Dias, Departamento de Pesquisa e Desenvolvimento, Belo Horizonte 30510010, Brazil; carolina.moreira@funed.mg.gov.br (C.P.d.S.M.); oliviamrv@gmail.com (O.M.d.S.R.V.)

* Correspondence: marcos_mol@yahoo.com.br; Tel.: +55-31-33144770

**Abstract:** Tenofovir disoproxil fumarate (TDF) is an antiretroviral drug extensively used by people living with HIV. The TDF molecule is hydrolysed in vivo and liberates tenofovir, the active part of the molecule. Tenofovir is a very stable drug and the discharge of its residues into the environment can potentially lead to risk for aquatic species. This study evaluated the TDF biodegradation and removal by cultures of *Microcystis novacekii* with the bacteria *Pseudomonas pseudoalcaligenes*. Concentrations of TDF of 12.5, 25.0, and 50.0 mg/L were used in this study. The process occurred in two stages. In the first 72 h, TDF was de-esterified, forming the tenofovir monoester intermediate by abiotic and enzymatic processes associated in an extracellular medium. In a second step, the monoester was removed from the culture medium by intracellular processes. The tenofovir or other by-products of TDF were not observed in the test conditions. At the end of the experiment, 88.7 to 94.1% of TDF and its monoester derivative were removed from the culture medium over 16 days. This process showed higher efficiency of TDF removal at the concentration of 25 mg/L. Tenofovir isoproxil monoester has partial antiviral activity and has shown to be persistent, maintaining a residual concentration after 16 days in the culture medium, therefore indicating the need to continue research on methods for total removal of this product from the aquatic environment.

**Keywords:** drug; *Microcystis novacekii; Pseudomonas pseudoalcaligenes*; by-products; tenofovir isoproxil monoester

## 1. Introduction

In recent decades, problems related to water contamination have become a global concern. In this context, the dispersion of pharmaceutical residues in the aquatic environment has been the target of numerous studies [1–3]. Drugs are biologically active molecules and their unpredictable effects on species exposed in the aquatic environment may represent a potential risk to aquatic ecosystems [4,5].

Studies have indicated the presence of drug residues in several environmental matrices, such as soil, sewage, surface, and treated waters [1–6]. Among the drugs, antiretrovirals (ARVs) are of particular concern, since some of them act in viral DNA synthesis and, potentially, may have effects on the replication process of other organisms.

According to United Nations data on human immunodeficiency virus (HIV) [7], currently 37.7 million people worldwide are living with HIV (PLHIV), and the control of the pandemic depends on continuous antiretroviral therapy. Thus, due to its widespread use, the presence of ARVs in the environment have been reported in different countries [8–11].

Tenofovir disoproxil fumarate (TDF) is an ARV of great importance because it is one of the first-line drugs, combined with other antiretroviral drugs, in HIV/AIDS treatment.

Tenofovir is an analogue of the nucleotide adenosine 5′-monophosphate and acts by inhibiting the reverse transcriptase enzyme necessary for viral replication in human cells. Due to the low lipid solubility [12] of tenofovir, it is administered as a prodrug, in esterified form, and it is de-esterified intracellularly to release the active portion of the molecule, tenofovir [13,14] (Figure 1).

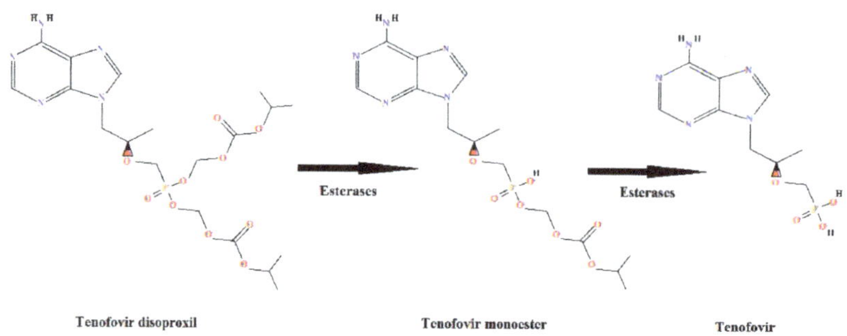


**Figure 1.** Metabolism of the prodrug tenofovir disoproxil with formation of the de-esterification products: tenofovir monoester and tenofovir.

About 80% of the dose of TDF administered to humans is eliminated in the active form, as a de-esterified product, tenofovir (Figure 1), without undergoing further metabolism [12]. The tenofovir molecule is very stable [15], which makes the treatment for its removal from the aquatic environment a challenge, both in the case of industrial effluents and domestic sewage.

Effluent treatment technologies that employ microorganisms are the most used processes for the degradation of organic compounds [16]. In recent years, the use of cyanobacteria and microalgae in waste treatment has aroused interest [16–18]. These are photosynthetic organisms and the presence of these groups in consortia makes the process more sustainable compared to technologies that use bacteria intercropping.

Cyanobacteria species also have the ability to degrade different types of chemical compounds by mixotrophic metabolisms acting together with bacteria in the degradation of chemicals in the environment [19,20].

Cyanobacteria of the genus Microcystis have presented interesting results in biodegradation studies of several substrates. These species show resistance and rapid responses to eutrophication [21], possibly by acclimation or adaptation to the adverse conditions [22].

Microcystis is one of the most commonly found genera in Brazil [23], which justifies evaluating the potential of Microcystis species locally for biodegradation processes. Cultures of *Microcystis novacekii* have been studied and have demonstrated their ability to remove different types of pollutants [24,25].

Cyanobacteria usually present themselves in the environment associated with bacteria through interactions considered mutualistic, where cyanobacteria provide heterotrophic bacteria with organic matter to sustain their growth and oxygen for greater efficiency of aerobic metabolism. At the same time, the bacteria provide the cyanobacteria with carbon dioxide and essential nutrients for their metabolism. Thus, these associations are fundamental for the cycling of carbon and other essential ions in the aquatic environment [26–28]. The mutualistic interactions between these organisms make it very difficult to maintain axenic cyanobacterial strains in biodegradation studies [27,29].

There are no studies in the literature on the removal of TDF from the aquatic environment by cyanobacteria culture, justifying the study of the potential of a *M. novacekii* culture to remove this antiviral. Therefore, the objective of this study was to evaluate the process of biodegradation of tenofovir disoproxil using a non-axenic culture of *M. novacekii* isolated from a Brazilian lake.

## 2. Materials and Methods

### 2.1. Reagents and Chemicals

The tenofovir disoproxil fumarate used in the experiments was obtained from Nortec Química (Duque de Caxias, Rio de Janeiro, Brazil) (lot 507034) as a white, amorphous solid. The drug was analysed and certified by the Department of Quality Control of the Ezequiel Dias Foundation (FUNED). The reagents, solvents, and other chemicals used were of analytical or high-performance liquid chromatography (HPLC) grade. All the solutions were prepared using purified water.

### 2.2. *Microcystis novacekii* Culture

A cyanobacterium strain, *M. novacekii* (Komárek) Compère, was isolated from water samples collected in Dom Helvécio Lake, in the Rio Doce State Park (42°35′595″ S; 19°46′419″ W), Minas Gerais, Southeastern Brazil) in May 2004. For the isolation of *M. novacekii* strains, pipetting and dilution series of lake water samples were used. The species was isolated in its pure form and cultivated in WC (Wright's cryptophytes) culture medium [30,31].

The procedure rendered a non-axenic unialgal culture which was kept in the culture with a WC medium at pH 7 and a controlled temperature of 25 ± 2 °C under a fluorescent light of 110.5 µmol photons $m^2/s$ and a 12 h photoperiod. The non-axenic *M. novacekii* strain was kept in culture in the algae and cyanobacteria bank of the Laboratory of Limnology, Ecotoxicology and Aquatic Ecology at the Institute of Biological Sciences of the Federal University of Minas Gerais (LIMNEA-ICB-UFMG).

The *Microcystis novacekii* strain was tested for the presence of microcystin to verify the toxigenic potential, using enzyme-linked immunosorbent assays and the polymerase chain reaction to amplify the mcyB gene in the DNA of this strain. Both results were negative for microcystin. During the experiments, the presence of *Pseudomonas pseudoalcaligenes* was observed in the culture. The bacterium associated with *M. novacekii* was identified by Neoprospecta Microbiome Technologies Company using next-generation amplicon sequencing (NGS) and Neobiome software.

### 2.3. *Microcystis novacekii* Culture Medium

ASM-1 medium [32] buffered by the addition of 750 mg/L of 3-(N-morpholino) propanesulfonic acid (MOPS), pKa 7.2, was used for *M. novacekii* culture. The pH was adjusted to 7.0 with 0.1 mol/L HCl or NaOH solution. The composition of the ASM-1 medium was (mg/L): $NaNO_3$ (170.00), $CaCl_2$ (29.00), $MgCl_{2.6}H_2O$ (41.00), $MgSO_4$ (49.00), $K_2HPO_4$ (8.70), $Na_2HPO_{4.12}H_2O$ (17.80), $CoCl_{2.6}H_2O$ (9.5 × $10^{-3}$), $CuCl_{2.2}H_2O$ (6.5 × $10^{-4}$), $MnCl_{2.4}H_2O$ (0.69), $ZnSO_{4.7}H2O$ (0.35), $H_3BO_3$ (1.24), $FeCl_{3.3}H_2O$ (0.54), and EDTA $Na_2$ (3.72).

### 2.4. TDF Biodegradation by *M. novacekii*

The pre-culture of *M. novacekii* was prepared by inoculation of *M. novacekii* to a flask containing ASM-1 medium. This flask was incubated under controlled temperature (25 ± 2 °C). Cell density was evaluated by optical density (680 nm) [33] and was monitored until the culture reached approximately $10^6$ cell/mL. The TDF biodegradation tests in culture of the cyanobacterium *M. novacekii* were carried out following the OECD standardized protocol (ready biodegradability tests) n° 301—Guidelines for testing of chemicals (2003) with adaptations [34]. The TDF solution (2.5 g/mL) was added to test flasks containing 100 mL of *M. novacekii* culture with a cell density of approximately 106 cell/mL to obtain final concentrations of 50.00, 25.00, and 12.50 mg/L. Cyanobacteria culture was used as growth control. To determine the stability of TDF in ASM-1 medium, an uninoculated sterile medium in the same concentrations of the test flasks was prepared. The test flasks and controls were transferred to a shaking table and were incubated for 16 days, under controlled temperature (25 ± 2 °C) and 5000 lx (c. 11 $W/m^2$) illumination from cool-white fluorescent lamps. All experiments were conducted in triplicate.

### 2.5. Extraction of TDF and Its Metabolites from Culture Medium

After 1, 3, 7, and 16 days, a 5.00 mL sample was removed from each test flask, filtered (0.45 µm—Millipore), and the aqueous portion was submitted to solid-phase extraction. The samples were subjected to solid-phase extraction using a Phenomenex Strata-X® cartridge (Phenomenex, Torrance, CA, USA). The cartridges were conditioned with 5 mL of purified water. The samples were transferred to the cartridge and eluted with 5 mL of 5% methanol solution, followed by 5 mL pure methanol. The eluate was filtered (Millipore Filter, 0.22 µm pore size), transferred to a vial with an insert, and injected into a liquid chromatography equipment coupled to a mass spectrophotometer for HPLC quadrupole time-of-flight mass spectrometry (HPLC/Q-TOF-MS) analysis.

### 2.6. HPLC/Q-TOF-MS HPLC Analysis

A high-pressure liquid chromatography coupled with quadrupole time-of-flight mass spectrometry (HPLC–QTOF-MS) system (6540 UHD Accurate Mass Q-TOF LC/MS equipped with Agilent Mass Hunter Workstation Data Acquisition software B.06.00) was used and a Zorbax Eclipse Plus C18 column (2.1 × 50 mm; particle size 1.8 mm), with the following experimental conditions: flow rate of 0.5 mL/min; mobile phase methanol:water, both with 0.1% formic acid in gradient elution (50% of methanol for 2 min and 50–100% methanol for 3 min, and then returned to 50% of methanol for 1.5 min, totalling 6.5 min); injection volume of 4 µL, temperature of 50 °C. ESI parameters were capillary voltage 3.5 kV for positive mode; gas temperature 325 °C; drying gas 8 L/min; fragmentor 175 V; skimmer 65 V; mass range from 100 to 1000 $m/z$ and no collision energy was used. The TDF calibration curve was prepared in triplicate in water, with concentrations ranging from 0.1 to 125 mg/L, followed by filtration (0.22 µm).

## 3. Results and Discussion

The analytical method for monitoring the experiments was developed using an aqueous solution of TDF (50.0 mg/L), analysed via HPLC-ESI-Q-TOF/MS. The presence of two peaks (Figure 2), a main peak (0.7 min) and a residual peak (0.3 min) was observed in the chromatogram. For peak characterization, the isolated ions in the mass spectrum were extracted and the corresponding chemical structure was proposed based on the exact mass obtained. Thus, the peak at 0.3 min, with $m/z$ 404.13, according to the mass fragmentation pattern described by Kurmi et al. [35], was assigned to tenofovir isoproxil protonated monoester (TMF). The peak at 0.7 min, with $m/z$ 520.18 was attributed to protonated tenofovir disoproxil TDF [M + H +]. In the final elution time (after 4 min), no TDF-derived fragment was observed, with the final peaks corresponding to the gradient elution.

According to Kurmi et al. [35], the tenofovir isoproxil monoester detected in the medium corresponds to the product of partial hydrolysis of TDF and is described as an impurity commonly present in the raw material of the drug [36].

To carry out the tests, the contribution of abiotic processes to the degradation of TDF in ASM-1 medium was evaluated, under controlled conditions (pH, temperature, and light radiation). It was observed in the controls (TDF concentrations 12.5, 25, and 50 mg/L) an increase in the peak attributed to the monoester throughout the experiment. At the end of the process (16 days), all the TDF was converted into tenofovir isoproxil monoester, showing that mono-de-esterification can occur spontaneously in the culture medium.

Hydrolysis was observed at all TDF concentrations used. ASM-1 medium is a mineral medium containing several salts, and the experiment was performed in buffered medium (pH 7–8). According to Silva [37], the hydrolysis of TDF occurs preferably at a neutral or alkaline pH, with the molecule being more stable at an acidic pH (2 to 3). Thus, the test conditions favored the de-esterification of the molecule.

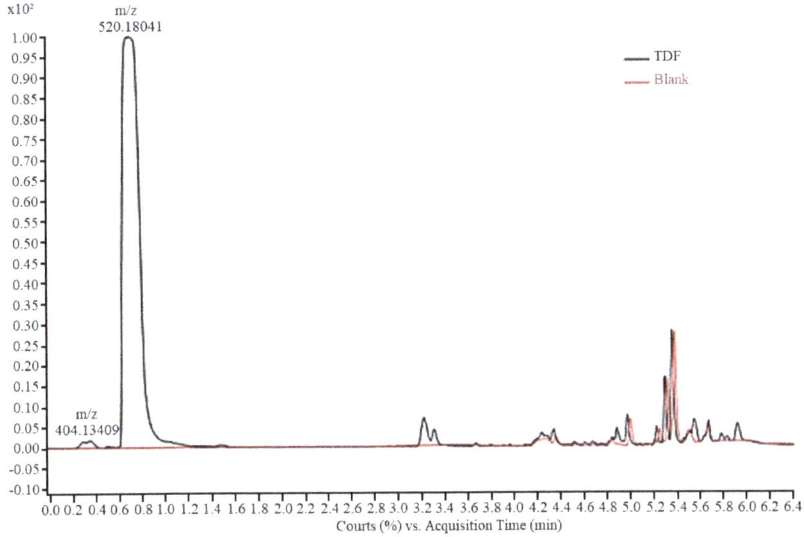


**Figure 2.** Chromatogram of TDF solution in ASM-1 medium (50 mg/L) showing peaks corresponding to tenofovir isoproxil monoester (*m*/*z* 404.13) and tenofovir disoproxil (*m*/*z* 520.18).

Biodegradation tests were performed using TDF concentrations of 12.5, 25, and 50 mg/L. It is important to highlight that this series of concentrations were defined considering that the ability of strains of microorganisms to degrade toxic compounds depends on their intrinsic ability to metabolise the xenobiotic. This property can be constitutive or acquired by microorganisms exposed to conditions considered stressful for the species [22,38].

Microbial biochemical pathways can be activated when microorganisms are exposed to critical conditions, such as high concentrations of pollutants, temperature, and pH variations [22,39]. In the case of *M. novacekii*, in a previous study [40], the strain tolerated high concentrations of TDF (EC50% 161.0 mg/L), allowing the use of drug concentrations of up to 50 mg/L in this study, configuring a stressful condition to evaluate the metabolization potential of this antiviral by the culture of *M. novacekii*.

During the TDF biodegradation experiments using the *M. novacekii* culture, the drug's mono-de-esterification was also verified; however, the hydrolysis in the tests was more intense and faster compared to controls (Figure 3). While in the control samples total mono-de-esterification occurred over 16 days, in cultures, after 72 h, only TDF residues were detected, indicating that although abiotic factors contribute to hydrolysis, metabolic pathways of microorganisms are probably responsible for accelerating the de-esterification process through extracellular enzymes.

It was verified during the test that the drug and its derivative were gradually extracted from the medium, with a percentage of TDF/TMF removal at the end of the experiment of 91.8% (12.5 mg/L), 94.1% (25 mg/L), and 88.7% (50 mg/L). In controls, no reduction in monoester concentration was observed (Figure 4). Neither tenofovir, nor any other metabolite besides TMF was detected in the culture medium, indicating that the removal of TMF is probably due to the direct action of microorganisms.

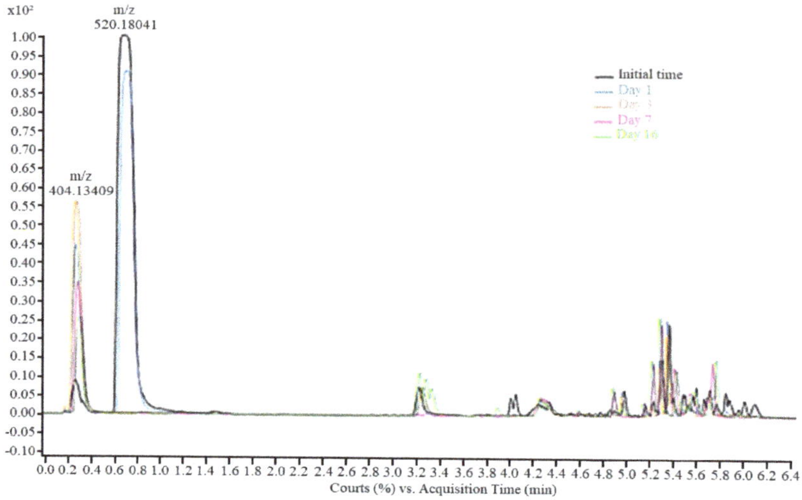


**Figure 3.** Tenofovir disoproxil and monoester of tenofovir removal during cyanobacteria degra-dation process in sample of 50 mg/L.

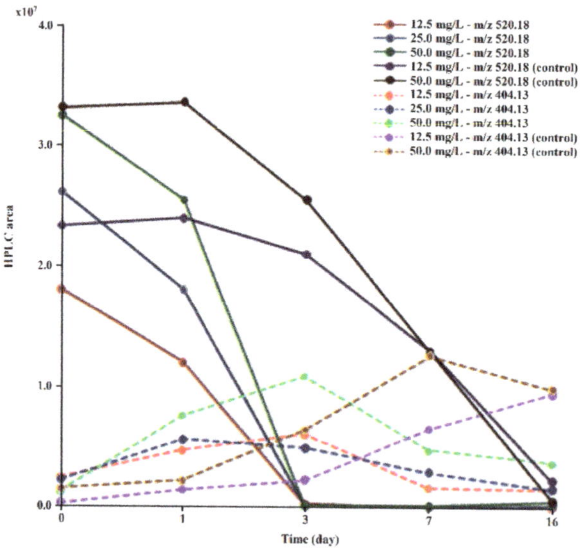


**Figure 4.** Evolution of the peak areas at $m/z$ 520.18 and $m/z$ 404.13 obtained via HPLC/MS during TDF (at concentrations of 12.5, 25.0, and 50.0 mg/L) biodegradation by *M. novacekii*.

To analyze these results, it is necessary to consider that the culture of *M. novacekii* employed is unialgal, but not axenic. According to the genetic sequencing performed, the bacterium *P. pseudoalcaligenes* was identified in the medium. This is an aerobic, Gram-negative species and its potential to metabolise toxic compounds has been described by several researchers [11–13]. Its production of esterases stands out, including potent arylesterases that give this species the ability to degrade various compounds, including polyesters [41,42]. The production of these esterases may be the accelerating factor of TDF hydrolysis.

The coexistence of cyanobacteria and microalgae with heterotrophic bacteria in the environment has aroused interest in the study of degradation of organic substances, leading to the inclusion of photosynthetic species in microbial consortia. The association of these groups of organisms can be advantageous for reducing energy expenditure due to in situ oxygen production, while also reducing $CO_2$ emissions and increasing the production of algal biomass that can be used in the production of various compounds of technological application [16,44]. Thus, several studies of the potential of these associations for the removal of different classes of organic substances have been carried out [45–47].

Phytoplanktonic species show similar behaviour in the face of stressors [48], which may increase the expression of enzymes degrading organic compounds and other compounds aimed at cellular protection [18].

In the study of microbial associations for the biodegradation of pollutants, the presence of species of the genus Microcystis is particularly interesting as they are very resistant to toxic agents due to protection mechanisms developed throughout the evolutionary process [38]. In the case of the genus Microcystis, resistance has been attributed to the characteristics of some species, including those that present a thick mucilaginous layer that surrounds the cells, with diverse functions such as nutrition and protection against dissection and against external agents, in addition to allowing the aggregation of cells in colonies that favor the formation of biofilm [17,38].

The role of cell protection by mucilage is highlighted by Pugnetti et al. [49] who reported that when exposed to adverse conditions such as the presence of toxic substances, some species intensify the production of the mucilaginous layer, which behaves like a sponge that absorbs xenobiotics [50]. In general, mucilage can biosorb xenobiotics through different interactions, usually weak bonds, without energy consumption and can retain metal ions, natural organic matter, and toxic organic substances [16,50]. Chan et al. [16] reported that microalgae can biosorb compounds rich in nitrogen, phosphorus, heavy metals, antibiotics, organochlorines, pesticides, and azo dyes from aqueous matrices.

The presence of mucilage is one of the factors that facilitates the association with bacteria, since the exopolymers that compose the mucilage can be used for fixation and as a nutritional source to bacteria, offering heterotrophic microorganisms an ideal microenvironment for their growth and metabolism [51]. Thus, in the associations of cyanobacteria and heterotrophic bacteria, two processes can act in the removal of pollutants, the immobilisation of toxic substances by mucilage, and microbial degradation by heterotrophic bacteria and by cyanobacteria themselves [47].

In the proposition of cyanobacteria/microalgae and bacteria consortia, although the immobilisation by adsorption to mucilage allows for the reduction of soluble organic carbon, it is preferable to associate the mechanisms of biodegradation and especially of mineralization of pollutants as they are more effective for the removal of organic compounds. In general, biodegradation occurs through metabolic pathways typical of various microorganisms (bacteria, fungi, and algae), which can be expressed under stressful conditions. Biodegradation can lead to partial decomposition of the molecule, generating by-products or total decomposition, reducing organic compounds through metabolic processes to their inorganic forms [16].

Microalgae are organisms capable of degrading various types of substances into carbon sources and removing them from water. Several studies indicate the dehydration capacity of different species of microalgae, such as *Chlorella* spp., *Scenedesmus* spp., and *Aphanocapsa* spp. [52,53] through processes of biosorption and/or bioconversion and/or biodegradation. In addition, phytomediation with microalgae species is an effective way to ensure environmental sustainability, as they are widely distributed around the world, are good at degrading substances, have fast metabolism, and are cost-effective [18]. Cyanobacteria are prokaryotic microorganisms that thrive in a variety of conditions and have the potential to degrade substances while in search of sources to dispense and then use in their growth, stimulating water quality [54].

Studies of microalgae/cyanobacterial and bacterial consortia for the purification of drugs from the aquatic environment have been reported with promising results [17,20]. The use of these associations for the biodegradation of antimicrobials is highlighted. This is an important characteristic of cyanobacteria and microalgae given the difficulty in obtaining bacterial species tolerant to the biocidal effects of these compounds. Wang et al. [47], using a consortium of microalgae-heterotrophic bacteria for the degradation of chlortetracycline, observed that this drug was initially removed by biosorption, followed by biodegradation.

At the end of the experiment, the authors concluded that biosorption alone had a negligible contribution to the drug removal process, which does not mean that biosorption does not have an initial action on drug retention, facilitating the action of enzymes retained in the mucilaginous layer. Biodegradation catalysed by enzymes secreted by the species present, under stress triggered by the antibiotic, has been identified as the main mechanism of chlortetracycline removal [47]. These authors observed that the use of microalgae/heterotrophic bacteria cultures for chlortetracycline biodegradation presented better results in terms of drug bioremoval than the respective axenic cultures.

Likewise, in a study of the removal of sulfamethoxazole from the aquatic environment by a consortium of algae and heterotrophic bacteria, Rodrigues et al. [46] found that the antibiotic was mainly removed by biodegradation. The bioaccumulation and biosorption of the drug by the microorganisms were negligible. The small contribution of drug biosorption on microalgae cell walls was attributed to the high water solubility of the sulfamethoxazole molecule [46].

In this study, using the association of *M. novacekii* and *P. pseudoalcaligenes*, it is likely that the removal of TDF occurs in a similar way to that reported in the degradation of chlortetracycline [47] and sulfamethoxazole [46]. Possibly in the initial step of the process, mono-de-esterification occurs by an abiotic process and by the action of extracellular hydrolases from both heterotrophic bacteria and cyanobacteria, since both groups of microorganisms can be esterase producers [27,55].

The rapid formation of TMF and its slow removal from the medium, without the presence of other metabolites, suggests that degradation occurs in the intracellular medium and that penetration into cells, whether bacterial or cyanobacterial, is the critical factor for the slow removal of TMF from the medium. Khan et al. [56] described that during the proliferation stage of bacterial and microalgae associations, both groups express enzymes such as phosphatase, sulfatase, glucosidase, and galactosidase that may be responsible for biodegradation processes of organic compounds. Thus, it is not possible to state which of the groups of microorganisms was responsible for the removal of TMF.

The role of *M. novacekii* in the degradation of TMF is reinforced by the presence of phosphate groups in the molecule, a limiting nutrient for the growth of cyanobacteria. Ren et al. [57] found that *Microcystis aeruginosa* can use dissolved organic phosphate from different chemical compounds to support its growth. In this way, the TMF molecule can provide a source of phosphate for the cyanobacteria.

Although the removal of TDF from the medium occurred with high yield, the presence of residual concentrations of TMF at the end of the experiment was observed for all concentrations tested. At the concentration of 25 mg/L, the highest removal was obtained, about 94% of the drug and its metabolite, which is a very promising result. Apparently, intra- and extracellular conjugated processes occur, requiring further studies to optimize the drug extraction process and its metabolite from the medium.

Although TDF has a high efficiency for biodegradation, there is the possibility of improving strains of *M. novacekii* through genetic engineering, increasing the number of active groups in the cell wall to stimulate the attraction of pollutants, increasing the rate of biodegradation [58–60]. However, there are some barriers to the application of genetic engineering, such as regulatory restrictions and ecological concerns [18].

The method for biodegradability assessment used in this study is based on the direct measurement of the disappearance of the original organic compound through specific chromatographic analysis to obtain information about primary biodegradation. This method is advantageous when compared to methods that use indirect measurements of bioconversion, such as $CO_2$, DOC, and BOD, as the use of only one indirect biodegradation parameter can provide misleading results [61].

Although biodegradation tests have been criticized, they are an important part of the information needed to predict the environmental fate of chemical products [62]. According to the OECD, TDF is classified as a readily biodegradable substance, as it achieved a sufficient extent of biodegradation in the OECD 301 test, respecting the 28-day time limit [34].

Thus, based on the OECD 301 series ready biodegradability tests, TDF is considered a "non-persistent" substance, as a positive result in a ready biodegradability test is sufficient to confirm non-persistence [62]. Additionally, taking into account the important observations made by Strotmann et al. (2023) [62], several important criteria were met to improve the quality of the biodegradation test, such as the assessment of the toxic effect of TDF on the cyanobacteria *M. novacekii* through the growth inhibition test [40].

Based on this criterion, it was possible to verify that there are no suspicions that TDF may have inhibitory effects on the inoculum in biodegradability tests, making it unnecessary to add inhibitory control vessels to avoid false negative conclusions [34,62,63].

Excluding the possible toxic effects of TDF on cyanobacteria [40], the reduction in biodegradation capacity with the increase in the concentration of the test substance may be a consequence of metabolic products formed during the biodegradation process and/or a change in the pH of the medium. Although traditional media used in OECD tests typically have a low phosphate concentration (3.7 mM), the culture medium used in this study was relatively high, approximately 18 mM. This increase in phosphate concentration results in good buffering properties, as demonstrated in a combined test system by Strotmann et al. (1995) [64], where the phosphate content was raised to 25.1 mM to allow stable buffering properties in the pH 7.0 range, which is necessary for biodegradation processes.

Thus, the production of metabolites, such as tenofovir and alkylphosphonate acids, during the biodegradation process can compromise the efficiency of the process [40]. However, many bacteria, such as *M. novacekii*, are capable of degrading not only the substances, but also their metabolites, forming smaller by-products, such as alcohols, oxygenated compounds, carbon, and water. During this biodegradation process, the breakdown of FTD causes a release of acids (such as carboxylic and phosphonic acids), which reduces the pH value.

However, even with the formation of possibly toxic by-products and the small change in pH, and even with the good results obtained in this study, it should be noted that tenofovir is an inhibitor of DNA synthesis, and its metabolite TMF is partially active [65], therefore its persistence in the environment can potentially lead to damage to the genetic heritage of exposed species. Thus, the persistence of TMF in culture medium for more than 15 days is worrying and points to the need for further studies on the biodegradation of this antiviral to prevent possible genotoxic actions to other aquatic organisms.

## 4. Conclusions

Through this study, excellent results were obtained in the removal of TDF from the culture medium using a culture of *M. novacekii* and *P. pseudoalkaligenes*. Approximately 94% of the drug and its metabolite were extracted at a concentration of 25 mg/L. After 16 days, residual concentrations of only one TDF metabolite, tenofovir isoproxil monoester, was detected. This result reinforces the potential of this association for studies on the removal of this drug from more complex matrices. The sustainability of the method, the ease of the technique and the good performance of the culture in removing TDF are advantages that justify the investigation of the potential of this association for environmental uses. The rapid de-esterification of TDF in the culture medium releasing the monoester—tenofovir isoproxil was one of the most important results of this study. TMF partially maintains the antiviral activity and the persistence of residual concentrations of this compound for more than 16 days in the culture medium is a worrying factor, as it may indicate that this metabolite can accumulate in the environment. In this sense, further studies aimed at removing this metabolite are necessary in order to prevent exposure of aquatic species to antiviral residues.

**Author Contributions:** Conceptualization, S.M.S.M. and F.A.R.B.; methodology, S.R.S., C.P.d.S.M., O.M.d.S.R.V., S.M.S.M., M.P.G.M. and F.A.R.B.; validation, S.R.S., G.S.-S., C.P.d.S.M., O.M.d.S.R.V., M.R.S., S.M.S.M., M.P.G.M. and F.A.R.B.; formal analysis, S.R.S., G.S.-S., C.P.d.S.M., O.M.d.S.R.V., S.M.S.M. and M.P.G.M.; investigation, S.R.S. and S.M.S.M.; resources, S.R.S.; data curation, S.R.S.; writing—original draft preparation, S.R.S., G.S.-S., C.P.d.S.M., O.M.d.S.R.V., M.R.S., S.M.S.M., M.P.G.M. and F.A.R.B.; writing—review and editing, G.S.-S., S.M.S.M., M.P.G.M. and F.A.R.B.; visualization S.R.S., G.S.-S., C.P.d.S.M., O.M.d.S.R.V., M.R.S., S.M.S.M., M.P.G.M. and F.A.R.B.; supervision, S.R.S.; project administration, S.R.S.; funding acquisition, S.R.S. All authors have read and agreed to the published version of the manuscript.

**Funding:** This study was funded by the Minas Gerais State Agency for Research and Development—FAPEMIG (process n. CEX-APQ-01626-14).

**Institutional Review Board Statement:** Not applicable.

**Informed Consent Statement:** Not applicable.

**Data Availability Statement:** Data are contained within the article.

**Acknowledgments:** This study was funded by the Minas Gerais State Agency for Research and Development—FAPEMIG (process n. CEX-APQ-01626-14). Grammar English review by Marília Ribeiro de Vasconcelos.

**Conflicts of Interest:** The authors declare no conflicts of interest.

## References


1. Ortúzar, M.; Esterhuizen, M.; Olicón-Hernández, D.R.; González-López, J.J.; Ballesteros, E.A. Pharmaceutical Pollution in Aquatic Environments: A Concise Review of Environmental Impacts and Bioremediation Systems. *Front. Microbiol.* **2022**, *26*, 869332. [CrossRef] [PubMed]
2. Mojiri, A.; Zhou, J.L.; Ratnaweera, H.; Rezania, S.; Nazari, M. Pharmaceuticals and personal care products in aquatic environments and their removal by algae-based systems. *Chemosphere* **2022**, *288*, 132580. [CrossRef]
3. Wilkinson, J.L.; Boxall, A.B.A.; Kolpin, D.W.; Leung, K.M.; Lai, R.W.; Galbán-Malagón, C.; Teta, C. Pharmaceutical pollution of the world's rivers. *Proc. Natl. Acad. Sci. USA* **2022**, *19*, e2113947119. [CrossRef]
4. Agunbiade, F.O.; Moodley, B. Occurrence and distribution pattern of acidic pharmaceuticals in surface water, wastewater, and sediment of the Msunduzi River, Kwazulu-Natal, South Africa. *Environ. Toxicol. Chem.* **2016**, *35*, 36–46. [CrossRef]
5. Rzymski, P.; Drewek, A.; Klimaszyk, P. Pharmaceutical pollution of aquatic environment: An emerging and enormous challenge. *Limnol. Rev.* **2017**, *17*, 97–107. [CrossRef]
6. Mahmood, A.R.; Al-Haideri, H.H.; Hassan, F.M. Detection of Antibiotics in Drinking Water Treatment Plants in Baghdad City, Iraq. *Adv. Public Health* **2019**, *2019*, 7851354. [CrossRef]
7. United Nations Programme on HIV/AIDS. *Confronting Inequalities: Lessons for Pandemic Responses from 40 Years of AIDS*; Report No.: UNAIDS/JC3020E. Global AIDS Update; Joint United Nations Programme on HIV/AIDS (UNAIDS): Geneva, Switzerland, 2021.
8. Wood, T.P.; Duvenage, C.S.J.; Rohwer, E. The occurrence of anti-retroviral compounds used for HIV treatment in South African surface water. *Environ. Pollut.* **2015**, *199*, 235–243. [CrossRef]
9. K'oreje, K.O.; Vergeynst, L.; Ombaka, D.; De Wispelaere, P.; Okoth, M.; Van Langenhove, H.; Demeestere, K. Occurrence patterns of pharmaceutical residues in wastewater, surface water and groundwater of Nairobi and Kisumu city, Kenya. *Chemosphere* **2016**, *149*, 238–244. [CrossRef]
10. Bottoni, P.; Caroli, S. Presence of residues and metabolites of pharmaceuticals in environmental compartments, food commodities and workplaces: A review spanning the three-year period 2014–2016. *Microchem. J.* **2018**, *136*, 2–24. [CrossRef]
11. Fekadu, S.; Alemayehu, E.; Dewil, R.; Van Der Bruggen, B. Pharmaceuticals in freshwater aquatic environments: A comparison of the African and European challenge. *Sci. Total Environ.* **2019**, *654*, 324–337. [CrossRef]
12. Kearney, B.P.; Flaherty, J.F.; Shah, J. Tenofovir disoproxil fumarate: Clinical pharmacology and pharmacokinetics. *Clin. Pharmacokinet.* **2004**, *43*, 595–612. [CrossRef]
13. Fung, H.B.; Stone, E.A.; Piacenti, F.J. Tenofovir disoproxil fumarate: A nucleotide reverse transcriptase inhibitor for the treatment of HIV infection. *Clin. Ther.* **2002**, *24*, 1515–1548. [CrossRef] [PubMed]
14. Leite, D.I.; Faria, J.V.; De Azevedo, L.D.; Figueiredo, Y.V.; Martins, W.A.; Maria Da Conceição, A.D.; Boechat, N. Tenofovir: Relação estrutura-atividade e métodos de síntese. *Rev. Virtual Quím.* **2015**, *7*, 2347–2376. [CrossRef]
15. Agrahari, V.; Putty, S.; Mathes, C.; Murowchick, J.B.; Youan, B.B.C. Evaluation of degradation kinetics and physicochemical stability of tenofovir. *Drug Test. Anal.* **2015**, *7*, 207–213. [CrossRef] [PubMed]
16. Chan, S.S.; Khoo, K.S.; Chew, K.W.; Ling, T.C.; Show, P.L. Recent advances biodegradation and biosorption of organic compounds from wastewater: Microalgae-bacteria consortium—A review. *Bioresour. Technol.* **2022**, *344*, 126159. [CrossRef]

17. Gonçalves, A.L.; Pires, J.C.; Simões, M. A review on the use of microalgal consortia for wastewater treatment. *Algal Res.* **2017**, *24*, 403–415. [CrossRef]
18. Touliabah, H.E.S.; El-Sheekh, M.M.; Ismail, M.M.; El-Kassas, H. A review of Microalgae- and Cyanobacteria-based biodegradation of organic pollutants. *Molecules* **2022**, *27*, 1141. [CrossRef]
19. Abinandan, S.; Shanthakumar, S. Challenges and opportunities in application of microalgae (Chlorophyta) for wastewater treatment: A review. *Renew. Sustain. Energy Rev.* **2015**, *52*, 123–132. [CrossRef]
20. Tolboom, S.N.; Carrillo-Nieves, D.; Rostro-Alanis, M.J.; Quiroz, R.C.; Barceló, D.; Iqbal, H.M.; Parra-Saldivar, R. Algal-based removal strategies for hazardous contaminants from the environment—A review. *Sci. Total Environ.* **2019**, *665*, 358–366. [CrossRef]
21. Xiao, M.; Li, M.; Reynolds, C.S. Colony formation in the cyanobacterium Microcystis. *Biol. Rev.* **2018**, *93*, 1399–1420. [CrossRef]
22. Zeng, J.; Yang, L.; Wang, W.X. Acclimation to and recovery from cadmium and zinc exposure by a freshwater cyanobacterium, Microcystis aeruginosa. *Aquat. Toxicol.* **2009**, *93*, 1–10. [CrossRef] [PubMed]
23. Bicudo, C.E.; Menezes, M. *Gêneros de Algas de Águas Continentais do Brasil. Chave para Identificação e Descrições*, 2nd ed.; Rima: São Paulo, Brazil, 2006.
24. Fioravante, I.A.; Albergaria, B.; Teodoro, T.S.; Magalhães, S.M.S.; Barbosa, F.; Augusti, R. Removal of 17α-ethinylestradiol from a sterile WC medium by the cyanobacteria Microcystis novacekii. *J. Environ. Monit.* **2012**, *14*, 2362–2366. [CrossRef]
25. Campos, M.M.C.; Faria, V.H.F.; Teodoro, T.S.; Barbosa, F.A.R.; Magalhães, S.M.S. Evaluation of the capacity of the cyanobacterium Microcystis novacekii to remove atrazine from a culture medium. *J. Environ. Sci. Health Part B* **2013**, *48*, 101–107. [CrossRef]
26. Lutzu, G.A.; Dunford, N.T. Interactions of microalgae and other microorganisms for enhanced production of high-value compounds. *Front. Biosci.* **2018**, *23*, 1487–1504. [CrossRef]
27. Zheng, Q.; Wang, Y.; Xie, R.; Lang, A.S.; Liu, Y.; Lu, J.; Zhang, X.; Sun, J.; Suttle, C.A.; Jiao, N. Dynamics of heterotrophic bacterial assemblages within Synechococcus cultures. *Appl. Environ. Microbiol.* **2018**, *84*, e01517-17. [CrossRef]
28. Ye, T.; Zhao, Z.; Bai, L.; Song, N.; Jiang, H. Characteristics and bacterial community dynamics during extracellular polymeric substance (EPS) degradation of cyanobacterial blooms. *Sci. Total Environ.* **2020**, *748*, 142309. [CrossRef] [PubMed]
29. Silva, S.B.; Pádua, R.M.; Barbosa, F.A.R.; Silva, M.A.N.; Azevedo, F.R.; Magalhães, S.M.S. Phytoplankton Cultures for Tannin Biodegradation. *Water Air Soil Pollut.* **2019**, *230*, 170. [CrossRef]
30. Guillard, R.R.L.; Lorenzen, C.J. Yellow-green algae with chlorophyllide C1,2. *J. Phycol.* **1972**, *8*, 10–14.
31. Andersen, R.A.; Berges, J.A.; Harrisons, P.J. Appendix A, Recipes for freshwater and seawater media. In *Algal Culturing Techniques*; Elsevier: Burlington, NJ, USA, 2005; pp. 429–474.
32. Carmichael, W.W.; Gorham, P.R. An Improved Method for Obtaining Axenic Clones of Planktonic Blue-Green Algae. *J. Phycol.* **1974**, *10*, 238–240.
33. Ma, J.; Lin, F.; Qin, W.; Wang, P. Differential response of four cyanobacterial and green algal species to triazophos, fentin acetate, and ethephon. *Bull. Environ. Contam. Toxicol.* **2004**, *73*, 890–897. [CrossRef]
34. Organization for Economic Co-Operation and Development—OECD. *Principles and Strategies Related to the Testing of Degradation of Organic Chemicals. Part 1*; OECD: Paris, France, 2003.
35. Kurmi, M.; Golla, V.M.; Kumar, S.; Sahu, A.; Singh, S. Stability behaviour of antiretroviral drugs and their combinations. 1: Characterization of tenofovir disoproxil fumarate degradation products by mass spectrometry. *RSC Adv.* **2015**, *5*, 96117–96129. [CrossRef]
36. World Health Organization. *Consultation Documents: The International Pharmacopoeia*; WHO Drug Information: Geneva, Switzerland, 2019.
37. Silva, J.P.A. Estudo de Estabilidade do Antirretroviral Tenofovir: Uma Abordagem Integrada para o Desenvolvimento de Medicamento. Master's Thesis, Universidade Federal de Pernambuco, Recife, Brazil, 2014.
38. Marsac, N.; Houmard, J. Adaptation of cyanobacteria to environmental stimuli: New steps towards molecular mechanisms. *FEMS Microbiol. Lett.* **1993**, *10*, 119–189. [CrossRef]
39. Żyszka-Haberecht, B.; Niemczyk, E.; Lipok, J. Metabolic relation of cyanobacteria to aromatic compounds. *Appl. Microbiol. Biotechnol.* **2019**, *103*, 1167–1178. [CrossRef] [PubMed]
40. Silva, S.R.; Barbosa, F.A.R.; Mol, M.P.G.; Magalhães, S.M.S. Toxicity for Aquatic Organisms of Antiretroviral Tenofovir Disoproxil. *J. Environ. Prot.* **2019**, *10*, 1565. [CrossRef]
41. Igeño, M.I.; Becerra, G.; Guijo, M.I.; Merchán, F.; Blasco, R. Metabolic adaptation of Pseudomonas pseudoalcaligenes CECT5344 to cyanide: Role of malate–quinone oxidoreductases, aconitase and fumarase isoenzymes. *Biochem. Soc. Trans.* **2011**, *39*, 1849–1853. [CrossRef]
42. Wallace, P.W.; Haernvall, K.; Ribitsch, D.; Zitzenbacher, S.; Schittmayer, M.; Steinkellner, G.; Birner-Gruenberger, R. PpEst is a novel PBAT degrading polyesterase identified by proteomic screening of Pseudomonas pseudoalcaligenes. *Appl. Microbiol. Biotechnol.* **2017**, *101*, 2291–2303. [CrossRef]
43. Safari, M.; Yakhchali, B. Comprehensive genomic analysis of an indigenous Pseudomonas pseudoalcaligenes degrading phenolic compounds. *Sci. Rep.* **2019**, *9*, 12736. [CrossRef]
44. Wang, Y.; Ho, S.H.; Cheng, C.L.; Guo, W.Q.; Nagarajan, D.; Ren, N.Q.; Lee, D.J.; Chang, J.S. Perspectives on the feasibility of using microalgae for industrial wastewater treatment. *Bioresour. Technol.* **2016**, *222*, 485–497. [CrossRef]

45. Maza-Márquez, P.; González-Martínez, A.; Martínez-Toledo, M.V.; Fenice, M.; Lasserrot, A.; González-López, J. Biotreatment of industrial olive washing water by synergetic association of microalgal-bacterial consortia in a photobioreactor. *Environ. Sci. Pollut. Res.* **2017**, *24*, 527–538. [CrossRef]
46. Rodrigues, D.A.S.; Da Cunha, C.C.R.F.; Freitas, M.G.; De Barros, A.L.C.; e Castro, P.B.N.; Pereira, A.R.; Silva, S.Q.; Santiago, A.F.; Afonso, R.J.D.C.F. Biodegradation of Sulfamethoxazole by microalgae-bacteria consortium in wastewater treatment plant effluents. *Sci. Total Environ.* **2020**, *749*, 141441. [CrossRef]
47. Wang, Y.; He, Y.; Li, X.; Nagarajan, D.; Chang, J.S. Enhanced biodegradation of chlortetracycline via a microalgae-bacteria consortium. *Bioresour. Technol.* **2022**, *343*, 126149. [CrossRef]
48. Wiśniewska, K.; Śliwińska-Wilczewska, S.; Lewandowska, A.; Konik, M. The effect of abiotic factors on abundance and photosynthetic performance of airborne cyanobacteria and microalgae isolated from the southern Baltic Sea region. *Cells* **2021**, *10*, 103. [CrossRef]
49. Pugnetti, A.; Armeni, M.; Camatti, E.; Crevatin, E.; Dell'anno, A.; Del Negro, P.; Danovaro, R. Imbalance between phytoplankton production and bacterial carbon demand in relation to mucilage formation in the Northern Adriatic Sea. *Sci. Total Environ.* **2005**, *353*, 162–177. [CrossRef]
50. Leppard, G.G. The characterization of algal and microbial mucilages and their aggregates in aquatic ecosystems. *Sci. Total Environ.* **1995**, *165*, 103–131. [CrossRef]
51. Shen, H.; Niu, Y.; Xie, P.; Tao, M.I.N.; Yang, X.I. Morphological and physiological changes in Microcystis aeruginosa as a result of interactions with heterotrophic bacteria. *Freshw. Biol.* **2011**, *56*, 1065–1080. [CrossRef]
52. Omar, H.H. Algal decolorization and degradation of monoazo and diazo dyes. *Pak. J. Biol. Sci.* **2008**, *11*, 1310–1316. [CrossRef]
53. El-Sheekh, M.M.; Abou-El-Souod, G.; El-Asrag, H. Biodegradation of some dyes by the green Alga Chlorella vulgaris and the Cyanobacterium Aphanocapsa elachista. *Egypt. J. Bot.* **2018**, *58*, 311–320. [CrossRef]
54. Kumar, R.; Ghosh, A.K.; Pal, P. Synergy of biofuel production with waste remediation along with value-added co-products recovery through microalgae cultivation: A review of membrane-integrated green approach. *Sci. Total Environ.* **2020**, *698*, 134169. [CrossRef]
55. Panda, T.; Gowrishankar, B.S. Production and applications of esterases. *Appl. Microbiol. Biotechnol.* **2005**, *67*, 160–169. [CrossRef]
56. Khan, M.I.; Shin, J.H.; Kim, J.D. The promising future of microalgae: Current status, challenges, and optimization of a sustainable and renewable industry for biofuels, feed, and other products. *Microb. Cell Fact.* **2018**, *17*, 36. [CrossRef]
57. Ren, L.; Wang, P.; Wang, C.; Chen, J.; Hou, J.; Qian, J. Algal growth and utilization of phosphorus studied by combined mono-culture and co-culture experiments. *Environ. Pollut.* **2017**, *22*, 274–285. [CrossRef]
58. Bilal, M.; Rasheed, T.; Sosa-Hernandez, J.E.; Raza, A.; Nabeel, F.; Iqbal, H.M.N. Biosorption: An Interplay between Marine Algae and Potentially Toxic Elements—A Review. *Mar. Drugs* **2018**, *16*, 65. [CrossRef]
59. Chekroun, K.B.; Sánchez, E.; Baghour, M. The role of algae in bioremediation of organic pollutants. *Int. Res. J. Public Environ. Health* **2014**, *1*, 19–32.
60. Kuritz, T.; Wolk, C.P. Use of Filamentous Cyanobacteria for Biodegradation of Organic Pollutants. *Appl. Environ. Microbiol.* **1995**, *61*, 234–238. [CrossRef]
61. Govind, R.; Gao, G.; Lai, L.; Tabak, H.H. Continuous automated and simultaneous measurement of oxygen uptake and carbon dioxide evolution in biological systems. *Water Environ. Res.* **1997**, *69*, 73–80. [CrossRef]
62. Strotmann, U.; Thouand, G.; Pagga, U.; Gartiser, S.; Heipieper, H.J. Toward the future of OECD/ISO biodegradability testing-new approaches and developments. *Appl. Microbiol. Biotechnol.* **2023**, *107*, 2073–2095. [CrossRef]
63. Strotmann, U.; Durand, M.J.; Thouand, G.; Eberlein, C.; Heipieper, H.J.; Gartiser, S.; Pagga, U. Microbiological toxicity tests using standardized ISO/OECD methods—Current state and outlook. *Appl. Microbiol. Biotechnol.* **2024**, *108*, 454. [CrossRef]
64. Strotmann, U.J.; Schwarz, H.; Pagga, U. The combined $CO_2$/DOC test—A new method to determine the biodegradability of organic compounds. *Chemosphere* **1995**, *30*, 525–538. [CrossRef]
65. Brooks, K.M.; Ibrahim, M.E.; Castillo-Mancilla, J.R.; MaWhinney, S.; Alexander, K.; Tilden, S.; Kerr, B.J.; Ellison, L.; McHugh, C.; Bushman, L.R.; et al. Pharmacokinetics of tenofovir monoester and association with intracellular tenofovir diphosphate following single-dose tenofovir disoproxil fumarate. *J. Antimicrob. Chemother.* **2019**, *74*, 2352–2359. [CrossRef]

toxics

MDPI

*Article*

# Temporal Variation and Industry-Specific Differences of the Use of Volatile Organic Compounds from 2018 to 2023 and Their Health Risks in a Typical Industrially Concentrated Area in South China

**Yijia Guo [1], Lihua Zhu [1], Liyin Zhang [1], Xinxin Tang [2], Xinjie Li [2], Yiming Ge [2], Feng Li [1], Jilong Yang [1,3], Shaoyou Lu [2], Jinru Chen [1,*] and Xiaotao Zhou [1,*]**

1 Public Health Service Center, Bao'an District, Shenzhen 518126, China; beckyguo1995@126.com (Y.G.); zhuivy810@126.com (L.Z.); 13798509754@126.com (L.Z.); lf13424232209@163.com (F.L.); yangjlong6@mail2.sysu.edu.cn (J.Y.)

2 School of Public Health (Shenzhen), Shenzhen Campus of Sun Yat-sen University, Shenzhen 518107, China; tangxx26@mail2.sysu.edu.cn (X.T.); lixj328@mail2.sysu.edu.cn (X.L.); geym5@mail2.sysu.edu.cn (Y.G.); lushy23@mail.sysu.edu.cn (S.L.)

3 School of Public Health, Sun Yat-Sen University, Guangzhou 510080, China

* Correspondence: sgzjk209@126.com (J.C.); cybersoul@126.com (X.Z.); Tel.: +86-139-2745-0193 (J.C.); +86-186-1718-7906 (X.Z.)

**Abstract:** The risk of occupational exposure to organic solvents varies across industries due to factors such as processing materials, ventilation conditions, and exposure duration. Given the dynamic nature of organic solvent use and occupational exposures, continuous monitoring and analysis are essential for identifying high-risk hazards and developing targeted prevention strategies. Therefore, this study aims to analyze the use of organic solvents and volatile organic compounds (VOCs) in different industries in Bao'an District, Shenzhen, China, from 2018 to 2023, to understand their temporal variation and industry-specific differences and to identify high-risk occupational hazards. This study includes 1335 organic solvent samples, used by 414 different industry enterprises, and 1554 air samples. The result shows that the usage of organic solvents in various industries decreased with the outbreak of the pandemic and, conversely, increased as the situation improved. The most frequently detected volatile components in organic solvents were alkanes, followed by aromatic hydrocarbons. The ratios of the detection frequency of VOCs to the total number of detected categories increased year by year after 2020, indicating a tendency towards reduction and concentration of the types of organic solvents used in industrial production. Among the 8 high-risk VOCs, toluene (22.5%), n-hexane (22.0%), xylene (16.1%), and ethylbenzene (15.3%) have relatively high detection rates, suggesting that they need to be focused on in occupational health. Through air samples, the results show that trichloroethylene and xylene pose a high risk to human health (HQ > 1). We recommend that industry should strengthen monitoring of these two VOCs.

**Keywords:** organic solvents; industries; occupational hazards; Shenzhen

## 1. Introduction

Organic solvents, integral to industrial processes [1], find widespread application across various sectors, including paints, coatings, adhesives, and cleaners [2–4]. They serve not only as solvents for dissolving raw materials and facilitating reactant cross-linking but also contribute to enhancing the performance, stability, and durability of the resulting products. Nonetheless, their extensive utilization carries inherent environmental implications that warrant attention [5]. During industrial operations, the volatilization of organic solvents releases harmful gases, contributing to atmospheric pollution. These emissions give rise to pollutants, such as ozone and nitrogen dioxide, detrimental to both human health [6–8] and vegetation [9]. Furthermore, organic solvents can infiltrate water bodies and soil through various pathways, posing risks to water resources and soil integrity, ultimately impacting biodiversity and ecological equilibrium [10]. In summary, organic solvents play an indispensable role in industrial production, but their potential impact on the environment should not be ignored.

Volatile organic compounds (VOCs) constitute the primary components of the organic solvents widely utilized in industrial processes [11]. However, their volatile nature renders VOCs a significant source of environmental contamination and occupational health hazards [12–14]. Through extensive testing of organic solvent samples, it has been observed that VOCs exhibit a diverse range of types and concentrations, underscoring their notable impact on both environmental quality and human health. Regarding occupational health, exposure to VOCs poses substantial risks to workers in relevant industries [15,16]. VOCs can infiltrate the human body through inhalation and dermal absorption, potentially leading to various health issues and causing long-term or irreversible harm [17]. Studies have revealed elevated hazard quotients for benzene and chloroform in electro-mechanical repair and automotive painting centers in Argentina [18], heightened exposure levels to benzene, toluene, and paraxylene among workers in a furniture factory in Colombia compared to control groups [19], and a significantly elevated non-carcinogenic risk of acrolein for employees in a furniture shop in Xi'an, China, surpassing acceptable levels by 283 times [20].

Guangdong Province, as an industrial and economic stronghold of China, has organic solvent poisoning cases accounting for more than 70% of its total cases of occupational poisoning [21], and this organic solvent poisoning demonstrates enterprise aggregation, commonly found in the manufacturing industry, especially in light industry and the electronic industry [22]. Bao'an District is an industrially important area in Shenzhen, Guangdong Province, with numerous frontline enterprises. It has been reported that the cases of occupational chemical poisoning accounted for a quarter of the incidence of emerging occupational diseases in this district during the period 2006–2021, most of which were due to benzene, n-hexane and trichloroethylene poisoning [23], with ensuing massive economic losses [24]. Additionally, in the United States, 12.1% of the workforce is occupationally exposed to organic solvents [25], with 87% of methylene chloride-related fatalities identified in occupational settings [26]. Therefore, occupational hazards caused by organic solvents need to be given more attention.

The risk of occupational exposure to organic solvents may vary due to operational site factors, such as differences in processing materials [27], ventilation conditions [28], exposure duration [29], and indoor and outdoor microenvironments [30]. Current studies have focused on occupational exposure to VOCs in individual industries, such as the electronics industry [31], wooden furniture manufacturing industry [32], painting industry [27], printing industry [33], and automobile repair industry [34], while there are comparatively fewer studies that have made cross-sectional comparisons between various categories of industries. In addition, considering the occupational disease-accelerating properties of organic solvents and the introduction of new materials and technologies, concern for the dynamics of organic solvent use and occupational exposure over specific time scales can generate more targeted recommendations for occupational disease prevention and control strategies. As mentioned above, it is imperative to carry out long-term continuous monitoring of organic solvent components in the workplaces of different industries, which can provide a more reliable basis for the selection of organic solvents and the supervision of occupational hazards in enterprises.

Therefore, this study aims to analyze the use of organic solvents and VOCs in different industries in Bao'an District, Shenzhen, China, from 2018 to 2023, to elucidate the temporal variation of volatile components in organic solvents and their industry-specific differences. At the same time, the HQ values of high-risk VOCs are calculated through air monitoring samples to estimate health risks to the human body and identify high-risk occupational hazards.

## 2. Materials and Methods

### 2.1. Samples Collection

The organic solvent samples were collected from routine testing and evaluation at the Yanluo Branch Center for Public Health Services in Bao'an District from 2018 to 2023, including 1335 samples used by 414 different industry enterprises. Among these enterprises, there are 142 enterprises from the electronics industry, 141 from the chemical industry, 92 from light industry, and 39 from the machinery industry.

During sampling, the staff randomly selected organic solvents from the industries' inventory and collected samples with 50 mL wide mouthed glass sample bottles. After collection, the bottles were sealed and transported back to the Yanluo Branch Center for Public Health Services and stored in the dark at room temperature for further analysis.

### 2.2. Analysis of VOCs

According to the Technical Guidelines of Guangdong Occupational Health Technology Quality Control Center (GDOHTOC 001–2020) [35], VOCs were qualitatively and semi-quantitatively analyzed using the Agilent 7890A-5975C gas chromatography–mass spectrometry (GC–MS) system and a DB-5MS column (60 m × 0.25 mm × 1.00 μm) with the headspace method. Five milliliters (or 5 g) of each sample were placed into a headspace vial and heated at 40 °C for 30 min for equilibration, then 100 μL of the liquid is injected for analysis. Chromatographic conditions: injection port temperature is at 260 °C, initial temperature is at 45 °C holding for 2 min, then increased at a rate of 10 °C/min to 230 °C, holding for 2 min. In the subsequent analysis, we will use peak area percentage (PAP) for comparison. PAP is an analytical method used to describe the relationship between peak size and concentration in chemical analysis. In chemical analysis, peak area refers to the area under the curve of a specific peak, while concentration refers to the concentration of solutes in a solution. By calculating the peak area percentage, the concentration of solutes in the solution can be determined.

### 2.3. Air Detection

Based on recent occupational disease occurrences related to organic solvents, this research regards benzene, 1,2-dichloroethane, n-hexane, trichloroethylene, toluene, ethylbenzene, xylene, and trichloromethane as high-risk occupational hazards [36–38] and calculates the health risks of these VOCs. The air samples from different industry enterprises were collected in 2023 for daily monitoring, with a total of 1554 samples, including 522 from the electronics industry, 504 from the chemical industry, 372 from light industry, and 156 from the machinery industry. Each sample was collected by an activated carbon tube with a flow rate of 50 mL/min for 2–8 h at the sampling point. The air samples were analyzed by 7890A gas chromatograph (Agilent, Santa Clara, CA, USA) equipped with flame ionization detector. The activated carbon in the front and rear sections was poured into two solvent desorption bottles and 1.0 mL of carbon disulfide added to each. Then, the bottles were sealed and desorbed for 30 min and shaken occasionally. This solution was later used for detection. The injection volume is 1 μL. The specific reference methods are shown in Table S1.

### 2.4. Human Health Risk Assessment

Health risk assessment can quantitatively describe the relationship between human exposure to VOCs and adverse health reactions. The level of VOC inhaled by the human body is calculated based on the environmental concentration of each type of VOC. Referring to previous literature [39], the exposure concentration $EC_i$ ($mg/m^3$) in this study was calculated as follows with Formula (1):

$$EC_i = C_i \frac{ET \times EF \times ED}{AT} \times 90\% \quad (1)$$

$C_i$ is the environmental concentration of $VOCs_i$ (mg/m$^3$); ET is the exposure time (8 h/day); EF is the exposure frequency (240 day/year); ED is the lifetime exposure time, which is 30 years; and AT is the average exposure time (70 years × 365 days/year × 24 h/day). The absorption coefficient in the formula is 90%.

The calculation formula for non-carcinogenic risk assessment is as follows using Formula (2):

$$HQ_i = \frac{EC_i}{REC_i} \tag{2}$$

The $HQ_i$ represents the hazard quotient associated with compound i, while $RFC_i$ is the reference concentration (mg/m$^3$) of compound i. Based on the non-carcinogenic risk assessment methodology employed by the U.S. Environmental Protection Agency (EPA), it is determined that a non-carcinogenic health risk is present when $HQ > 1$. When $HQ < 1$, it indicates that the level of risk is within an acceptable range. Since the EPA did not calculate the RFC of 1,2-dichloroethane and trichloromethane, the calculation of HQ values does not involve these two VOCs.

### *2.5. Statistical Analysis*

Statistical analysis is conducted using IBM SPSS Statistics 26. The chi-square test is employed to examine differences in the composition ratio of organic solvent samples across different industries and the detection rates of VOCs in organic solvents. For the volume fraction of VOCs in organic solvents, if homogeneity of variance is satisfied, One-Way ANOVA is applied; if not satisfied, Welch's ANOVA is used. The significance level for testing is set at $\alpha = 0.05$.

## **3. Results and Discussion**

### *3.1. Overview of VOC Testing Industry and Sample Analysis*

#### 3.1.1. Numbers of Organic Solvent Samples in Four Industries

According to the industrial classification for national economic activities (GB/T 4754-2017) [40] and the table of business scope of occupational health technical service organizations [2022 Edition], a total of 414 enterprises in the Yanluo and Songgang Street areas of Bao'an District conducted organic solvent testing from 2018 to 2023. These enterprises represent 16 specific industries, which are further categorized into four major categories: electronics industry, chemical metallurgy and building materials industry (referred to as "chemical industry"), light and textile industry (referred to as "light industry"), and machinery industry. Detailed information regarding these industries is provided in Table S2.

Between 2018 and 2023, a total of 1335 organic solvent samples were analyzed within the jurisdiction. Significantly, the number of organic solvent samples detected notably increased in 2023, particularly within the electronics and light industry sectors (Figure 1A). Conversely, there was minimal change observed in the chemical and mechanical industries (Table S3). This discrepancy may stem from the traditional nature of these industries, which often employ specific types of organic solvents with extended service life and stability, thereby requiring less frequent replacement or testing [41]. Conversely, since the onset of the epidemic, the electronics and light industry sectors have experienced rapid expansion [42,43]. These sectors extensively utilize organic solvents in their production processes, resulting in heightened demand for solvent testing [44]. Regarding the composition ratio of samples from various industries, the majority of organic solvent testing samples in the jurisdiction originated from the electronics industry (10.95–45.52%) and the chemical industry (18.82–36.09%) (Figure 1B, Table S3).

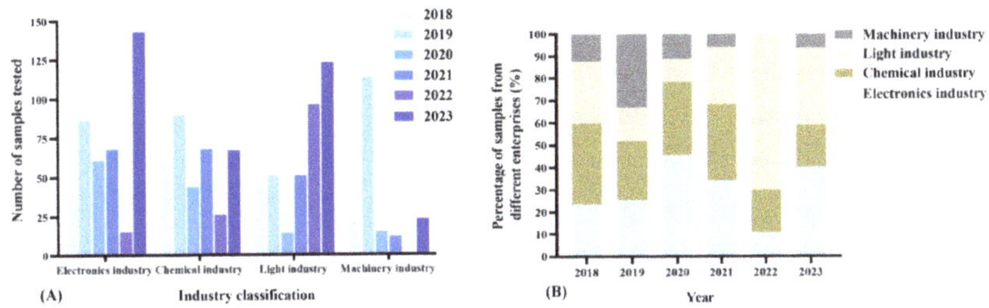


**Figure 1.** The number (**A**) and proportion (**B**) of organic solvent samples across different industries from 2018 to 2023.

3.1.2. Composition of Organic Solvent Samples from Four Industries

Among all the samples analyzed, the top five compositions were adhesive (18.28%), cleaner (16.78%), ink (9.74%), alcohol (7.04%), and diluent (5.69%) (Figure 2). The common VOC composition of these five organic solvents is shown in Table S4. Adhesive and cleaner find extensive applications across various industries. Adhesives serve as crucial joining materials in the assembly and manufacturing of diverse products [45], thus accounting for their significant presence in the samples. Cleaners play a vital role in equipment and product cleaning processes [46], underscoring their indispensable nature in manufacturing operations, which contributes to their high representation in the sample composition.

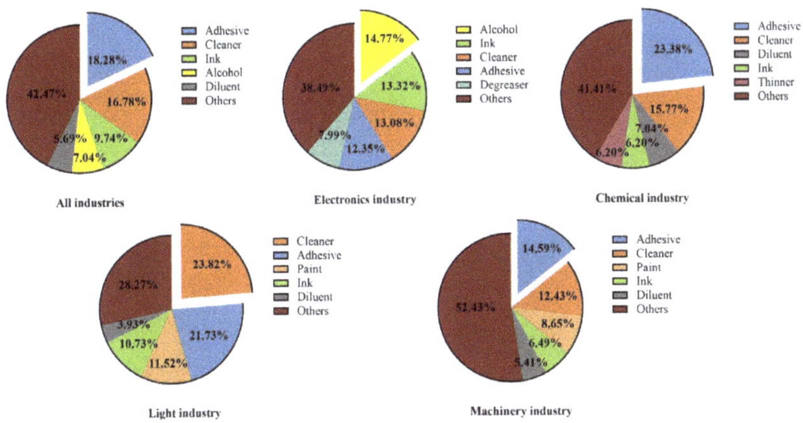


**Figure 2.** Composition of organic solvents in different industries' samples.

In the electronics industry, alcohol comprises the largest percentage in the sample composition (14.77%), followed by ink (13.32%) (Figure 2). Alcohol, characterized by high volatility and low surface tension, facilitates rapid evaporation [47], effectively removing surface oils and impurities to enhance the cleanliness and reliability of components. Hence, it is utilized extensively for cleaning electronic equipment. On the other hand, ink is employed in the fabrication of printed circuit boards (PCBs) to delineate circuit lines and conductive patterns [48]. Additionally, it finds application in the production of flexible circuits, touchscreens, and other electronic components owing to its conductivity, rendering it indispensable in the electronics industry [49].

*3.2. Analysis of VOCs in Organic Solvents*

3.2.1. Categories of VOCs in Organic Solvents

During 2018–2023, a total of 400 VOCs were detected in 1335 organic solvent samples, including 142 alkanes, 53 aromatic hydrocarbons, 48 esters, 38 alcohols, 11 ketones,

20 halogenated hydrocarbons, 20 ethers, and 68 other species, of which 66 were listed in the Occupational exposure limits for hazardous agents in the workplace (GBZ 2.1-2019) [50]. As shown in Figure 3 and Table S5, VOCs were detected 5719 times, and the composition ratios of different components was in the following order: alkanes (33.6%) > aromatic hydrocarbons (19.3%) > alcohols (16.1%) > esters (12.7%) > ketones (7.3%) > halogenated hydrocarbons (6.0%) > others (3.2%) > ethers (1.8%). Overall, the number of alkanes detected (F = 7.346, $p < 0.01$) and their detection frequency (F = 11.925, $p < 0.01$) were significantly higher than those of other VOCs, followed by aromatic hydrocarbons, which was similar to the organic solvent determinations in the other districts of Shenzhen [51–53].

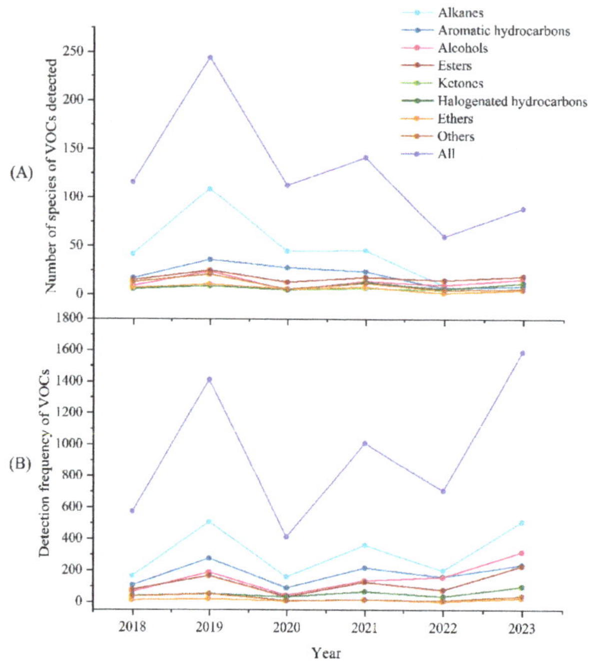


**Figure 3.** The number (**A**) and detection frequency (**B**) of VOCs in various organic solvents from 2018–2023.

As shown in Figure 3, the number of detected VOCs and the detection frequency reached a peak in 2019 and a trough in 2020, which is attributed to the outbreak of the COVID-19 pandemic in 2020, causing the shutdown of many industrial operations, which reduced the demand for organic solvents. Notably, after 2020, there is an overall downward trend in the number of detected VOC categories, but an upward trend in the detection frequency. The calculated results show that the ratios of the detection frequency of VOCs to the total number of detected categories for each year are as follows: 17.9 in 2023 > 11.89 in 2022 > 7.14 in 2021 > 5.79 in 2019 > 4.96 in 2018 > 3.67 in 2020, suggesting that the ratio is increasing year by year after 2020. Evidently, there is a trend towards homogenization with the gradual reduction and centralization of the types of organic solvents applied in industrial production. This reduces the potential for serious risks associated with mixed exposures, while the introduction of new substitutes may raise new health concerns and require further research into the hazards of organic solvents and their occupational exposure risks.

### 3.2.2. Detection of VOCs in Organic Solvents

The ten most frequently detected VOCs in the 1335 organic solvent samples during 2018–2023 are displayed in Table 1, and they have all been listed in the Occupational exposure limits for hazardous agents in the workplace (GBZ 2.1-2019) [50]. The five substances with the highest detection rates were in the following order: methanol (31.1%) > toluene (22.5%) > n-hexane (22.0%) > xylene (16.1%) > ethylbenzene (15.3%). Among them, toluene (22.1%) exhibited the highest average peak area percentage, followed by methanol (16.9%), n-hexane (4.96%), xylene (0.25%), and ethylbenzene (0.33%). They are all high-risk occupational hazards of organic solvents, especially benzene derivatives with carcinogenic risk, and their detection rate is the main concern of the many studies on organic solvent analysis in Shenzhen [52,54–56], requiring key attention in occupational health monitoring.

**Table 1.** Top ten volatile organic compounds detected in 1335 organic solvent samples.

| NO. | Components | DF | DR (%) | PAP (%) | |
|---|---|---|---|---|---|
| | | | | Mean | Range |
| 1 | Methanol | 415 | 31.1 | 16.9 | 0.10–100 |
| 2 | Toluene | 300 | 22.5 | 22.1 | 0.10–100 |
| 3 | n-Hexane | 294 | 22.0 | 4.96 | 0.01–41.4 |
| 4 | Xylene | 215 | 16.1 | 0.25 | 0.10–99.1 |
| 5 | Ethylbenzene | 204 | 15.3 | 0.33 | 0.10–56.1 |
| 6 | Ethyl acetate | 196 | 14.7 | 12.9 | 0.10–100 |
| 7 | Dichloromethane | 182 | 13.6 | 25.8 | 0.10–100 |
| 8 | Methyl acetate | 174 | 13.0 | 9.57 | 0.10–92.5 |
| 9 | Dimethoxymethane | 169 | 12.7 | 13.3 | 0.10–82.9 |
| 10 | Acetone | 168 | 12.6 | 11.1 | 0.10–100 |

DF: detection frequency, the times that VOCs were detected; DR: detection rates, the proportion of VOCs detected in all samples; PAP: peak area percentage, the proportion of the peak area to the total peak area.

In the present study, methanol was the most frequently detected and was also the major VOC of various organic samples, including inks, alcohols, diluents, degreaser, thinner, and screen wash. It has been reported that methanol poisoning has a high mortality rate among survivors and can lead to long-term visual sequelae and severe brain damage [57–59]. Since methanol can be absorbed by humans through various routes, such as dermal contact, ingestion and inhalation, corresponding protective measures must be taken, such as ventilation and wearing appropriate protective equipment, to avoid prolonged exposure to methanol vapors. It is noteworthy that a relatively low detection rate (6.67%) and mean peak area percentage (0.36%) of benzene were observed in this study. The almost complete cessation of benzene's industrial use as a solvent has allowed its congeners, such as toluene and xylene, to be introduced into industrial production as low-toxicity alternatives, but benzene remains as a raw ingredient in other materials, such as styrene [60]. However, chronic exposure to low ambient concentrations of benzene still significantly increases the risk of death in the population [61]. Furthermore, beyond the well-known carcinogenicity and hematologic toxicity of benzene, toluene, ethylbenzene, and xylene (BTEX) [27], epidemiological studies have demonstrated associations between BTEX exposure and cardiovascular disease [62], decreased lung function [63], liver damage [64], and neurological symptoms [65]. n-Hexane is frequently applied as an auxiliary production material in small and medium-sized enterprises, especially in the Pearl River Delta region where the electronics, hardware and printing industries are concentrated, and it has been associated with peripheral neuropathy [29]. More remarkably, workers are subject to combined exposures to various hazards that may exacerbate occupational injuries when laboring in workshops. It has been reported that exposure to mixtures of organic solvents is related to the prevalence of hypertension [66], and that the combined effects of organic solvents and noise have a synergistic effect on hearing damage [67–69].

### 3.3. *Detection and Analysis of High-Risk Occupational Disease Hazards*

#### 3.3.1. Detection of High-Risk Occupational Hazards in Organic Solvents from 2018 to 2023

Table 2 shows the detection rates and average peak area percentages of high-risk VOCs of organic solvents from 2018 to 2023. Significant differences were found in the detection rates of 8 VOCs ($\chi^2 = 741.120$, $p < 0.001$). Over the past six years, toluene (22.47%), n-hexane (22.02%), ethylbenzene (15.28%), and xylene (16.10%) had higher detection rates, while the detection rates of trichloromethane (0.60%), 1,2-dichloroethane (1.05%), trichloroethylene (5.54%), and benzene (6.67%) were lower. Additionally, the average peak area percentages of toluene (14.48%) and xylene (12.93%) were significantly higher than other VOCs ($F = 10.119$, $p < 0.001$) (Table 2), as substitutes for benzene, toluene and xylene have been widely used in various industries in recent years [70,71]. Although the toxicity of toluene and xylene is lower than that of benzene, studies have also shown that long-term exposure to toluene and xylene is associated with cardiovascular diseases and can lead to premature births [62,72]. Therefore, the use of toluene and xylene needs to be cautious, and efforts should continue to find alternatives with lower toxicity.

**Table 2.** High-risk volatile organic compounds from 2018 to 2023.

| Years | Benzene | | 1,2-Dichloroethane | | n-Hexane | | Trichloroethylene | |
|---|---|---|---|---|---|---|---|---|
| | DR (%) | APAP (%) | DR (%) | APAP (%) | DR (%) | APAP (%) | DR (%) | APAP (%) |
| 2018 ($n$ = 169) | 5.92 | 0.09 | 0.59 | 1.39 | 9.47 | 3.92 | 9.47 | 13.82 |
| 2019 ($n$ = 340) | 4.41 | 0.63 | 1.18 | 1.59 | 13.53 | 6.05 | 5 | 9.78 |
| 2020 ($n$ = 134) | 2.99 | 0.02 | 0.75 | 0.34 | 15.67 | 2.08 | 8.21 | 4.99 |
| 2021 ($n$ = 199) | 6.53 | 0.04 | 1.51 | 0.58 | 27.64 | 1.77 | 8.04 | 5.13 |
| 2022 ($n$ = 137) | 12.41 | 0.03 | 1.46 | 0.01 | 30.66 | 1.53 | 3.65 | 1.24 |
| 2023 ($n$ = 356) | 8.43 | 0.11 | 0.84 | 0.01 | 32.02 | 2.42 | 2.53 | 1.76 |
| all samples ($n$ = 1335) | 6.67 $^{a}$ | 0.16 $^{1}$ | 1.05 $^{b}$ | 0.70 $^{1}$ | 22.02 $^{c}$ | 2.80 $^{1}$ | 5.54 $^{a}$ | 7.38 $^{1}$ |
| **Years** | **Toluene** | | **Ethylbenzene** | | **Xylene** | | **Trichloromethane** | |
| | DR (%) | APAP (%) | DR (%) | APAP (%) | DR (%) | APAP (%) | DR (%) | APAP (%) |
| 2018 ($n$ = 169) | 17.75 | 15.46 | 10.65 | 1.91 | 15.98 | 21.5 | 0 | 0 |
| 2019 ($n$ = 340) | 21.47 | 25.45 | 16.18 | 4.91 | 12.94 | 22.01 | 0 | 0 |
| 2020 ($n$ = 134) | 18.66 | 13.58 | 7.46 | 0.62 | 7.46 | 2.89 | 0.75 | 1.53 |
| 2021 ($n$ = 199) | 27.14 | 9.63 | 21.11 | 2.75 | 21.11 | 10.52 | 1.01 | 1.93 |
| 2022 ($n$ = 137) | 37.23 | 12.09 | 28.47 | 2.51 | 29.2 | 8.29 | 0.73 | 4.17 |
| 2023 ($n$ = 356) | 18.82 | 8.16 | 11.24 | 3.02 | 14.61 | 8.26 | 1.12 | 0.43 |
| all samples ($n$ = 1335) | 22.47 $^{c}$ | 14.48 $^{2}$ | 15.28 $^{d}$ | 3.16 $^{1}$ | 16.10 $^{d}$ | 12.93 $^{2}$ | 0.60 $^{b}$ | 1.41 $^{1}$ |

DR: detection rates; APAP: average peak area percentage; $^{a, b, c, d}$: Different superscript letters indicate $p < 0.001$ (Chi-square test); $^{1, 2}$: Different superscript numbers indicate $p < 0.001$ (One-Way ANOVA).

Between 2018 and 2023, the annual detection rates of benzene, 1,2-dichloroethane, toluene, ethylbenzene, xylene, and trichloromethane fluctuated within a certain range. Benzene (12.41%), toluene (37.23%), ethylbenzene (28.47%), and xylene (29.20%) all saw their highest detection rates in 2022 (Table 2). It is worth noting that the detection rate of n-hexane increased from 9.47% in 2018 to 32.02% in 2023, showing an upward trend over the past six years. n-Hexane is an industrial organic solvent, and its toxicity is mainly neurotoxic. Prolonged exposure to n-hexane can induce severe peripheral neuropathy, and it is also associated with central-peripheral axonopathy [73,74]. Meanwhile, the detection rate of trichloroethylene decreased from 9.47% in 2018 to 2.53% in 2023, showing a downward trend over the past six years. Trichloroethylene is also a commonly used industrial solvent, its toxicity may lead to congenital defects and neurodegenerative diseases, and it is considered a potential risk factor for the development of Parkinson's disease (PD) [75,76]. The gradual decrease in the detection rate of trichloroethylene in recent years indicates that other solvents have gradually been used to replace it in industry. Although the usage rate has declined, its toxicity still cannot be ignored.

#### 3.3.2. Detection of High-Risk Occupational Hazards in Organic Solvents Used in Different Industries

Table 3 shows the detection rates and average peak area percentages of high-risk VOCs in different industries. Among the 414 enterprises, samples from the electronics industry, chemical industry and light industry were able to detect all eight high-risk VOCs. In the machinery industry, all VOCs except trichloromethane were detected. The detection rates of n-hexane (13.11~24.94%), toluene (12.02~32.2%), ethylbenzene (7.26~21.99%), and xylene (8.23~22.25%) are higher than for other VOCs.

**Table 3.** High-risk volatile organic compounds in different industries.

| Industry Classification | Benzene | | 1,2-Dichloroethane | | n-Hexane | | Trichloroethylene | |
|---|---|---|---|---|---|---|---|---|
| | DR (%) | APAP (%) | DR (%) | APAP (%) | DR (%) | APAP (%) | DR (%) | APAP (%) |
| Electronics industry ($n$ = 413) | 4.84 [a] | 0.19 | 0.48 [b] | 0.09 | 24.94 [c] | 4.38 | 8.96 [a,d] | 27.47 |
| Chemical industry ($n$ = 355) | 6.76 [a] | 0.22 | 0.85 [b] | 7.48 | 24.23 [c,d] | 6.44 | 5.92 [a] | 54.89 |
| Light industry ($n$ = 382) | 8.90 [a] | 0.1 | 2.09 [b] | 2.17 | 21.47 [c] | 2.86 | 1.31 [b] | 0.48 |
| Machinery industry ($n$ = 185) | 6.01 [a,b] | 1.32 | 0.55 [a] | 19.5 | 13.11 [b] | 8.2 | 2.73 [a] | 55.61 |
| **Industry classification** | **Toluene** | | **Ethylbenzene** | | **Xylene** | | **Trichloromethane** | |
| | **DR (%)** | **APAP (%)** | **DR (%)** | **APAP (%)** | **DR (%)** | **APAP (%)** | **DR (%)** | **APAP (%)** |
| Electronics industry ($n$ = 413) | 12.35 [d] | 14.96 | 7.26 [a,d] | 2.55 | 8.23 [a,d] | 10.97 | 0.48 [b] | 3.48 |
| Chemical industry ($n$ = 355) | 28.73 [d] | 24.69 | 18.59 [c] | 5.91 | 21.69 [c,d] | 21.9 | 0.85 [b] | 8.98 |
| Light industry ($n$ = 382) | 32.2 [d] | 24.15 | 21.99 [c] | 3.66 | 22.25 [c,d] | 16.27 | 0.79 [b] | 35.71 |
| Machinery industry ($n$ = 185) | 12.02 [b] | 21.12 | 12.57 [b] | 11.71 | 10.93 [b] | 43.76 | 0.00 [a] | 0 |

DR: detection rates; APAP: average peak area percentage; [a, b, c, d]: Different superscript letters indicate $p < 0.001$ (Chi-square test).

The average peak area percentage of VOCs represents the volumetric fraction of VOCs. According to Table 3, we can see that the volumetric fraction of toluene (14.96~24.69%) is relatively high in each industry, while the volumetric fraction of benzene (0.1~1.32%) is lower than other VOCs. The volumetric fractions of 1,2-dichloroethane (19.5%), trichloroethylene (55.61%), and xylene (43.76%) in the mechanical industry are higher than in other industries. In addition, the volumetric fraction of trichloroethylene (54.89%) in the chemical industry is also higher than in other industries. The volumetric fraction of trichloromethane (35.71%) in the light industry is higher than in other industries. This study detected the content of VOCs in organic solvents used in industries, while most other literature detected the content of VOCs in factory air. Although the samples are different, the detection levels also have certain reference value. Xie et al. detected the types and levels of VOCs in the air from the petrochemical industry in China. The results indicate that benzene and n-hexane are characteristic VOCs of the petrochemical industry (mass content: 14.54% and 4.24%) [77]. Another piece of research from Jinan, China shows that, besides benzene (average mixing ratio: 1.35 ± 1.32 ppbv) and n-hexane (0.92 ± 0.49 ppbv), there are also higher levels of toluene (1.46 ± 1.34 ppbv) and xylene (1.51 ± 1.57 ppbv) in the air from the petrochemical industry [78]. Based on the results of this study and other related research, we suggest that, in order to control the hazards of VOCs to human health, it is necessary for various industries to reduce the usage of benzene, toluene, xylene, and ethylbenzene, and continue to search for low-toxicity alternatives for these compounds.

### *3.4. The Health Risk of High-Risk Occupational Hazards*

#### 3.4.1. Detection of High-Risk Occupational Hazards in Air Samples

Table 4 shows the detection rate and mean concentration of high-risk VOCs in air samples from different enterprises in 2023. Except for 1,2-dichloroethane in the chemical industry and 1,2-dichloroethane and trichloromethane in the machinery industry, all

VOCs have been detected in all industries. According to Table 4, the DRs of benzene, 1,2-dichloroethane, and trichloromethane are relatively low, while the DRs of n-hexane, toluene, and xylene are all around 50%. The detection rate of toluene in air samples from the light industry is as high as 71.15%, and the detection rate of toluene in the organic solvent samples from the light industry is also the highest (32.2%), indicating that the use of toluene in light industry is relatively high. It is necessary to provide targeted protection for workers in light industry to reduce the harm of toluene to them.

**Table 4.** High-risk VOCs in air samples from different industries in 2023.

| Industry Classification | Benzene | | 1,2-Dichloroethane | | n-Hexane | | Trichloroethylene | |
|---|---|---|---|---|---|---|---|---|
| | DR (%) | Mean (mg/m$^3$) | DR (%) | Mean (mg/m$^3$) | DR (%) | Mean (mg/m$^3$) | DR (%) | Mean (mg/m$^3$) |
| Electronics industry ($n$ = 522) | 10.59 | 0.18 | 7.84 | 0.13 | 40.23 | 2.82 | 7.59 | 105.89 |
| Chemical industry ($n$ = 504) | 11.54 | 0.12 | 0 | - | 43.04 | 4.21 | 24.20 | 145.02 |
| Light industry ($n$ = 372) | 16.67 | 0.03 | 3.85 | 0.03 | 56.67 | 2.50 | 5.13 | 54.56 |
| Machinery industry ($n$ = 156) | 16.67 | 0.03 | 0 | - | 50.00 | 10.73 | 16.67 | 175.68 |
| **Industry classification** | **Toluene** | | **Ethylbenzene** | | **Xylene** | | **Trichloromethane** | |
| | DR (%) | Mean (mg/m$^3$) | DR (%) | Mean (mg/m$^3$) | DR (%) | Mean (mg/m$^3$) | DR (%) | Mean (mg/m$^3$) |
| Electronics industry ($n$ = 522) | 34.85 | 4.70 | 25.42 | 2.00 | 42.30 | 6.64 | 13.04 | 6.54 |
| Chemical industry ($n$ = 504) | 53.62 | 24.52 | 43.55 | 3.27 | 60.00 | 13.66 | 5.88 | 0.20 |
| Light industry ($n$ = 372) | 71.15 | 18.60 | 45.83 | 2.04 | 50.00 | 3.87 | 6.25 | 0.04 |
| Machinery industry ($n$ = 156) | 50.00 | 10.04 | 40.00 | 3.36 | 50.00 | 30.85 | 0 | - |

DR: detection rates.

In air samples, the mean concentration of trichloroethylene is significantly higher than that of other VOCs, but its detection rate in organic solvent samples is not the highest. This indicates that the volatility of trichloroethylene is relatively high in all industries, and it may also be due to improper ventilation and exhaust protection measures in workshops using trichloroethylene. In addition to trichloroethylene, the mean concentrations of n-hexane, toluene, and xylene in some industries are also relatively high. A study of the petrochemical industry in Map Ta Phut, Thailand, found that the highest annual ambient concentration of xylene (57.9 μg/m$^3$) in the air is approximately 6 to 20 times higher than benzene's (9 μg/m$^3$) and toluene's (2.8 μg/m$^3$) [79]. Compared to research in Thailand, the exposure to VOCs in Shenzhen is significantly more severe. The research on the pharmaceutical industry in the Yangtze River Delta in China [80], the manufacturing industries in Dalian, Shenyang, and Kaifeng in China [81], and the paint production plants in Iran [82] all show that benzene, xylene, and ethylbenzene are the most prominent VOCs. These types of VOCs have a certain toxicity to the human body, especially neurotoxicity [75,83–85]. Therefore, industries should enhance health monitoring of workers and, if workers have early symptoms, immediate action should be taken to avoid further exacerbation of poisoning.

### 3.4.2. The Health Risk of High-Risk Occupational Hazards in Different Industries

Table 5 shows the health risks of high-risk volatile organic compounds in different industries in 2023. For all industries, exposure to benzene, n-hexane, toluene, and ethylbenzene poses no risk to human health (HQ < 1). Xylene poses a certain risk to the health of workers in the Electronics and Light industries (3 < HQ < 10), while its harm in the other two enterprises is several times greater (HQ > 10). The health hazards of trichloroethylene for all industrial workers are significant (HQ > 10), indicating that Shenzhen industries need to impose stricter restrictions on the use of trichloroethylene.

**Table 5.** The HQ of high-risk VOCs in different industries in 2023.

| Industry Classification | Benzene | n-Hexane | Trichloroethylene | Toluene | Ethylbenzene | Xylene |
|---|---|---|---|---|---|---|
| Electronics industry | 0.51 | 0.09 | >10 [a] | 0.08 | 0.06 | 5.61 |
| Chemical industry | 0.34 | 0.14 | >10 [a] | 0.41 | 0.10 | >10 [a] |
| Light industry | 0.08 | 0.08 | >10 [a] | 0.31 | 0.06 | 3.27 |
| Machinery industry | 0.08 | 0.35 | >10 [a] | 0.17 | 0.10 | >10 [a] |

[a]: When the HQ value is much greater than 10, use >10 instead of the true value.

Previous studies have shown that xylene poses almost no health risk to humans in petrochemical plants in Iran [86] and industrial zones in Chongqing [87] and Hefei [88], China (HQ < 1). However, in other industries, the results of benzene, toluene, ethylbenzene, and xylenes (BTEX) are completely different. In Iranian refineries, BTEX in the air poses a health risk to humans (1 < HQ < 5) [89,90]. In printing factories in Beijing, the health risks of benzene and xylene cannot also be ignored (1 < HQ < 5) [91]. However, compared with other studies, the health risks of trichloroethylene and xylene in this study are still higher, indicating that the usage of trichloroethylene and xylene in Shenzhen industry is relatively high, and the prevention and control measures are not timely. At present, the permissible concentration-time weighted average (PC-TWA) of xylene is 50 mg/m$^3$ in China, while the PC–TWA of trichloroethylene is 30 mg/m$^3$. We can see that industries in Shenzhen all follow the national limit for the use of xylene, but the usage of trichloroethylene is over the limit. We suggest that factories should strengthen ventilation measures and reduce the use of these two types of VOCs. For trichloroethylene, factories should reduce its usage below the national limit, while the limit value of xylene should be appropriately lowered, as the current limit value cannot effectively reduce the health hazards of xylene for workers. Factories should regularly monitor the concentration of these in the air and conduct timely physical examinations of workers to minimize the adverse health effects of trichloroethylene and xylene.

## 4. Conclusions

This study detected 1335 organic solvent samples used by 414 different industry enterprises in Bao'an District from 2018 to 2023. The result shows that, during the outbreak of the pandemic, most industries reduced their use of organic solvents. However, after the end of the pandemic, rapid expansion due to economic demands led to an increased demand for organic solvents. At the same time, with a trend towards the homogenization of organic solvent types in industrial production, this has reduced the associated risks of mixed exposure. Among all detected VOCs, methanol has the highest detection rate. However, among the eight high-risk VOCs, toluene, ethylbenzene, xylene, and n-hexane have relatively high detection rates. Trichloroethylene and xylene pose high health risks to workers (HQ > 1), and industries should strengthen monitoring of these two VOCs.

This study also has certain limitations, the most important one is that no biological samples were collected from workers for detection, making it impossible to directly determine the exposure effects of VOCs on workers. We only use HQ values to reflect the potential hazards of VOCs to workers. The hazards of VOCs and their occupational exposure risks should continue to be given attention to, with ongoing research efforts aimed at gaining a deeper understanding.

**Supplementary Materials:** The following supporting information can be downloaded at https://www.mdpi.com/article/10.3390/toxics12090634/s1, Table S1. Detection-related information for high-risk VOCs; Table S2: Industry classification and sectors; Table S3: Enterprise and sample situation of organic solvent detection in different industries from 2018 to 2023; Table S4: The top ten VOCs with high detection rates in common organic solvent samples; Table S5: Detection of volatile components in different categories of organic solvents from 2018–2023.

**Author Contributions:** Methodology, Y.G. (Yijia Guo); validation, L.Z. (Lihua Zhu) and L.Z. (Liyin Zhang); formal analysis, F.L. and J.Y.; investigation, X.L.; writing—original draft, X.T. and Y.G. (Yiming Ge); J.C. data curation; S.L. funding acquisition; X.Z. project administration. All authors have read and agreed to the published version of the manuscript.

**Funding:** Please add: This research was funded by the Bao'an District Medical and Health Research Project (2023JD143) and the National Natural Science Foundation of China (No. 42277424).

**Institutional Review Board Statement:** Not applicable.

**Informed Consent Statement:** Not applicable.

**Data Availability Statement:** Data are available from the corresponding author by request.

**Acknowledgments:** Acknowledgments are due to those who contributed to this research.

**Conflicts of Interest:** The authors declare that they have no known competing financial interests or personal relationships that could have appeared to influence the work reported in this paper.

## References

1. Joshi, D.R.; Adhikari, N. An Overview on Common Organic Solvents and Their Toxicity. *J. Pharm. Res. Int.* **2019**, *28*, 1–18. [CrossRef]
2. Durrani, T.; Clapp, R.; Harrison, R.; Shusterman, D. Solvent-Based Paint and Varnish Removers: A Focused Toxicologic Review of Existing and Alternative Constituents. *J. Appl. Toxicol.* **2020**, *40*, 1325–1341. [CrossRef]
3. Mo, Z.; Cui, R.; Yuan, B.; Cai, H.; McDonald, B.C.; Li, M.; Zheng, J.; Shao, M. A Mass-Balance-Based Emission Inventory of Non-Methane Volatile Organic Compounds (NMVOCs) for Solvent Use in China. *Atmos. Chem. Phys.* **2021**, *21*, 13655–13666. [CrossRef]
4. Stockwell, C.E.; Coggon, M.M.; Gkatzelis, G.I.; Ortega, J.; McDonald, B.C.; Peischl, J.; Aikin, K.; Gilman, J.B.; Trainer, M.; Warneke, C. Volatile Organic Compound Emissions from Solvent- and Water-Borne Coatings—Compositional Differences and Tracer Compound Identifications. *Atmos. Chem. Phys.* **2021**, *21*, 6005–6022. [CrossRef]
5. Sheldon, R.A. Green Solvents for Sustainable Organic Synthesis: State of the Art. *Green Chem.* **2005**, *7*, 267–278. [CrossRef]
6. Karami, S.; Bassig, B.; Stewart, P.A.; Lee, K.-M.; Rothman, N.; Moore, L.E.; Lan, Q. Occupational Trichloroethylene Exposure and Risk of Lymphatic and Haematopoietic Cancers: A Meta-Analysis. *Occup. Environ. Med.* **2013**, *70*, 591–599. [CrossRef]
7. Spencer, P.S.; Schaumburg, H.H. Organic Solvent Neurotoxicity. Facts and Research Needs. *Scand. J. Work Environ. Health* **1985**, *11* (Suppl. S1), 53–60.
8. Vlaanderen, J.; Straif, K.; Pukkala, E.; Kauppinen, T.; Kyyrönen, P.; Martinsen, J.I.; Kjaerheim, K.; Tryggvadottir, L.; Hansen, J.; Sparén, P.; et al. Occupational Exposure to Trichloroethylene and Perchloroethylene and the Risk of Lymphoma, Liver, and Kidney Cancer in Four Nordic Countries. *Occup. Environ. Med.* **2013**, *70*, 393–401. [CrossRef] [PubMed]
9. Baker, E.L. A Review of Recent Research on Health Effects of Human Occupational Exposure to Organic Solvents. A Critical Review. *J. Occup. Med.* **1994**, *36*, 1079–1092. [CrossRef]
10. Chao, K.-P.; Wang, P.; Lin, C.-H. Estimation of Diffusion Coefficients and Solubilities for Organic Solvents Permeation through High-Density Polyethylene Geomembrane. *J. Environ. Eng.* **2006**, *132*, 519–526. [CrossRef]
11. Lee, S.C.; Chiu, M.Y.; Ho, K.F.; Zou, S.C.; Wang, X. Volatile Organic Compounds (VOCs) in Urban Atmosphere of Hong Kong. *Chemosphere* **2002**, *48*, 375–382. [CrossRef]
12. Cheng, L.; Wei, W.; Guo, A.; Zhang, C.; Sha, K.; Wang, R.; Wang, K.; Cheng, S. Health Risk Assessment of Hazardous VOCs and Its Associations with Exposure Duration and Protection Measures for Coking Industry Workers. *J. Clean. Prod.* **2022**, *379*, 134919. [CrossRef]
13. Liu, Y.; Li, S.; Wang, Q.; Zheng, X.; Zhao, Y.; Lu, W. Occupational Health Risks of VOCs Emitted from the Working Face of Municipal Solid Waste Landfill: Temporal Variation and Influencing Factors. *Waste Manag.* **2023**, *160*, 173–181. [CrossRef] [PubMed]
14. Yan, Y.; Peng, L.; Cheng, N.; Bai, H.; Mu, L. Health Risk Assessment of Toxic VOCs Species for the Coal Fire Well Drillers. *Environ. Sci. Pollut. Res. Int.* **2015**, *22*, 15132–15144. [CrossRef]
15. Lamplugh, A.; Harries, M.; Xiang, F.; Trinh, J.; Hecobian, A.; Montoya, L.D. Occupational Exposure to Volatile Organic Compounds and Health Risks in Colorado Nail Salons. *Environ. Pollut.* **2019**, *249*, 518–526. [CrossRef]
16. Lin, N.; Rosemberg, M.-A.; Li, W.; Meza-Wilson, E.; Godwin, C.; Batterman, S. Occupational Exposure and Health Risks of Volatile Organic Compounds of Hotel Housekeepers: Field Measurements of Exposure and Health Risks. *Indoor Air* **2021**, *31*, 26–39. [CrossRef]
17. Wolkoff, P.; Wilkins, C.K.; Clausen, P.A.; Nielsen, G.D. Organic Compounds in Office Environments—Sensory Irritation, Odor, Measurements and the Role of Reactive Chemistry. *Indoor Air* **2006**, *16*, 7–19. [CrossRef] [PubMed]
18. Colman Lerner, J.E.; Sanchez, E.Y.; Sambeth, J.E.; Porta, A.A. Characterization and Health Risk Assessment of VOCs in Occupational Environments in Buenos Aires, Argentina. *Atmos. Environ.* **2012**, *55*, 440–447. [CrossRef]

19. Vargas-Ramos, Y.E.; Marrugo-Negrete, J.L. Exposure to VOCs in furniture factories in two populations in northern Colombia. *Rev. Salud Publica* **2014**, *16*, 834–846.
20. Xu, H.; Li, Y.; Feng, R.; He, K.; Ho, S.S.H.; Wang, Z.; Ho, K.F.; Sun, J.; Chen, J.; Wang, Y.; et al. Comprehensive Characterization and Health Assessment of Occupational Exposures to Volatile Organic Compounds (VOCs) in Xi'an, a Major City of Northwestern China. *Atmos. Environ.* **2021**, *246*, 118085. [CrossRef]
21. Li, X.; Tu, H.; Chen, J.; Yu, H.; Zhang, R.; Chen, P.; Hu, S. Trends of 7 Organic Solvent-Induced Occupational Diseases in Guangdong, from 2006 to 2015. *Biomed. Environ. Sci.* **2020**, *33*, 862–866. [CrossRef] [PubMed]
22. Li, X.; Qu, H.; Hu, S.; Chen, J.; Tu, H.; Wen, X.; Yu, H.; Zhou, S.; Qi, Y. Study on the Epidemic Characteristics and Trends of Occupational Chemical Poisoning in Guangdong Province. *China Occup. Med.* **2018**, *45*, 436–442.
23. Wang, L.; Weng, S.; Zhu, Z.; Chen, Z.; Dai, Z.; Feng, J. Characteristic and trend analysis of occupational diseases from 2006 to 2021 in Bao'anDistrict of Shenzhen. *Ind. Health Occup. Dis.* **2023**, *49*, 215–219. [CrossRef]
24. Chen, S.; Wang, S.; Lin, B. Study on the Economic Burden of Occupational Poisoning Induced by Organic Solvents in Baoan District of Shenzhen City. *Chin. Occup. Med.* **2005**, *32*, 15–17. [CrossRef]
25. Stephan-Recaido, S.C.; Peckham, T.K.; Lavoué, J.; Baker, M.G. Characterizing the Burden of Occupational Chemical Exposures by Sociodemographic Groups in the United States, 2021. *Am. J. Public Health* **2024**, *114*, 57–67. [CrossRef]
26. Hoang, A.; Fagan, K.; Cannon, D.L.; Rayasam, S.D.G.; Harrison, R.; Shusterman, D.; Singla, V. Assessment of Methylene Chloride-Related Fatalities in the United States, 1980–2018. *JAMA Intern. Med.* **2021**, *181*, 797–805. [CrossRef] [PubMed]
27. Mo, Z.; Lu, S.; Shao, M. Volatile Organic Compound (VOC) Emissions and Health Risk Assessment in Paint and Coatings Industry in the Yangtze River Delta, China. *Environ. Pollut.* **2021**, *269*, 115740. [CrossRef]
28. Şahin, Ü.A.; Oğur, N.E.; Ayvaz, C.; Dumanoğlu, Y.; Onat, B.; Uzun, B.; Özkaya, F.; Akın, Ö. Volatile Organic Compound Concentrations under Two Different Ventilation Structures and Their Health Risks in the Adhesive Tape Manufacturing Workplace. *Air Qual. Atmos. Health* **2023**, *16*, 2177–2191. [CrossRef]
29. Zhu, J.; Su, S.; Wen, C.; Wang, T.; Xu, H.; Liu, M. Application of Multiple Occupational Health Risk Assessment Models in the Prediction of Occupational Health Risks of N-Hexane in the Air-Conditioned Closed Workshop. *Front. Public Health* **2022**, *10*, 1017718. [CrossRef]
30. Mai, J.-L.; Yang, W.-W.; Zeng, Y.; Guan, Y.-F.; Chen, S.-J. Volatile Organic Compounds (VOCs) in Residential Indoor Air during Interior Finish Period: Sources, Variations, and Health Risks. *Hyg. Environ. Health Adv.* **2024**, *9*, 100087. [CrossRef]
31. Huang, Q.; Su, S.; Zhang, X.; Li, X.; Zhu, J.; Wang, T.; Wen, C. Occupational Health Risk Assessment of Workplace Solvents and Noise in the Electronics Industry Using Three Comprehensive Risk Assessment Models. *Front. Public Health* **2023**, *11*, 1063488. [CrossRef] [PubMed]
32. Xiang, Y.; Huang, Y.; Zhong, X.; Zhou, W. Present status of occupational hazards by organic solvents and its risk assessment in wood furniture manufacturing industry in Shenzhen city. *Chin. J. Ind. Med.* **2023**, *36*, 442–445. [CrossRef]
33. Sutton, P.; Wolf, K.; Quint, J. Implementing Safer Alternatives to Lithographic Cleanup Solvents to Protect the Health of Workers and the Environment. *J. Occup. Environ. Hyg.* **2009**, *6*, 174–187. [CrossRef]
34. Bates, M.N.; Reed, B.R.; Liu, S.; Eisen, E.A.; Hammond, S.K. Solvent Exposure and Cognitive Function in Automotive Technicians. *Neurotoxicology* **2016**, *57*, 22–30. [CrossRef] [PubMed]
35. Guangdong Occupational Health Technical Quality Control Center. *Technical Guidelines of Guangdong Occupational Health Technology Quality Control Center*; Guangdong Occupational Health Technical Quality Control Center: Guangdong, China, 2020. Available online: https://zyjk.xinanli.com/d/file/2024-03-19/GDOHTQC%20003%E2%80%942021%20%E5%B7%A5%E4%BD%9C%E5%9C%BA%E6%89%80%E7%A9%BA%E6%B0%94%E4%B8%AD%E4%B8%99%E7%83%AF%E9%85%B8%E7%94%B2%E9%85%AF%E7%9A%84%E6%BA%B6%E5%89%82%E8%A7%A3%E5%90%B8-%E6%B0%94%E7%9B%B8%E8%89%B2%E8%B0%B1%E6%B3%95.pdf (accessed on 28 July 2024).
36. Chen, H.; Qiu, X. Analysis on morbidity status of occupational diseases in Shenzhen, 2006–2012. *China Occup. Med.* **2014**, *41*, 478–480.
37. Li, Y.; Xie, X.; Wang, J.; Du, W.; Liu, Y. Analysis on volatile organic ingredients in adhesives and solvents in 14 cases of 1,2-dichloroethane poisoning accidents. *China Occup. Med.* **2014**, *41*, 602–604.
38. Yu, X.; Qiu, X.; Bian, H.; Zhang, S.; Zhu, Z.; Wu, J. Analysis on occupational disease in Shajing Street Baoan District of Shenzhen from 2003-2012. *Occup. Health* **2014**, *30*, 2303–2305. [CrossRef]
39. Zhu, L.; Guan, X.; Peng, Y.; Li, J.; Zhang, X.; Gong, A.; Li, M.; Xie, H.; Chen, S.; Li, J.; et al. Characterization of VOCs Emissions and Associated Health Risks Inherent to the Packaging and Printing Industries in Shandong Province, China. *Sci. Total Environ.* **2024**, *946*, 174108. [CrossRef]
40. National Bureau of Statistics of China. *Industrial Classification for National Economic Activities*; National Bureau of Statistics of China: Beijing, China, 2017. Available online: https://www.stats.gov.cn/xxgk/tjbz/gjtjbz/201710/t20171017_1758922.html (accessed on 25 July 2024).
41. Constable, D.J.C.; Jimenez-Gonzalez, C.; Henderson, R.K. Perspective on Solvent Use in the Pharmaceutical Industry. *Org. Process Res. Dev.* **2007**, *11*, 133–137. [CrossRef]
42. Chaniago, Y.D.; Minh, L.Q.; Khan, M.S.; Koo, K.-K.; Bahadori, A.; Lee, M. Optimal Design of Advanced Distillation Configuration for Enhanced Energy Efficiency of Waste Solvent Recovery Process in Semiconductor Industry. *Energy Convers. Manag.* **2015**, *102*, 92–103. [CrossRef]

43. Nayak, J.; Mishra, M.; Naik, B.; Swapnarekha, H.; Cengiz, K.; Shanmuganathan, V. An Impact Study of COVID-19 on Six Different Industries: Automobile, Energy and Power, Agriculture, Education, Travel and Tourism and Consumer Electronics. *Expert. Syst.* **2022**, *39*, e12677. [CrossRef] [PubMed]
44. Chaniago, Y.D.; Harvianto, G.R.; Bahadori, A.; Lee, M. Enhanced Recovery of PGME and PGMEA from Waste Photoresistor Thinners by Heterogeneous Azeotropic Dividing-Wall Column. *Process Saf. Environ. Prot.* **2016**, *103*, 413–423. [CrossRef]
45. Geldermann, J.; Peters, N.-H.; Nunge, S.; Rentz, O. Best Available Techniques in the Sector of Adhesives Application. *Int. J. Adhes. Adhes.* **2004**, *24*, 85–91. [CrossRef]
46. Giagnorio, M.; Amelio, A.; Grüttner, H.; Tiraferri, A. Environmental Impacts of Detergents and Benefits of Their Recovery in the Laundering Industry. *J. Clean. Prod.* **2017**, *154*, 593–601. [CrossRef]
47. Baker, H.R.; Leach, P.B.; Singleterry, C.R.; Zisman, W.A. Cleaning by surface displacement of water and oils. *Ind. Eng. Chem.* **1967**, *59*, 29–40. [CrossRef]
48. Joe Lopes, A.; MacDonald, E.; Wicker, R.B. Integrating Stereolithography and Direct Print Technologies for 3D Structural Electronics Fabrication. *Rapid Prototyp. J.* **2012**, *18*, 129–143. [CrossRef]
49. Beltrão, M.; Duarte, F.M.; Viana, J.C.; Paulo, V. A Review on In-Mold Electronics Technology. *Polym. Eng. Sci.* **2022**, *62*, 967–990. [CrossRef]
50. NHS. *Occupational Exposure Limits for Hazardous Agents in the Workplace—Part 1: Chemical Hazardous Agents*; NHS: London, UK, 2019.
51. Huang, L.; Peng, Z.; Chi, H.; Zhang, X.; Yin, Q. Analysis of harmful volatile components of organic solvents in workplace of Luohu district of Shenzhen. *Chin. J. Public Health Manag.* **2014**, *30*, 593–594. [CrossRef]
52. Huang, Z.; Liang, N.; Cai, Z.; Zhang, N.; Chen, Z.; Zheng, Y.; Luo, X.; Yuan, L.; Chen, T. Determination of organic solvents in small and medium-sized enterprises of Longgang District, Shenzhen in 2019. *Hainan Med. J.* **2021**, *32*, 225–227.
53. Zhong, W.; Xu, X.; Yin, J. Analysis of main volatile organic compounds in chemicals used by industrial enterprises in Shenzhen. *Occup. Health Emerg. Rescue* **2020**, *38*, 478–481. [CrossRef]
54. Ji, L.; Tan, C.; Yu, B. Analyzing the prevalence of key occupational hazards in 56 printing enterprises in Shenzhen City. *China Occup. Med.* **2022**, *49*, 596–600. [CrossRef]
55. Wang, L.; Zhu, Z.; Dai, Z.; Feng, J.; Weng, S. Analysis of volatile organic components of organic solvents used in Bao'an District of Shenzhen. *China J. Ind. Hyg. Occup. Dis.* **2022**, *40*, 867–871. [CrossRef]
56. Yang, G.; Xiang, Y.; Zhu, X.; Zhou, W. Status survey on occupational disease hazards of organic solvents in 39 electronic enterprises in Shenzhen City. *China Occup. Med.* **2019**, *46*, 403–406.
57. Chan, A.P.L.; Chan, T.Y.K. Methanol as an Unlisted Ingredient in Supposedly Alcohol-Based Hand Rub Can Pose Serious Health Risk. *Int. J. Environ. Res. Public Health* **2018**, *15*, 1440. [CrossRef]
58. Dear, K.; Grayson, L.; Nixon, R. Potential Methanol Toxicity and the Importance of Using a Standardised Alcohol-Based Hand Rub Formulation in the Era of COVID-19. *Antimicrob. Resist. Infect. Control* **2020**, *9*, 129. [CrossRef]
59. Zakharov, S.; Hlusicka, J.; Nurieva, O.; Kotikova, K.; Lischkova, L.; Kacer, P.; Kacerova, T.; Urban, P.; Vaneckova, M.; Seidl, Z.; et al. Neuroinflammation Markers and Methyl Alcohol Induced Toxic Brain Damage. *Toxicol. Lett.* **2018**, *298*, 60–69. [CrossRef] [PubMed]
60. Boogaard, P.J. Human Biomonitoring of Low-Level Benzene Exposures. *Crit. Rev. Toxicol.* **2022**, *52*, 799–810. [CrossRef] [PubMed]
61. Wang, J.; Ma, Y.; Tang, L.; Li, D.; Xie, J.; Sun, Y.; Tian, Y. Long-Term Exposure to Low Concentrations of Ambient Benzene and Mortality in a National English Cohort. *Am. J. Respir. Crit. Care Med.* **2024**, *209*, 987–994. [CrossRef]
62. He, Y.; Qiu, H.; Wang, W.; Lin, Y.; Ho, K.F. Exposure to BTEX Is Associated with Cardiovascular Disease, Dyslipidemia and Leukocytosis in National US Population. *Sci. Total Environ.* **2024**, *919*, 170639. [CrossRef]
63. He, Y.; Lin, Y.; Qiu, H.; Wu, L.; Ho, K.F. Low-Dose Blood BTEX Are Associated with Pulmonary Function through Changes in Inflammatory Markers among US Adults: NHANES 2007–2012. *Environ. Sci. Pollut. Res. Int.* **2023**, *30*, 69064–69079. [CrossRef]
64. Werder, E.J.; Beier, J.I.; Sandler, D.P.; Falkner, K.C.; Gripshover, T.; Wahlang, B.; Engel, L.S.; Cave, M.C. Blood BTEXS and Heavy Metal Levels Are Associated with Liver Injury and Systemic Inflammation in Gulf States Residents. *Food Chem. Toxicol.* **2020**, *139*, 111242. [CrossRef]
65. Werder, E.J.; Engel, L.S.; Blair, A.; Kwok, R.K.; McGrath, J.A.; Sandler, D.P. Blood BTEX Levels and Neurologic Symptoms in Gulf States Residents. *Environ. Res.* **2019**, *175*, 100–107. [CrossRef] [PubMed]
66. Attarchi, M.; Golabadi, M.; Labbafinejad, Y.; Mohammadi, S. Combined Effects of Exposure to Occupational Noise and Mixed Organic Solvents on Blood Pressure in Car Manufacturing Company Workers. *Am. J. Ind. Med.* **2013**, *56*, 243–251. [CrossRef] [PubMed]
67. Choi, Y.-H.; Kim, K. Noise-Induced Hearing Loss in Korean Workers: Co-Exposure to Organic Solvents and Heavy Metals in Nationwide Industries. *PLoS ONE* **2014**, *9*, e97538. [CrossRef] [PubMed]
68. Jiang, W.; Zheng, D.; Qu, L. Effect of benzene series combined with noise on hearing. *Occup. Health* **2016**, *32*, 305–307. [CrossRef]
69. Li, X.; Dong, Q.; Wei, C.; Ding, L.; Song, H. Effect of noise combined with high temperature and n-hexane on human hearing in an engine manufacturer in 2015. *Occup. Health* **2016**, *32*, 308–310. [CrossRef]
70. Jin, J.; Xie, C.; Gao, J.; Wang, H.; Zhang, J.; Zhao, Y.; Gao, M.; Ma, J.; Wang, Z.; Guan, J. Elucidating the Toluene Formation Mechanism in the Reaction of Propargyl Radical with 1,3-Butadiene. *Phys. Chem. Chem. Phys.* **2023**, *25*, 13136–13144. [CrossRef]

71. Pan, S.; Li, X.; Xu, X.; Zhang, D.; Xu, Z. Synthesis and Application of Quaternary Amine-Functionalized Core-Shell-Shell Magnetic Polymers for Determination of Metabolites of Benzene, Toluene and Xylene in Human Urine Samples and Study of Exposure Assessment. *J. Chromatogr. A* **2023**, *1708*, 464320. [CrossRef]
72. Partha, D.B.; Cassidy-Bushrow, A.E.; Huang, Y. Global Preterm Births Attributable to BTEX (Benzene, Toluene, Ethylbenzene, and Xylene) Exposure. *Sci. Total Environ.* **2022**, *838*, 156390. [CrossRef]
73. Li, X.; Yu, T.; Wang, S.; Wang, Q.; Li, M.; Liu, Z.; Xie, K. Diallyl Sulfide-Induced Attenuation of n-Hexane-Induced Peripheral Nerve Impairment Is Associated with Metabolic Inhibition of n-Hexane. *Food Chem. Toxicol.* **2020**, *137*, 111167. [CrossRef]
74. Piao, F.; Chen, Y.; Yu, L.; Shi, X.; Liu, X.; Jiang, L.; Yang, G.; Wang, N.; Gao, B.; Zhang, C. 2,5-Hexanedione-Induced Deregulation of Axon-Related microRNA Expression in Rat Nerve Tissues. *Toxicol. Lett.* **2020**, *320*, 95–102. [CrossRef]
75. Br, D.M.; Jt, G. Trichloroethylene, a Ubiquitous Environmental Contaminant in the Risk for Parkinson's Disease. *Environ. Sci. Process. Impacts* **2020**, *22*, 543–554. [CrossRef]
76. Horzmann, K.A.; Portales, A.M.; Batcho, K.G.; Freeman, J.L. Developmental Toxicity of Trichloroethylene in Zebrafish (*Danio rerio*). *Environ. Sci. Process. Impacts* **2020**, *22*, 728–739. [CrossRef] [PubMed]
77. Xie, H.; Gao, W.; Zhao, W.; Han, Y.; Gao, Y.; Liu, B.; Han, Y. Source Profile Study of VOCs Unorganized Emissions from Typical Aromatic Devices in Petrochemical Industry. *Sci. Total Environ.* **2023**, *889*, 164098. [CrossRef] [PubMed]
78. Mu, J.; Zhang, Y.; Xia, Z.; Fan, G.; Zhao, M.; Sun, X.; Liu, Y.; Chen, T.; Shen, H.; Zhang, Z.; et al. Two-Year Online Measurements of Volatile Organic Compounds (VOCs) at Four Sites in a Chinese City: Significant Impact of Petrochemical Industry. *Sci. Total Environ.* **2023**, *858*, 159951. [CrossRef] [PubMed]
79. Keawboonchu, J.; Thepanondh, S.; Kultan, V.; Pinthong, N.; Malakan, W.; Robson, M.G. Integrated Sustainable Management of Petrochemical Industrial Air Pollution. *Int. J. Environ. Res. Public Health* **2023**, *20*, 2280. [CrossRef] [PubMed]
80. Cheng, N.; Jing, D.; Zhang, C.; Chen, Z.; Li, W.; Li, S.; Wang, Q. Process-Based VOCs Source Profiles and Contributions to Ozone Formation and Carcinogenic Risk in a Typical Chemical Synthesis Pharmaceutical Industry in China. *Sci. Total Environ.* **2021**, *752*, 141899. [CrossRef]
81. You, G.; Jin, Z.; Lu, S.; Ren, J.; Zhang, Y.; Hu, K.; Xie, S. Emission Factors and Source Profiles of Volatile Organic Compounds from the Automobile Manufacturing Industry. *Sci. Total Environ.* **2024**, *927*, 172183. [CrossRef]
82. Ghobakhloo, S.; Khoshakhlagh, A.H.; Morais, S.; Mazaheri Tehrani, A. Exposure to Volatile Organic Compounds in Paint Production Plants: Levels and Potential Human Health Risks. *Toxics* **2023**, *11*, 111. [CrossRef]
83. Cruz, S.L.; Rivera-García, M.T.; Woodward, J.J. Review of Toluene Action: Clinical Evidence, Animal Studies and Molecular Targets. *J. Drug Alcohol. Res.* **2014**, *3*, 235840. [CrossRef]
84. Druzhinina, E.S.; Kozyreva, A.A.; Bembeeva, R.T.; Kozlovsky, A.S.; Sokolova, V.E.; Isaev, I.V.; Narbutov, A.G.; Zavadenko, N.N.; Tikhonova, O.A. Neuropathy in N-Hexane Poisoning. *S.S. Korsakov J. Neurol. Psychiatry* **2024**, *124*. [CrossRef] [PubMed]
85. Mokammel, A.; Rostami, R.; Niazi, S.; Asgari, A.; Fazlzadeh, M. BTEX Levels in Rural Households: Heating System, Building Characteristic Impacts and Lifetime Excess Cancer Risk Assessment. *Environ. Pollut.* **2022**, *298*, 118845. [CrossRef]
86. Shanh, F.G.; Rahimnejad, S.; Bahrami, A.; Farhadian, M. Risk Assessment of Workers' Exposure to Volatile Organic Compounds in the Air of a Petrochemical Complex in Iran. *Indian. J. Occup. Environ. Med.* **2017**, *21*, 121–127. [CrossRef]
87. Chen, R.; Li, T.; Huang, C.; Yu, Y.; Zhou, L.; Hu, G.; Yang, F.; Zhang, L. Characteristics and Health Risks of Benzene Series and Halocarbons near a Typical Chemical Industrial Park. *Environ. Pollut.* **2021**, *289*, 117893. [CrossRef]
88. Hu, R.; Liu, G.; Zhang, H.; Xue, H.; Wang, X. Levels, Characteristics and Health Risk Assessment of VOCs in Different Functional Zones of Hefei. *Ecotoxicol. Environ. Saf.* **2018**, *160*, 301–307. [CrossRef] [PubMed]
89. Khajeh Hoseini, L.; Jalilzadeh Yengejeh, R.; Mohammadi Rouzbehani, M.; Sabzalipour, S. Health Risk Assessment of Volatile Organic Compounds (VOCs) in a Refinery in the Southwest of Iran Using SQRA Method. *Front. Public Health* **2022**, *10*, 978354. [CrossRef] [PubMed]
90. Khoshakhlagh, A.H.; Yazdanirad, S.; Mousavi, M.; Gruszecka-Kosowska, A.; Shahriyari, M.; Rajabi-Vardanjani, H. Summer and Winter Variations of BTEX Concentrations in an Oil Refinery Complex and Health Risk Assessment Based on Monte-Carlo Simulations. *Sci. Rep.* **2023**, *13*, 10670. [CrossRef]
91. Su, M.; Sun, R.; Zhang, X.; Wang, S.; Zhang, P.; Yuan, Z.; Liu, C.; Wang, Q. Assessment of the Inhalation Risks Associated with Working in Printing Rooms: A Study on the Staff of Eight Printing Rooms in Beijing, China. *Environ. Sci. Pollut. Res. Int.* **2018**, *25*, 17137–17143. [CrossRef]

*toxics*

*Article*

# Bisphenol Chemicals in Surface Soil from E-Waste Dismantling Facilities and the Surrounding Areas: Spatial Distribution and Health Risk

**Lei Zhao [1], Fengli Zhou [1], Shuyue Wang [1], Yan Yang [2,3,4], Haojia Chen [2,3,4], Xufang Ma [1] and Xiaotu Liu [1,*]**

1 Guangdong Key Laboratory of Environmental Pollution and Health, College of Environment and Climate, Jinan University, Guangzhou 510632, China; lei_zhao1998@163.com (L.Z.); zhoufengli@stu2020.jnu.edu.cn (F.Z.); wangshuyue@stu2022.jnu.edu.cn (S.W.); xufang_ma2002@163.com (X.M.)
2 School of Environmental Science and Engineering, Institute of Environmental Health and Pollution Control, Guangdong University of Technology, Guangzhou 510006, China; yangyan1209@gdut.edu.cn (Y.Y.); chenhaojia_gdut@163.com (H.C.)
3 Synergy Innovation Institute of Guangdong University of Technology, Shantou 515041, China
4 Chemistry and Chemical Engineering Guangdong Laboratory, Shantou 515041, China
* Correspondence: liuxiaotu@jnu.edu.cn

**Abstract:** Electronic waste (e-waste) dismantling facilities are well-known bisphenol chemical (BP) sources. In this study, non-targeted screening combined with targeted analysis of BPs in surface soil from e-waste dismantling facilities and their surroundings revealed their presence, distribution, and exposure risk. A total of 14 BPs were identified including bisphenol A (BPA) and its novel structural analogs and halogenated BPs. The total concentrations of BPs ranged from 963 to 47,160 ng/g (median: 6970 ng/g) in e-waste soil, higher than those measured in surface soil from surrounding areas, i.e., 10–7750 ng/g (median 197 ng/g). BPA, tetrabromobisphenol A (TBBPA), and bisphenol F (BPF) were the dominant ones from the two areas. Concentrations of TBBPA and its debromination product from the surrounding area significantly decreased with increasing distances from the e-waste dismantling facilities. Estimation of daily intake via oral ingestion of soil suggests that current contamination scenarios are unlikely to pose health risks for e-waste dismantling workers and adults and toddlers living in the surrounding areas, with their intakes generally well below the tolerable daily intakes proposed for several BPs. However, the BPA intakes of workers exceeded the more strict tolerable daily intake for BPA established recently, which merits continuous environmental surveillance.

**Keywords:** bisphenol chemicals; e-waste dismantling facilities; surface soil; daily intake

## 1. Introduction

Bisphenol chemicals (BPs) refer generically to compounds that possess two phenol groups in their structure [1]. Among the BPs documented for use in industrial applications, bisphenol A (BPA) stands out as one of the most recognized compounds globally [2]. BPA is widely used in the synthesis of epoxy resin and polycarbonate plastic as an organic synthetic additive, as well as in plastic items including toys and drinking containers [3]. Increasing demand for plastic materials has resulted in ubiquitous environmental distributions of BPA [4]. Increasing evidence suggests that BPA could potentially have harmful effects on humans, ranging from endocrine disruption and developmental toxicity to carcinogenicity, obesity, and reproductive toxicity [5–7]. Therefore, the use of BPA is restricted in many countries due to concerns about health and ecological risks from exposure [1,7].

The regulation of BPA has led to the use of an increasing number of alternatives. Over 200 bisphenol alternatives with structural similarities to BPA have been documented [8]. Due to structural similarities, some of those alternatives, such as bisphenol F (BPF) and bisphenol S (BPS), have also been found to have endocrine-disrupting effects [9]. This sparked concerns regarding the possible environmental exposures and health risks of the diverse range of these alternatives. In addition, halogenated BPs, such as tetrabromobisphenol A (TBBPA), are also an important class of BPs and have raised concerns recently [10]. Although there have been studies focusing on the environmental presence and human exposure to bisphenols other than BPA [1], due to the complexity of BPs, there might be a large number of unknown BPs that have not yet been identified.

Electronic waste (e-waste) disassembly is a significant source of many additive-based substances [11]. E-waste contains up to 30% plastic materials and epoxy resins (by weight), which are the main uses of BPs [12]. The e-waste dismantling process might result in the uncontrolled release of large quantities of these substances into the environment [13]. Previous studies have found high levels of BPA and its analog alternatives and halogenated derivatives of BPA in various environmental matrices around e-waste dismantling areas [14–16]. However, no study has systematically screened BPs in e-waste dismantling areas, and information remains limited other than BPA and TBBPA. In addition, as a significant source, the influences of e-waste disassembly activities on the concentrations of BPs in the surrounding environment have rarely been studied.

Recently, we have developed a non-targeted screening strategy D-ISF for the identification of BPs based on dansyl chloride (DnsCl) derivatization and positive electrospray ionization high-resolution mass spectrometry (HRMS) in-source fragmentation [17]. The strategy largely enhanced the detection sensitivities and accuracy by introducing easily ionizable functional groups to BPs and generating characteristic fragments. In light of D-ISF, this study aims to (1) screen BPs in surface soil samples collected from two typical e-waste dismantling sites in South China in a non-targeted manner; (2) characterize the concentrations and spatial distribution of all identified BPs in surface soil from e-waste dismantling facilities and the surrounding areas; and (3) estimate the exposure risks of BPs for occupation workers and residents living around the dismantling sites.

## 2. Materials and Methods

### 2.1. Sample Collection

Two typical large-scale e-waste dismantling parks in Qingyuan and Shantou, South China, were selected, and a total of 24 surface soil samples were gathered from various locations within the two industrial parks, including e-waste storage areas, dismantling operation zones, internal roadways, etc. Surface soil samples from the surroundings (including residential areas, commercial areas, schools, etc.) within a radius of 8 km of the e-waste dismantling park of Shantou were also collected (n = 34). Detailed information on sampling sites is listed in Table S1 and Figure S1 of Supplementary Materials. At each sampling site, soil was collected using precleaned brushes and wrapped with clean aluminum foil on consistently sunny days in 2021. Soil samples were sieved by passage through a 125 μm stainless sieve (Hogentogler & Co., Inc., Columbia, MD, USA) to remove some large stones and kept at −20 °C until analysis at the analytical laboratory.

### 2.2. Chemicals and Sample Preparation

All the standards and reagents in this study were purchased commercially, of which the purity of the standards was above 95% and the reagents were of high-performance liquid chromatography purity (details are provided in the Supplementary Materials). A previously established method for deriving dansyl chloride was applied for sample preparation. Approximately 50 mg of the soil sample was weighted and then spiked with 10 μL surrogate standards and vortexed (L600, Hunan Xiangyi Laboratory Instrument Development Co., Ltd., Changsha, China). Then, 3 mL of acetonitrile (ACN) was added for extraction. After sonication for 15 min (EFAA-DC24, ANPEL Laboratory Technologies Shanghai Inc., Shanghai, China), the extract was centrifuged at 2500 rpm for 5 min. After being repeated twice, the combined extract was concentrated to about 100 μL. The concentrated extract was then derivatized with DnsCl using the procedures described previously [17].

### 2.3. Instrumental Analysis

The screening of derivatized BPs in pooled e-waste soil using the D-ISF strategy was carried out on an ultra-performance liquid chromatography (UPLC)-HRMS device (Vanquish Flex, Thermo Fisher Scientific, Pleasanton, CA, USA) equipped with an Orbitrap Exploris 240 (Thermo Fisher Scientific, CA, USA). Detailed information was provided in our previous study [17]. The determination of identified derivatized BPs was performed on an Agilent 1290 Infinity LC coupled to a 6470 triple quadrupole MS equipped with Jet Stream Technology Ion Source electrospray ionization (AJS-ESI) (Agilent Technologies, Santa Clara, CA, USA). The conditions in the LC section, including column and mobile phases as well as the gradients, were as the same as those in the UPLC-Orbitrap Exploris 240 MS [17]. Derivatized BPs were detected under positive AJS-ESI in the multiple reaction monitoring (MRM) mode. The settings of the AJS-ESI source were as follows: the nozzle and capillary voltage were 1500 and 3000 V, respectively, and the sheath gas flow rate and temperature were 11 L/min and 250 °C, respectively.

### 2.4. Quality Assurance and Quality Control (QA/QC)

The performance of the DnSCl derivatization methods was evaluated with satisfactory results. For example, the derivatization efficiencies of target BPs ranged from 91.8% to 99.8% in solvents [17]. In this study, BPA, BPF, BPS, bisphenol E (BPE), bisphenol B (BPB), bisphenol G (BPG), and bisphenol Z (BPZ) were used as target chemicals for QA/QC. The matrix effects of derivatized BPs in soil samples ranged from 65.4 ± 5.6% to 114 ± 3.8%. The method recoveries were measured by spiking 20 ng of each of the target BPs into pooled soil samples collected from the university campus and processed in five replicates. The recoveries of the spiked BPs ranged from 65.6 ± 3.1% to 85.6 ± 3.6% after subtracting the original concentrations measured in soil. Two blanks were processed along with every ten soil samples. No target bisphenol compounds were detected in the procedural blanks. The calibration curves for the derivatized bisphenol standards exhibited linear regression coefficients > 0.99 with a range of 10 to 200 ng/mL. Additionally, the limits of detection (LODs), defined as a response 3 times the standard deviation of the noise, and the limits of quantification (LOQs), defined as a response 10 times the standard deviation, are provided in Table S2.

### 2.5. Exposure Assessment

The estimated daily intake of BPs (*EDI*, ng/kg BW/day) via oral ingestion of soil was calculated using the following equation [18]:

$$EDI = \frac{C \times DIR \times EF}{BW}$$

where C (ng/g) is the concentration of BPs in the soil, *DIR* represents the ingestion rate (g/day), *EF* indicates the exposure fraction (unitless, hours spent outdoors over a day), and *BW* stands for body weight (kg). The values of these parameters used for the assessment of e-waste dismantling workers and residents (adults and toddlers) living in the surrounding area are summarized in Table S3.

The hazard quotient (*HQ*) approach was used to estimate the potential health risk of exposure to BPs via oral ingestion:

$$HQ = \frac{EDI}{tolerable\ daily\ intakes}$$

A value of 4 μg/kg bw/day for the tolerable daily intake (TDI) of BPA was used in the study [19]. The TDI values for BPS, BPF, and TBBPA were 4.4, 3.5, and 1000 μg/kg bw/day, respectively [20,21], while TDIs for other BPs are not available.

### 2.6. Data Analysis

For BPs with a detection frequency exceeding 60%, if their measured values are below the LOQs, a value of $LOQ/\sqrt{2}$ will be assigned for statistical analysis. We used Spearman's correlation analysis (two-tailed) to determine the associations between the concentrations of individual BPs and between the concentrations and the distances from the e-waste dismantling facilities to the sampling sites (PASW Statistics 18.0, IBM Inc., Armonk, NY, USA). Additionally, for the comparisons of concentrations of BPs between soil samples from the dismantling area and the surrounding area, the Mann–Whitney U test was utilized.

## 3. Results and Discussion

### 3.1. Screening of Bisphenol Chemicals in E-Waste Soil

A total of 24 potential BP candidates were screened out of the pooled samples of e-waste soil. After manually checking and confirming MS/MS spectra, 14 BPs were identified (Table S2). Among these, BPA, BPE, BPF, BPB, BPG, BPS, BPZ, bisphenol TMC (BP-TMC), 3-monobromobisphenol A (monoBBPA), and TBBPA were confirmed and quantified using reference standards. The other four BPs whose commercial standards are not available, including monochlorobisphenol A (monoClBPA), dibromobisphenol A (DiBBPA), tetramethyl bisphenol A (TMBPA), and 3,3′,5-Tribromobisphenol A (TriBBPA), were identified as follows: their measured MS spectrum matched with the theoretical isotopologue distributions and some characteristic fragments were observed in their MS/MS spectrum (Figures S2–S5). The identified BPs were analyzed in individual samples using MRM mode in LC-MS. A pseudo-MRM method was developed for the four BPs without standards and they were semi-quantified using standards of other compounds (Table S4) [22].

### 3.2. Concentrations and Profiles of Bisphenol Chemicals in Surface Soil

The concentrations and detection frequencies (DFs) of BPs in soil samples from e-waste dismantling facilities and the surrounding areas are shown in Table 1. A total of nine BPs had DFs of 100% in e-waste samples, including BPA, BPF, BPB, TMBPA, TBBPA, monoClBPA, monoBBPA, DiBBPA, and TriBBPA, while only BPA and BPF were detected in all samples from surrounding areas. The total concentrations of all identified BPs (referred to as ΣBPs) ranged from 962 to 47,165 ng/g with a median of 6968 ng/g in e-waste soil. These were significantly elevated compared to those detected in soil samples from the surroundings, i.e., 10.3–7751 ng/g (median: 197 ng/g) ($p < 0.001$). The data suggest that BPs are widely present in the e-waste dismantling environment and that e-waste dismantling activities may be an important source of bisphenols.

BPA dominated over other BPs present in both e-waste soil and surrounding area surface soil, accounting for 53 ± 9.0% and 54 ± 10.5% of ΣBPs, respectively (Figure S6), indicating that BPA remains the most widely used bisphenol in China. The concentrations of BPA in e-waste soil ranged from 531 to 167,00 ng/g, with a median of 3350 ng/g, which were significantly higher than those observed in surrounding area samples, i.e., a median of 43 ng/g ($p < 0.001$). Our BPA concentrations were also much higher compared to previous studies which reported the concentrations in dust or soil from e-waste dismantling areas. For example, the mean concentration was 4.59 ng/g in soil samples from Taizhou [16], 49 ng/g from Xian [23], and 140 ng/g from India [14]. However, our concentration was comparable to those in indoor dust, also collected from South China (mean, 19,200 ng/g) [15]. This difference may be related to the type of waste handled by the e-waste dismantling workshop. In comparison with studies conducted in non-electronic-waste dismantling areas, the detection frequencies and concentrations in our studies were much higher. For example, our concentrations in the surrounding area soil far exceed those in soil samples collected from 21 provinces in China [24]. The notably high BPA concentrations in e-waste soil might be attributed to the burning of BPA-containing computer-printed circuit boards [25].

**Table 1.** Concentrations and detection frequencies (DF, %) of bisphenol chemicals in surface soil (ng/g) from e-waste dismantling facilities and surrounding areas in South China.

| BPs | E-Waste Dismantling Facilities (n = 24) | | | | Surrounding Areas (n = 34) | | | | p Value [3] |
|---|---|---|---|---|---|---|---|---|---|
| | DF | Mean | Median | Range | DF | Mean | Median | Range | |
| BPA | 100 | 4760 | 3350 | 532–16,740 | 100 | 460 | 84.1 | 7.20–4050 | <0.001 |
| BPF | 100 | 1170 | 920 | 30.3–4580 | 100 | 26.4 | 12.3 | 2.17–238 | <0.001 |
| BPS | 75 | 11.9 | 9.86 | <0.062–25.0 | 9 | 0.75 | <0.062 | <0.062–14.3 | — [4] |
| BP-TMC | 79 | 0.83 | 0.59 | <0.058–2.54 | 21 | 0.20 | <0.058 | <0.058–4.79 | — |
| BPG | 75 | 0.84 | 0.40 | <0.0031–3.29 | 62 | 0.021 | 0.005 | <0.0031–0.30 | <0.001 |
| BPE | 67 | 194 | 42.5 | <0.055–1060 | 21 | 0.99 | <0.055 | <0.055–13.7 | — |
| BPB | 100 | 89.8 | 11.9 | 1.05–655 | 59 | 1.08 | 0.17 | <0.0045–12.0 | — |
| BPZ | 75 | 20.3 | 3.75 | <0.064–83.4 | 65 | 0.66 | 0.21 | <0.064–9.24 | 0.005 |
| TMBPA [1] | 100 | 12.8 | 2.96 | 0.2–92.5 | 21 | <0.0087 | <0.0087 | <0.0087–4.06 | — |
| TBBPA | 100 | 2250 | 1850 | 268–6390 | 82 | 420 | 55.9 | <0.48–3850 | <0.001 |
| monoClBPA [2] | 100 | 136 | 13.7 | 0.52–991 | 12 | 0.43 | <0.48 | <0.48–7.77 | — |
| monoBBPA | 100 | 2620 | 129 | 12.2–17,140 | 74 | 28.7 | 3.74 | <0.48–320 | <0.001 |
| DiBBPA [2] | 100 | 181 | 6.85 | 1.11–1020 | 35 | 2.96 | <0.48 | <0.48–49.2 | — |
| TriBBPA [2] | 100 | 65.6 | 8.21 | 0.71–387 | 38 | 4.61 | <0.48 | <0.48–80.4 | — |

[1] semi-quantified using a reference standard of BPA; [2] semi-quantified using a reference standard of TBBPA; [3] comparison between samples from the dismantling facilities and the surrounding area; [4] detection rate below 60% without statistical analysis.

Seven structural analogs of BPA were widely detected, including BPF, BPE, BPB, BPS, BPZ, BP-TMC, and BPG. Among these, BPF was the dominant one with the highest concentrations (median, 920 ng/g and 12.3 ng/g for e-waste and surrounding area, respectively). In contrast, BPS was only detected in 75% of e-waste samples and 9% of surrounding area samples. The concentrations of BPS in e-waste soil (median, 9.86 ng/g) were much lower than those in household dust collected from Guangzhou (median, 320 ng/g) [26]. The low concentrations and detection rates of BPS in the e-waste dismantling areas were in line with previous studies [15]. A biomonitoring study showed that people living near e-waste recycling facilities contained significantly higher levels of BPF in their urine than those living in rural areas, but there was no such trend for BPS [13]. The high concentration of BPF in e-waste soil could be ascribed to the utilization of BPF as the primary alternative to BPA for epoxy resin and plastic synthesis [1]. By contrast, BPS is used in a wider range of consumer products, including paper and personal care products [27]. In addition to BPF and BPS, BP-TMC, BPG, BPE, BPB, and BPZ were rarely detected in previous studies. Their widespread detection suggests that they are also widely used in products related to e-waste.

As a structurally similar substance to BPA, TMBPA is primarily used as an intermediate in synthesizing polycarbonate resin [28]. A series of studies have compared the toxicity of BPA and TMBPA. TMBPA has been reported to be more potent than BPA in inducing lipid deposition in HepG2 cells [29], and similar to BPA in inducing estrogenic activity in human breast cancer cells [28]. However, to date, no study has investigated the environmental occurrence of TMBPA. Our data showed that TMBPA was detected in all e-waste samples with semi-quantitative concentrations of 0.2–92.5 ng/g and in 21% of samples from surrounding areas (<0.0087–4.06 ng/g). The concentrations of TMBPA exceeded those of the commonly used BPA alternatives, such as BPG and BPZ, suggesting that this emerging BP merits continuous environmental surveillance.

E-waste dismantling facilities were one of the main emission sources of halogenated BPs [30]. TBBPA and other four halogenated BPs (monoClBPA, monoBBPA, DiBBPA, TriBBPA) were 100% detected in e-waste soil. TBBPA and monoBBPA also had detection rates higher than 80% from the surrounding area. The concentrations of TBBPA in e-waste soil in our study (mean, 2250 ng/g) were comparable with those detected in e-waste outdoor dust from Taizhou (mean, 1998 ng/g) [31]. Although the median concentrations of TBBPA were higher than those of its debromination products, including monoBBPA,

DiBBPA, and TriBBPA, they exhibited extremely high concentrations in some e-waste samples. For example, the concentration of monoBBPA (17,100 ng/g) collected from a sampling site in the dismantling operation area was approximately three times higher than that of TBBPA (6390 ng/g). This may be due to the fact that debromination is likely to occur during the disposal of e-waste by processes such as heating [10]. However, in the surrounding area, the concentrations of TBBPA in surface soil were consistently higher than those of monoBBPA.

### 3.3. Spatial Distribution of BPs around E-Waste Dismantling Area and Source Implications

Spatial distributions of BPs with detection frequencies higher than 60% were analyzed to investigate the influences of e-waste dismantling activities on the BPs in the surrounding environment. Samples from the surroundings were divided into two groups according to the distances between the e-waste dismantling facilities: one was ≤3.5 km and the other was >3.5 km. Except for BPF and BPG, all BPs showed lower concentrations at sampling points with distances > 3.5 km (Figure 1a). However, only the results of monoBBPA were significant. Further, correlations between the concentrations of individual BPs and the distances were conducted. Concentrations of TBBPA and monoBBPA show significantly negative correlations with the distances, and the Spearman coefficients ($r_s$) were −0.363 ($p$ = 0.035) and −0.384 ($p$ = 0.025), respectively (Figure 1b). Although not significant, the concentrations of BPA and BPZ also decreased as the distance between the sampling sites and e-waste dismantling parks increased. Similar distribution trends with decreasing concentrations and increasing distances have also been observed for halogenated flame retardants in samples from the same e-waste recycling areas [32]. These findings further illustrate the importance of e-waste dismantling activities for the emission of BPs to the environment. Following emissions from e-waste dismantling, these BPs may enter the atmosphere and be deposited in the surrounding area through dry and wet deposition of atmospheric particulate matter. The absence of such trends for some bisphenols may be attributed to the fact that they have more diversified sources in the surrounding areas, such as consumer products.

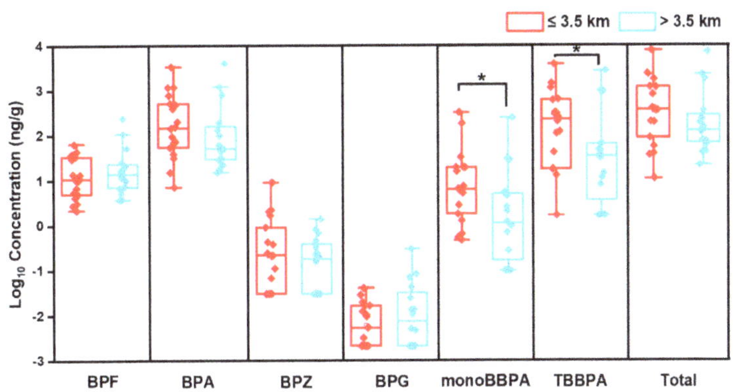


(a)

**Figure 1.** *Cont.*

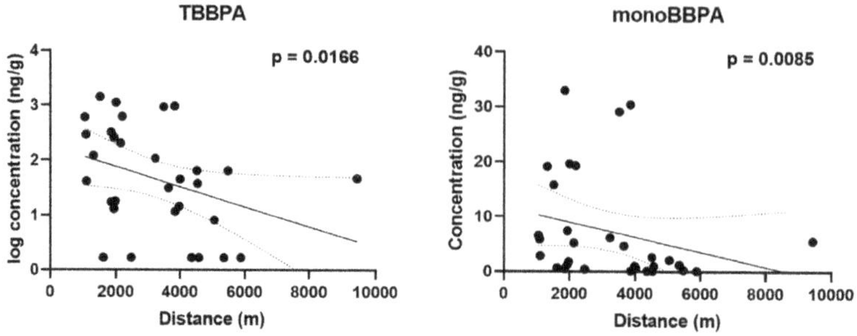


(**b**)

**Figure 1.** (**a**) Concentrations of BPs in the surface soil from surrounding areas within and beyond the distance of 3.5 km from the e-waste dismantling facilities; * indicates $p < 0.05$; (**b**) the relationship between concentrations of TBBPA and monoBBPA in the surface soil from surrounding areas and the distance from the e-waste dismantling facilities to the sampling sites.

To further explore the potential sources of BPs in the e-waste dismantling facilities and the surrounding areas, correlations between individual chemicals in soil from two regions were determined. There were significant positive correlations among all identified BPs in e-waste soil except for BPZ, with the correlation coefficient ranging from 0.44 to 0.95 (Figure S7). These findings suggest common applications of a diverse array of BPs in the materials of e-waste. The results of a recent study validate this speculation. The study showed the simultaneous detection of high concentrations of nine BP analogs in e-waste samples [33]. The absence of correlations for BPZ suggests that it has other unique sources that merit further attention. There were also significant correlations between all chemicals including BPZ in the surrounding area samples (Figure S8), suggesting they had similar sources. As the previous results show, the emissions from e-waste dismantling facilities might be the important ones.

*3.4. Risk Assessment and Health Implications*

The EDIs of BPs for e-waste dismantling workers and residents (adults and toddlers) via oral ingestions of soil were estimated (Figure 2, Table S5). Occupational workers have the highest exposure, far more than general toddlers and adults living in the surrounding areas. Taking BPA as an example, the median EDI was 2.09 ng/kg bw/day for workers, and it was almost 200 times the EDI of the adult residents (0.01 ng/kg bw/day) and 20 times the EDI of toddler residents (0.14 ng/kg bw/day). BPA and TBBPA contributed the majority of the total EDIs for the three populations. Notably, in the highest exposure scenarios, monoBBPA surpassed BPA and TBBPA as the largest contributor to the total exposure of workers, accounting for approximately 36.7% of total EDIs. However, currently, no studies have indicated the toxicity of monoBBPA, which warrants attention for future research.

In both the average and highest exposure scenarios, the EDIs of BPA were far lower than the temporary TDI (4 μg/kg·bw/day). The HQs of other BPs with available TDIs were also thousands of times lower than one (Table 2). Therefore, no considerable health risks are expected from the exposure to BPs via soil ingestion both for workers and residents. However, the EFSA recently proposed a draft TDI for BPA of 0.2 ng/kg bw/day [34]. Even in the lowest exposure scenarios, the EDI of workers (0.3 ng/kg bw/day) exceeded this draft TDI. The median EDI of toddlers (0.13 ng/kg bw/day) was also close to the new TDI, indicating potential health concerns.

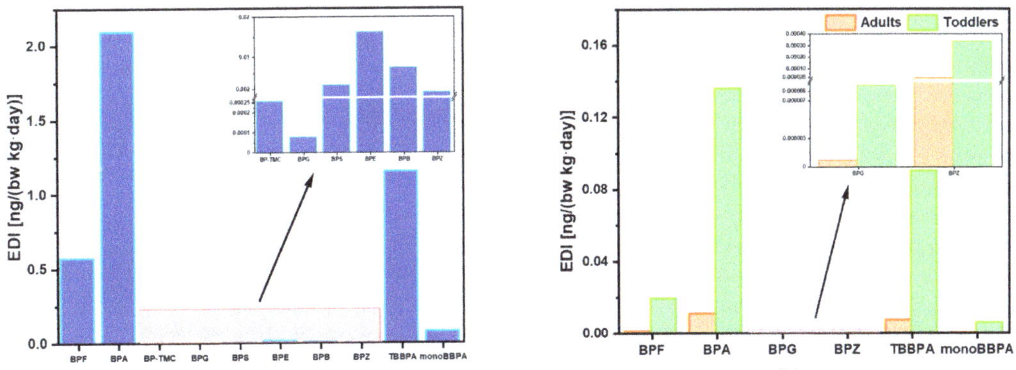


(a) (b)

**Figure 2.** Estimated daily intake (EDI; ng/kg bw/day) of BPs via oral ingestion for (**a**) e-waste dismantling workers, and (**b**) toddlers and adults living in the surrounding areas.

**Table 2.** The hazard quotients of BPs via oral ingestion for e-waste dismantling workers and toddlers and adults living in the surrounding areas.

| | E-Waste Dismantling Workers | | | Adults | | | Toddlers | | |
|---|---|---|---|---|---|---|---|---|---|
| | Min | Median | Max | Min | Median | Max | Min | Median | Max |
| BPF | $5.4 \times 10^{-6}$ | $1.6 \times 10^{-4}$ | $8.2 \times 10^{-4}$ | $8.3 \times 10^{-8}$ | $4.7 \times 10^{-7}$ | $9.1 \times 10^{-6}$ | $1.0 \times 10^{-6}$ | $5.7 \times 10^{-6}$ | $1.1 \times 10^{-4}$ |
| BPA | $8.3 \times 10^{-5}$ | $5.2 \times 10^{-4}$ | $2.6 \times 10^{-3}$ | $2.4 \times 10^{-7}$ | $2.8 \times 10^{-6}$ | $1.4 \times 10^{-4}$ | $2.9 \times 10^{-6}$ | $3.4 \times 10^{-5}$ | $1.6 \times 10^{-3}$ |
| TBBPA | $1.7 \times 10^{-7}$ | $1.2 \times 10^{-6}$ | $4.0 \times 10^{-6}$ | $2.2 \times 10^{-10}$ | $7.5 \times 10^{-9}$ | $5.1 \times 10^{-7}$ | $2.7 \times 10^{-9}$ | $9.0 \times 10^{-8}$ | $6.2 \times 10^{-6}$ |
| BPS | $6.0 \times 10^{-9}$ | $7.4 \times 10^{-7}$ | $3.6 \times 10^{-6}$ | | | | | | |

The risk of exposure to BPs estimated in this study may be much lower than the actual situation. Firstly, other exposure sources besides soil ingestion, such as inhalation, dermal absorption, and diet, were found to contribute more to internal exposure to BPs [35]. Secondly, due to the lack of corresponding reference doses for most BPs identified in this study, neglecting their contribution to risk may result in underestimation. Furthermore, the mixed toxic effects of exposure to multiple BPs may differ significantly from those of individual BPs. For example, BPA exposure along with seven other estrogenic chemicals exhibited significant estrogenic activity while each chemical was at a concentration below its effect threshold [36]. Therefore, future research should be more focused on improving the accuracy of risk evaluations and the continued regulation of occupational exposure to BPs.

## 4. Conclusions

In the present study, we performed suspect screening of BPs in e-waste soil, and a total of 14 BPs were identified. Our results demonstrated broad occurrences of BPs in surface soil from the e-waste dismantling facilities and the surrounding areas, suggesting their wide applications. Among all identified BPs, BPA was still the dominant one. The concentrations of TBBPA were also among the highest, and its debromination product, monoBBPA, exhibited extremely high concentrations in some e-waste samples. An emerging alternative of BPA, TMBPA, was identified and showed wide distribution. The spatial distributions of BPs showed that the e-waste dismantling facilities were important sources to the surrounding areas, especially for TBBPA and monoBBPA. The estimation of daily intake via oral ingestion suggests that current contamination scenarios are unlikely to produce considerable exposure risks for e-waste dismantlers and residents. However, the health risk of dismantling workers should not be overlooked given the additional exposure routes and possible mixed exposure effects from co-existing BPs.

**Supplementary Materials:** The following supporting information can be downloaded at https://www.mdpi.com/article/10.3390/toxics12060379/s1: References [37,38] are cited in the Supplementary Material. Table S1: Detailed information on sampling sites; Table S2: Information of identified bisphenol chemicals; Table S3: Parameters used for the estimation of daily intake via dust ingestion; Table S4: Multiple reaction monitoring (MRM) ions of each bisphenol chemical; Table S5: The EDIs of BPs for e-waste dismantling workers and residents (adults and toddlers) via soil ingestion; Figure S1: Distribution of sampling sites for the surface soil samples; Figure S2: Identification of 3,3′,5-tribromobisphenol A in e-waste soil; Figure S3: Identification of monochlorobisphenol A in e-waste soil; Figure S4: Identification of dibromobisphenol A in e-waste soil; Figure S5: Identification of tetramethyl bisphenol A in e-waste soil; Figure S6: Compositions of bisphenol chemicals in surface soil from e-waste dismantling facilities and surrounding areas in South China; Figure S7: Spearman correlations between individual bisphenol chemicals in surface soil from e-waste dismantling facilities; Figure S8: Spearman correlations between individual bisphenol chemicals in surface soil from e-waste dismantling park surrounding areas.

**Author Contributions:** L.Z.: writing—original draft preparation, investigation; F.Z.: methodology; S.W.: investigation; Y.Y.: resources; H.C.: resources; X.M.: investigation; X.L.: methodology, writing—reviewing and editing, supervision. All authors have read and agreed to the published version of the manuscript.

**Funding:** The present study was financially supported by the National Key R&D Program of China (2023YFC3706600) and the National Natural Science Foundation of China (No. 42177407).

**Institutional Review Board Statement:** Not applicable.

**Data Availability Statement:** The original contributions presented in the study are included in the article/supplementary material, further inquiries can be directed to the corresponding author.

**Conflicts of Interest:** The authors declare no conflicts of interest.

## References

1. Chen, D.; Kannan, K.; Tan, H.; Zheng, Z.; Feng, Y.-L.; Wu, Y.; Widelka, M. Bisphenol Analogues Other Than BPA: Environmental Occurrence, Human Exposure, and Toxicity—A Review. *Environ. Sci. Technol.* **2016**, *50*, 5438–5453. [CrossRef]
2. Liao, C.; Kannan, K. A survey of bisphenol A and other bisphenol analogues in foodstuffs from nine cities in China. *Food Addit. Contam. A* **2014**, *31*, 319–329. [CrossRef] [PubMed]
3. Vandenberg, L.N.; Hauser, R.; Marcus, M.; Olea, N.; Welshons, W.V. Human exposure to bisphenol A (BPA). *Reprod. Toxicol.* **2007**, *24*, 139–177. [CrossRef]
4. Hahladakis, J.N.; Iacovidou, E.; Gerassimidou, S. An overview of the occurrence, fate, and human risks of the bisphenol—A present in plastic materials, components, and products. *Integr. Environ. Assess. Manag.* **2023**, *19*, 45–62. [CrossRef] [PubMed]
5. Bonefeld-Jørgensen, E.C.; Long, M.; Hofmeister, M.V.; Vinggaard, A.M. Endocrine-disrupting potential of bisphenol A, bisphenol A dimethacrylate, 4-n-nonylphenol, and 4-n-octylphenol in vitro: New data and a brief review. *Environ. Health Perspect.* **2007**, *115*, 69–76. [CrossRef] [PubMed]
6. Naveira, C.; Rodrigues, N.; Santos, F.S.; Santos, L.N.; Neves, R.A.F. Acute toxicity of bisphenol a (BPA) to tropical marine and estuarine species from different trophic groups. *Environ. Pollut.* **2021**, *268*, 115911. [CrossRef]
7. Wang, X.; Nag, R.; Brunton, N.P.; Siddique, M.A.B.; Harrison, S.M.; Monahan, F.J.; Cummins, E. Human health risk assessment of bisphenol A (BPA) through meat products. *Environ. Res.* **2022**, *213*, 113734. [CrossRef]
8. Kemikalieinspektionen. Bisfenoler—En Kartläggning Och Analys. 2017. Available online: https://www.kemi.se/global/rapporter/2017/rapport-5-17-bisfenoler-en-kartlaggning-och-analys.pdf (accessed on 14 March 2024).
9. Catenza, C.J.; Farooq, A.; Shubear, N.S.; Donkor, K.K. A targeted review on fate, occurrence, risk and health implications of bisphenol analogues. *Chemosphere* **2021**, *268*, 129273. [CrossRef]
10. Liu, J.; Ma, S.; Lin, M.; Tang, J.; Yue, C.; Zhang, Z.; Yu, Y.; An, T. New Mixed Bromine/Chlorine Transformation Products of Tetrabromobisphenol A: Synthesis and Identification in Dust Samples from an E-Waste Dismantling Site. *Environ. Sci. Technol.* **2020**, *54*, 12235–12244. [CrossRef]
11. Xu, X.; Zeng, X.; Boezen, H.M.; Huo, X. E-waste environmental contamination and harm to public health in China. *Front. Med.* **2015**, *9*, 220–228. [CrossRef]
12. Papaoikonomou, K.; Emmanouil, C.; Vasilato, V.; Diapouli, E.; Grigoratos, T.; Zafirakou, A.; Kungolos, A. PM10 and elemental concentrations in a dismantling plant for waste of electrical and electronic equipment in Greece. *Aerosol Air Qual. Res.* **2018**, *18*, 1457–1469. [CrossRef]
13. Zhang, T.; Xue, J.; Gao, C.-z.; Qiu, R.-l.; Li, Y.-x.; Li, X.; Huang, M.-z.; Kannan, K. Urinary concentrations of bisphenols and their association with biomarkers of oxidative stress in people living near e-waste recycling facilities in China. *Environ. Sci. Technol.* **2016**, *50*, 4045–4053. [CrossRef]

14. Chakraborty, P.; Sampath, S.; Mukhopadhyay, M.; Selvaraj, S.; Bharat, G.K.; Nizzetto, L. Baseline investigation on plasticizers, bisphenol A, polycyclic aromatic hydrocarbons and heavy metals in the surface soil of the informal electronic waste recycling workshops and nearby open dumpsites in Indian metropolitan cities. *Environ. Pollut.* **2019**, *248*, 1036–1045. [CrossRef] [PubMed]
15. Pan, Y.; Xie, R.; Wei, X.; Li, A.J.; Zeng, L. Bisphenol and analogues in indoor dust from E-waste recycling sites, neighboring residential homes, and urban residential homes: Implications for human exposure. *Sci. Total Environ.* **2024**, *907*, 168012. [CrossRef] [PubMed]
16. Wei, D.; Yuan, K.; Ai, F.; Li, M.; Zhu, N.; Wang, Y.; Zeng, K.; Yin, D.; Bu, Y.; Zhang, Z. Occurrence, spatial distributions, and temporal trends of bisphenol analogues in an E-waste dismantling area: Implications for risk assessment. *Sci. Total Environ.* **2023**, *867*, 161498. [CrossRef]
17. Liu, X.; Lv, Q.; Song, X.; Chen, Y.; Zhao, L.; Yan, M.; Hu, B.; Chen, D. Screening for bisphenol chemicals: A strategy based on dansyl chloride derivatizatio n coupled with in-source fragmentation by high-resolution mass spectrometry. *Anal. Chem.* **2023**, *95*, 6227–6234. [CrossRef] [PubMed]
18. Liao, C.; Liu, F.; Guo, Y.; Moon, H.-B.; Nakata, H.; Wu, Q.; Kannan, K. Occurrence of eight bisphenol analogues in indoor dust from the United States and several Asian countries: Implications for human exposure. *Environ. Sci. Technol.* **2012**, *46*, 9138–9145. [CrossRef] [PubMed]
19. EFSA Panel on Food Contact Materials, Enzymes, Flavourings and Processing Aids (CEF). Scientific opinion on the risks to public health related to the presence of bisphenol A (BPA) in foodstuffs. *EFSA J.* **2015**, *13*, 3978. [CrossRef]
20. Barghi, M.; Shin, E.-S.; Kim, J.-C.; Choi, S.-D.; Chang, Y.-S. Human exposure to HBCD and TBBPA via indoor dust in Korea: Estimation of external exposure and body burden. *Sci. Total Environ.* **2017**, *593*, 779–786. [CrossRef]
21. Lyu, Z.; Harada, K.H.; Kim, S.; Fujitani, T.; Hitomi, T.; Pan, R.; Park, N.; Fujii, Y.; Kho, Y.; Choi, K. Temporal trends in bisphenol exposures and associated health risk among Japanese women living in the Kyoto area from 1993 to 2016. *Chemosphere* **2023**, *316*, 137867. [CrossRef]
22. Zheng, J.; Yang, J.; Zhao, F.; Peng, B.; Wang, Y.; Fang, M. CIL-ExPMRM: An Ultrasensitive Chemical Isotope Labeling Assisted Pseudo-MRM Platform to Accelerate Exposomic Suspect Screening. *Environ. Sci. Technol.* **2023**, *57*, 10962–10973. [CrossRef] [PubMed]
23. Qi, Y.; He, J.; Xiu, F.-R.; Yu, X.; Gao, X.; Li, Y.; Lu, Y.; Song, Z. A convenient chemiluminescence detection for bisphenol A in E-waste dismantling site based on surface charge change of cationic gold nanoparticles. *Microchem. J.* **2019**, *147*, 789–796. [CrossRef]
24. Xu, Y.; Hu, A.; Li, Y.; He, Y.; Xu, J.; Lu, Z. Determination and occurrence of bisphenol A and thirteen structural analogs in soil. *Chemosphere* **2021**, *277*, 130232. [CrossRef] [PubMed]
25. Vasiljevic, T.; Harner, T. Bisphenol A and its analogues in outdoor and indoor air: Properties, sources and global levels. *Sci. Total Environ.* **2021**, *789*, 148013. [CrossRef]
26. Yang, Y.; Shi, Y.; Chen, D.; Chen, H.; Liu, X. Bisphenol A and its analogues in paired urine and house dust from South China and implications for children's exposure. *Chemosphere* **2022**, *294*, 133701. [CrossRef]
27. Qiu, W.; Zhan, H.; Hu, J.; Zhang, T.; Xu, H.; Wong, M.; Xu, B.; Zheng, C. The occurrence, potential toxicity, and toxicity mechanism of bisphenol S, a substitute of bisphenol A: A critical review of recent progress. *Ecotoxicol. Environ. Saf.* **2019**, *173*, 192–202. [CrossRef] [PubMed]
28. Kitamura, S.; Suzuki, T.; Sanoh, S.; Kohta, R.; Jinno, N.; Sugihara, K.; Yoshihara, S.i.; Fujimoto, N.; Watanabe, H.; Ohta, S. Comparative study of the endocrine-disrupting activity of bisphenol A and 19 related compounds. *Toxicol. Sci.* **2005**, *84*, 249–259. [CrossRef] [PubMed]
29. Liu, Q.; Shao, W.; Weng, Z.; Zhang, X.; Ding, G.; Xu, C.; Xu, J.; Jiang, Z.; Gu, A. In vitro evaluation of the hepatic lipid accumulation of bisphenol analogs: A high-content screening assay. *Toxicol. Vitr.* **2020**, *68*, 104959. [CrossRef] [PubMed]
30. Malkoske, T.; Tang, Y.; Xu, W.; Yu, S.; Wang, H. A review of the environmental distribution, fate, and control of tetrabromobisphenol A released from sources. *Sci. Total Environ.* **2016**, *569*, 1608–1617. [CrossRef]
31. Wu, Y.; Li, Y.; Kang, D.; Wang, J.; Zhang, Y.; Du, D.; Pan, B.; Lin, Z.; Huang, C.; Dong, Q. Tetrabromobisphenol A and heavy metal exposure via dust ingestion in an e-waste recycling region in Southeast China. *Sci. Total Environ.* **2016**, *541*, 356–364. [CrossRef]
32. Ge, X.; Ma, S.; Zhang, X.; Yang, Y.; Li, G.; Yu, Y. Halogenated and organophosphorous flame retardants in surface soils from an e-waste dismantling park and its surrounding area: Distributions, sources, and human health risks. *Environ. Int.* **2020**, *139*, 105741. [CrossRef] [PubMed]
33. Runde, K.; Castro, G.; Vike-Jonas, K.; González, S.V.; Asimakopoulos, A.G.; Arp, H.P.H. Occurrence and sorption behaviour of bisphenols and benzophenone UV-filters in e-waste plastic and vehicle fluff. *J. Hazard. Mater.* **2022**, *426*, 127814. [CrossRef] [PubMed]
34. EFSA Panel on Food Contact Materials, Enzymes and Processing Aids (CEP); Lambré, C.; Barat Baviera, J.M.; Bolognesi, C.; Chesson, A.; Cocconcelli, P.S.; Crebelli, R.; Gott, D.M.; Grob, K.; Lampi, E.; et al. Re-evaluation of the risks to public health related to the presence of bisphenol A (BPA) in foodstuffs. *EFSA J.* **2023**, *21*, e06857. [PubMed]
35. Geens, T.; Aerts, D.; Berthot, C.; Bourguignon, J.-P.; Goeyens, L.; Lecomte, P.; Maghuin-Rogister, G.; Pironnet, A.-M.; Pussemier, L.; Scippo, M.-L. A review of dietary and non-dietary exposure to bisphenol-A. *Food Chem. Toxicol.* **2012**, *50*, 3725–3740. [CrossRef]
36. Silva, E.; Rajapakse, N.; Kortenkamp, A. Something from "nothing"—Eight weak estrogenic chemicals combined at concentrations below NOECs produce significant mixture effects. *Environ. Sci. Technol.* **2002**, *36*, 1751–1756. [CrossRef]

37. China Environmental Protection Agency (China EPA). *Exposure Factors Handbook of Chinese Population (Adults)*; China Environmental Press: Beijing, China, 2013; p. 948.
38. China Environmental Protection Agency (China EPA). *Exposure Factors Handbook of Chinese Population (0–5 Years)*; China Environmental Press: Beijing, China, 2013; p. 1002.

toxics

MDPI

Article

# Emerging Contaminants in the Effluent of Wastewater Should Be Regulated: Which and to What Extent?

**Weiwei Yang [1], Qingwei Bu [1,*], Qianhui Shi [1], Ruiqing Zhao [1], Haitao Huang [1], Lei Yang [2], Jianfeng Tang [3] and Yuning Ma [4,*]**

1 School of Chemical & Environmental Engineering, China University of Mining & Technology-Beijing, Beijing 100083, China; 2110340107@student.cumtb.edu.cn (Q.S.)
2 State Key Laboratory of Urban and Regional Ecology, Research Center for Eco-Environmental Sciences, Chinese Academy of Sciences, Beijing 100085, China
3 Key Laboratory of Urban Environment and Health, Institute of Urban Environment, Chinese Academy of Sciences, Xiamen 361021, China
4 College of Environmental and Resource Sciences, Zhejiang University, Hangzhou 310058, China
* Correspondence: qingwei.bu@cumtb.edu.cn (Q.B.); julius.yuningma@gmail.com (Y.M.)

**Abstract:** Effluent discharged from urban wastewater treatment plants (WWTPs) is a major source of emerging contaminants (ECs) requiring effective regulation. To this end, we collected discharge datasets of pharmaceuticals (PHACs) and endocrine-disrupting chemicals (EDCs), representing two primary categories of ECs, from Chinese WWTP effluent from 2012 to 2022 to establish an exposure database. Moreover, high-risk ECs' long-term water quality criteria (LWQC) were derived using the species sensitivity distribution (SSD) method. A total of 140 ECs (124 PHACs and 16 EDCs) were identified, with concentrations ranging from N.D. (not detected) to 706 μg/L. Most data were concentrated in coastal regions and Gansu, with high ecological risk observed in Gansu, Hebei, Shandong, Guangdong, and Hong Kong. Using the assessment factor (AF) method, 18 high-risk ECs requiring regulation were identified. However, only three of them, namely carbamazepine, ibuprofen, and bisphenol-A, met the derivation requirements of the SSD method. The LWQC for these three ECs were determined as 96.4, 1010, and 288 ng/L, respectively. Exposure data for carbamazepine and bisphenol-A surpassed their derived LWQC, indicating a need for heightened attention to these contaminants. This study elucidates the occurrence and risks of ECs in Chinese WWTPs and provides theoretical and data foundations for EC management in urban sewage facilities.

**Keywords:** China's WWTPs; emerging contaminants; risk; SSD; water quality criteria

## 1. Introduction

Emerging contaminants (ECs) encompass newly discovered or recognized pollutants that pose risks to ecological environments and human health yet lack effective regulatory measures [1]. They mainly consist of PHACs, EDCs, persistent organic pollutants, and microplastics [2,3], among others. Advances in science and monitoring technology have led to increased detection of ECs in aquatic ecosystems such as rivers, lakes, and groundwater [4]. Due to the limited treatment performance of traditional water treatment processes, urban sewage still contains high concentrations of ECs even after treatment, making urban wastewater discharge a significant source of ECs [5–8]. Continuous discharge of ECs may accumulate in aquatic environments, endangering ecosystems [8]. Moreover, utilizing treated wastewater for landscaping [9], agricultural irrigation [10], and industrial reuse [11] may expose humans to ECs, thereby posing potential health risks [12,13]. Consequently, controlling EC discharge from WWTP effluent is crucial for aquatic ecosystems and human health. While governments prioritize identifying and regulating high-risk ECs [4], specific regulatory guidance is lacking. Variation in EC detection, discharge, risk, and regional characteristics necessitates determining which ECs to regulate and to what extent in WWTP effluent discharge.

Despite numerous reports on EC occurrence and risks in WWTP effluent, most focus on limited ECs in small-scale areas, leaving nationwide or larger-scale data gaps. Deep mining of nationwide data is crucial for identifying high-risk areas and EC species, informing regulation in WWTP effluent discharge. Water quality criteria (WQC) guide water quality assessment and pollution prevention. Generally, WQC derivation methods include statistical extrapolation [14], the AF method [15], and the biotic ligand model method [16]. The AF method, while simple, is subjective, leading to conservative outcomes, primarily used for risk assessment. The biotic ligand model's incomplete theory hampers accurate baseline value derivation. In contrast, the statistical extrapolation method, with its rigorous logic and ample data, yields more objective and accurate WQC. Specifically, the statistical extrapolation method can be subdivided into species sensitivity distribution (SSD) and toxicity percentile rank (TPR) methods [14]. The SSD method, based on species sensitivity distribution theory, establishes a probability distribution model based on dose–response relationships, representing the sensitivity differences among different species and deducing the hazardous concentration for a given percentage of species, thereby protecting the ecosystem [14]. The TPR method, recommended by the US Environmental Protection Agency (US EPA) [17], considers water quality characteristics and the bioaccumulation effect of organisms but only utilizes the mean toxicity values of the four most sensitive genera. When deriving the WQC for pollutants unrelated to water quality characteristics and bioaccumulation effects, the SSD method is more capable of reflecting the overall ecological toxicity of chemicals, especially when considering the differences in species sensitivity. Therefore, WQC derived based on the SSD method hold promise in providing effective solutions for establishing discharge limits for high-risk ECs in WWTP effluent.

In this study, PHACs and EDCs were chosen as the focus, aiming to identify EC species, concentrations, risks, and high-risk areas in WWTP effluent across China, using nationwide data. High-risk ECs needing specific attention are selected, with SSD employed to calculate discharge limits. The objectives of this study include the following: (1) analyze PHACs and EDCs reported from Chinese WWTP effluent (2012–2022) to identify ECs needing attention; (2) screen high-risk ECs using the AF method; and (3) derive WQC for ECs in WWTP effluent using the SSD method.

## 2. Methodology

### 2.1. Methods of Data Collection

A comprehensive search was conducted utilizing the ISI Web of Science (https://webofscience.clarivate.cn, accessed on 30 December 2022) and China National Knowledge Infrastructure (CNKI) (https://kns.cnki.net/, accessed on 30 December 2022) to collect the relevant literature between 2012 and 2022 concerning PHACs and EDCs in WWTP effluent. The collected data focused solely on urban domestic wastewater, excluding pharmaceutical and industrial WWTPs, among others. This approach aimed to create provincial-level discharge maps specifically for PHACs and EDCs. Publications lacking clear location information were excluded from the analysis. Additionally, data from different wastewater treatment plants within the same study and data from the same wastewater treatment plant across different seasons were documented separately.

Furthermore, toxicological data on PHACs and EDCs were sourced from the ECOTOX database (https://cfpub.epa.gov/ecotox/) of the US EPA, accessed on 15 June 2023.

### 2.2. Risk Assessment Methods

To assess the ecological risk of a specific pollutant to aquatic organisms, the risk quotient (RQ) is computed by dividing the reported measured environmental concentration (MEC) by the predicted no-effect concentration (PNEC) [18]. PNEC is derived by dividing the most sensitive biological toxicity value by an AF. The selection criteria for assessment factors in the AF method are outlined in Table S4, while the PNEC values for PHACs and

EDCs are provided in Tables S5 and S6. To accurately evaluate the exposure risk of regional ECs, values such as "N.D." (not detected), "<LOD" (below the limit of detection), and "<LOQ" (below the limit of quantitation) are considered as zero in the calculation. The risk exceeding rate (RER) is then calculated by determining the ratio of the number of RQs with values greater than 0.1 to the total number of RQs.

*2.3. Water Quality Criteria Derivation Method*

The derivation of WQC using the SSD method was conducted according to the procedures outlined in the Technical Guidelines for Deriving Water Quality Criteria for Freshwater Organisms published by the Ministry of Ecology and Environment of the People's Republic of China [19]. Initially, the SSD method was applied following the minimum data requirements specified in the freshwater biological water quality criteria derivation method. These requirements include covering at least three different trophic levels, including producers, and encompassing at least 10 species representing diverse biological groups. These groups include one species each of cyprinid fish, non-cyprinid fish, zooplankton, benthic invertebrates (e.g., mollusks, benthic crustaceans), amphibians, or other aquatic organisms belonging to different phyla from the aforementioned animals, and one species of phytoplankton or aquatic vascular plant.

For acute toxicity data, animals require exposure times of approximately 24 h for rotifers, 48 h for daphnids and midges, and 96 h for other species, while plants need an exposure time of approximately 96 h. For chronic toxicity data, animals need exposure durations of 48 h or longer for nematodes, 21 days or longer for other species, or exposure encompassing a sensitive life stage. Plants require exposure durations of 21 days or longer, or spanning at least one generation. Subsequently, the acute value for the same effect (AVE) and chronic value for the same effect (CVE) for all species of a certain EC were calculated following the Technical Guidelines for Deriving Water Quality Criteria for Freshwater Organisms [19].

The obtained AVE and CVE values were fitted using the National Ecological Environment Criteria Calculation Software [20], and four models (normal distribution, log-normal distribution, logistic distribution, and log-logistic distribution) were fitted to obtain the hazardous concentration for 5% of species derived from the toxicity data ($HC_5$), root mean square error (RSME), and P(A-D) value. P(A-D) represents the Anderson–Darling test value, where a $p$-value > 0.05 indicates that the fit passes the A-D test and the model conforms to the theoretical distribution; RSME represents the root mean square error, with a smaller RSME indicating a higher accuracy of model fitting. The model fitting result ($HC_5$) with $p > 0.05$ and the smallest RSME was selected to derive the WQC.

Short-term water quality criteria (SWQC) for aquatic organisms were calculated by dividing the hazardous concentration for 5% of species derived from the acute toxicity data ($SHC_5$) by the short-term assessment factor (SAF). Long-term water quality criteria (LWQC) were derived by dividing the hazardous concentration for 5% of species derived from chronic toxicity data ($LHC_5$) by the long-term assessment factor (LAF). The value of SAF or LAF was determined based on the number of data used for deriving the WQC. When the number of species included in the effective toxicity data exceeds 15, SAF or LAF is set to 2; when the number of species included in the effective toxicity data is less than or equal to 15, SAF or LAF is generally set to 3 [19].

## 3. Exposure and Risk of ECs

*3.1. Exposure*

A comprehensive dataset comprising 1178 data points related to 140 ECs was extracted from 43 available studies. Of these, 1116 data points were associated with 124 different PHACs, accounting for 94.74% of the total, and were sourced from 29 reports (Table S1). Additionally, 62 data points were related to EDCs, covering 16 EDCs, making up 5.26% of the total, and were obtained from 14 reports (Table S2). Notably, no studies reported both PHACs and EDCs simultaneously. Detailed information on both PHACs and EDCs can be found in Table S3. The geographical coverage of these data spanned 18 provinces in China, with significant concentrations observed in the Bohai Sea Rim region, the Yangtze River Delta, and the Pearl River Delta basins (Figure S1). Among the provinces, Beijing, Jiangsu, Shanghai, Fujian, Guangdong, and Hong Kong reported both PHACs and EDCs, while others reported only one type of EC. Moreover, EDC-related data points were relatively scarce in these provinces, accounting for only 1.75% to 27.59% of the total. The concentrations of the 140 ECs in WWTP effluent ranged widely, from 0 ng/L to 706 μg/L. Specifically, the mean concentrations of pharmaceuticals varied from 0 ng/L to 5.09 μg/L, while those of EDCs ranged from 0.84 ng/L to 2.45 μg/L.

In terms of individual ECs, sulfamethoxazole exhibited the highest detection number with 46 data points and a mean detected concentration of 219 ng/L (Figure 1a). Notably, a recent study by Guo et al. highlighted sulfamethoxazole as the most frequently detected EC in Chinese surface water, with a detection concentration of 45 ng/L [4]. Moreover, sulfamethoxazole has been detected in WWTP effluent in Germany (22.9~34.9 ng/L) [21], the United States (1640 ng/L) [22], and other countries. In addition to sulfamethoxazole, ofloxacin, erythromycin, and salicylic acid also exhibited a high detection number (exceeded the overall detection frequency, 10.8), with 45, 38, and 12 data points, respectively. Their mean detected concentrations of 1151 ng/L, 1415 ng/L, and 2819 ng/L, respectively, exceeded the mean detected concentration (934 ng/L) of all ECs. Furthermore, their mean detected concentrations generally exceeded those found in Chinese surface water [18] by tenfold, as well as being higher than those in WWTP effluent in India (ofloxacin: 0~212 ng/L; erythromycin: 1~12 ng/L) [23] and the United States (erythromycin: 230 ng/L [24]; salicylic acid: 630 ng/L [25]).

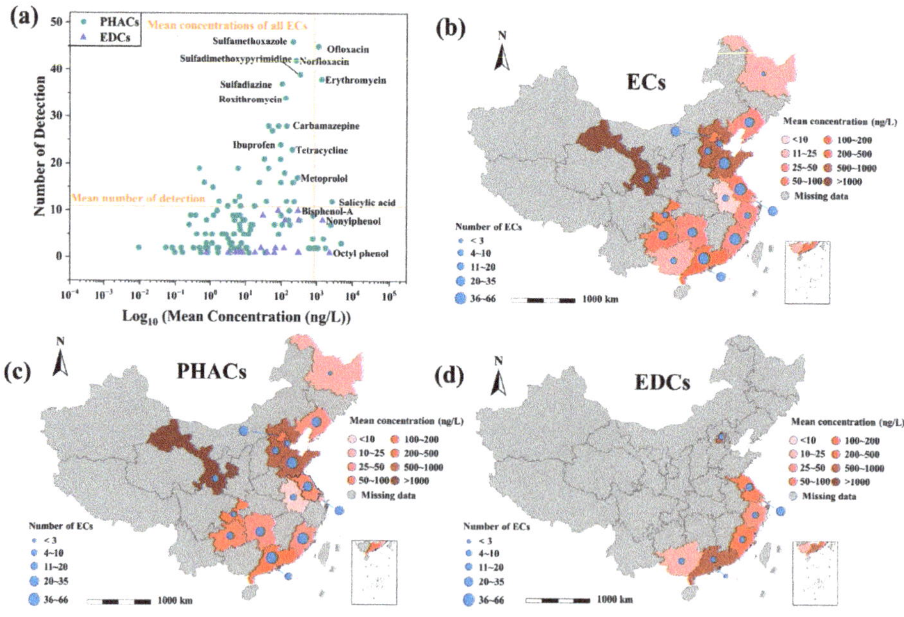


**Figure 1.** (**a**) Number of detections and mean concentration of individual ECs in the exposure database. Distribution of ECs (**b**), PHACs (**c**), and EDCs (**d**) by province recorded in the exposure.

The distribution of EC exposure, in terms of composition and concentration, varied significantly among different provinces' WWTP effluent (Figure 1b). Coastal provinces, except for Zhejiang, exhibited a relatively high number of EC exposure, with Guangdong Province having the highest number of exposed ECs at 66, followed by Fujian with 53 ECs, while Heilongjiang reported only one PHAC, caffeine. This variation could be attributed to the extensive production and consumption of chemicals in economically advanced regions, alongside heightened scientific research driven by their developed economies [4]. Regarding EC concentration, Gansu, the Bohai Sea region, and the Pearl River Delta region were identified as the most severely polluted areas. The mean EC concentrations in Gansu, Hebei, Shandong, and Guangdong were 1116 ng/L, 816 ng/L, 519 ng/L, and 354 ng/L, respectively, all exceeding the mean concentration of all ECs (258 ng/L). The high concentration of ECs in Gansu was primarily due to the high concentrations of ofloxacin (mean concentration of 9240 ng/L) and sulfapyridine (mean concentration of 1150 ng/L) among PHACs, making Gansu the province with the most severe PHAC pollution (Figure 1c). Hebei and Shandong ranked second and third in terms of PHAC pollution, with mean PHAC concentrations of 862 ng/L and 586 ng/L, respectively. The PHACs in Hebei were mainly attributed to sulfonamide (mean concentration: sulfadimethoxypyrimidine, 552.5 ng/L; sulfamethoxazole, 777.5 ng/L; sulfapyridine, 442.5 ng/L; sulfadimethoxine, 745 ng/L; sulfamerazine, 1200 ng/L; sulfathiazole, 797.5 ng/L) and beta-lactam antibiotics (mean concentration: cefazolin, 797.5 ng/L; procaine hydrochloride, 1010 ng/L; cefotaxime, 672.5 ng/L; ceftriaxone, 1867.5 ng/L; cefaclor, 890 ng/L), while in Shandong, in addition to sulfonamide (mean concentration: sulfadimethoxypyrimidine, 3967.76 ng/L; sulfamethoxypyridazine, 5086.95 ng/L; sulfamethoxazole, 267.07 ng/L) and quinolone antibiotics (mean concentration: ciprofloxacin, 390.05; enoxacin, 3264.5 ng/L; sarafloxacin, 1597.2 ng/L; oxilinic acid, 622.75 ng/L), there were also non-antibiotic PHACs such as diclofenac (mean concentration of 568 ng/L), ibuprofen (mean concentration of 525 ng/L), and carbamazepine (mean concentration of 1117 ng/L). Research on EDCs was relatively scarce, mainly concentrated in the Yangtze River Delta and the Pearl River Delta regions (Figure 1d). Hong Kong had the most severe EDC pollution, with a mean concentration of 1043 ng/L, mainly composed of nonylphenol (mean concentration of 1546 ng/L) and bisphenol A (mean concentration of 540 ng/L), followed by Guangdong Province, where major EDC species included nonylphenol (mean concentration of 2512 ng/L) and bisphenol A (mean concentration of 485 ng/L), as well as octylphenol (mean concentration of 2454 ng/L).

### 3.2. Risk

Due to the limited availability of toxicity data for ECs, RQs were calculated for only 79 PHACs and 13 EDCs. As shown in Figure 2a, there was a significant difference in the overall risk between PHACs and EDCs, with median RQ values of 0.11 and 67.9, respectively, both exceeding 0.1, indicating unacceptable ecological risks. Specifically, a total of 479 data points of PHACs had risk values exceeding 0.1, accounting for 50.8% of the total data points for PHACs with calculable risks (Figure S2a). For EDCs, only five data points had RQ values < 0.1, while the remaining 54 data points had RQ values exceeding 0.1, accounting for 91.5% of the total data points for EDCs with calculable risks (Figure S2b).

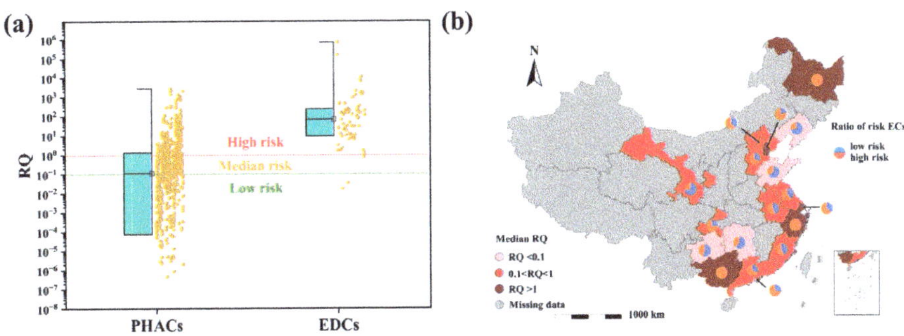


**Figure 2.** (**a**) Distribution of risk of PHACs and EDCs. (**b**) Distribution of risk by province, which was evaluated by median RQ.

Regarding the risk for individual provinces, except for Liaoning, Shandong, Guizhou, and Hunan, the median risk values for the other provinces were all higher than 0.1, indicating unacceptable risks. Nonetheless, the aforementioned four provinces still had 28.6%, 35.7%, 45.0%, and 38.7% of data points with an RQ > 0.1, signifying unacceptable risks (Figure 2b). Shandong Province reported an RQ for carbamazepine and ibuprofen of 1117.33 and 525.33, respectively, exceeding the critical value for unacceptable risks by more than three orders of magnitude, warranting special attention. Heilongjiang had the highest median RQ value, mainly due to the small data size (only two data points of caffeine). The median RQ of Zhejiang and Guangxi was also higher than 1, ranking among the highest in the country. This was attributed to the fact that only concentrations of EDCs were reported in these two regions, and since the PNEC for EDCs is generally low, this resulted in significantly higher median RQ values compared to other provinces. Furthermore, we also calculated RQ values for all data points of ECs (Figure S2). For PHACs, the median RQ values of 19 PHACs including sulfamethoxypyridazine, trimethoprim, ofloxacin, erythromycin, roxithromycin, spiramycin, tetracycline, ibuprofen, gemfibrozil, carbamazepine, bezafibrate, metoprolol, ketoprofen, caffeine, venlafaxine, fluoxetine, diazepam, oxazepam, and acetaminophen were all greater than 1, indicating high risk levels. For EDCs, except for chloroform and diethylstilbestrol, all other ECs had median RQ values exceeding 1, placing them in the high-risk category. These ECs pose a serious threat to aquatic organisms upon discharge after wastewater treatment and require close attention.

The identified high-risk ECs also exhibit a significant trend of risk exceeding rate (RER > 50%) and are detected frequently (exceeding the mean detection number). As illustrated in Figure 3, a total of 16 high-risk ECs were identified, comprising 14 PHACs and 2 EDCs. Among the PHACs are ofloxacin, norfloxacin, sulfamethoxazole, erythromycin, carbamazepine, roxithromycin, diclofenac, ciprofloxacin, caffeine, ibuprofen, metoprolol, bezafibrate, salicylic acid, and naproxen, while the EDCs include estrone and bisphenol A. The high ecological risks of these ECs have also been extensively documented in other studies (e.g., ofloxacin [26–29], norfloxacin [26,28,30,31], sulfamethoxazole [32–34], erythromycin [35–37], carbamazepine [32,38–41], roxithromycin [42–45], diclofenac [46–48], ciprofloxacin [49,50], caffeine [51–54], ibuprofen [38,55–57], metoprolol [58,59], bezafibrate [60,61], salicylic acid [62], naproxen [63–65], estrone [66–69], and bisphenol-A [70–75]). Moreover, these contaminants have been widely documented in surface waters across China [4,76]. Nevertheless, the current discharge standards of WWTPs in China fail to address these ECs, resulting in a lack of effective regulation over ECs with significant ecological risks.

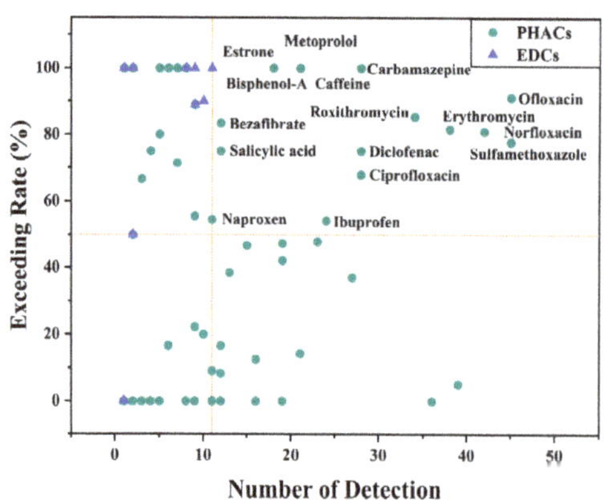


**Figure 3.** Risk exceeding rate (RER) and number of detections of ECs based on risk.

## 4. The Derivation Water Quality Criteria of ECs

Taking into account the detection frequency, mean detection concentration, high-risk exceedance rate, and the presence of locally severe ECs in WWTP effluent in China, a total of 18 ECs including ofloxacin, norfloxacin, sulfamethoxazole, sulfapyridine, erythromycin, carbamazepine, roxithromycin, diclofenac, ciprofloxacin, caffeine, ibuprofen, metoprolol, bezafibrate, salicylic acid, bisphenol-A, estrone, nonylphenol, and octylphenol were selected for deriving WQC using the SSD method. However, only carbamazepine, ibuprofen, and bisphenol-A met the data requirements for SSD derivation. The toxicity data for these three ECs were processed according to the SSD method's requirements for deriving short-term and long-term water quality standards for freshwater organisms. The processed toxicity data are detailed in Tables S7–S12.

The toxicity data were inputted into the National Ecological Environment Standard Calculation Software and fitted using four models: normal distribution, log-normal distribution, logistic, and log-logistic. The resulting $HC_5$, RSME, and P(A-D) values are shown in Table S13. As shown in Figure 4, the logistic model demonstrated excellent fits for the acute toxicity data of carbamazepine and ibuprofen, as well as the chronic toxicity data of ibuprofen. The log-logistic model exhibited excellent fits for the acute toxicity data of bisphenol A, the chronic toxicity data of carbamazepine, and the chronic toxicity data of bisphenol A. Further calculations revealed that the SWQC for carbamazepine, ibuprofen, and bisphenol A were 3.40 mg/L, 1.86 mg/L, and 0.89 mg/L (Table 1), respectively, which are higher than the actual concentrations of these ECs in water bodies by two to three orders of magnitude, indicating limited guidance for controlling the discharge of ECs in wastewater. However, the LWQC for carbamazepine, ibuprofen, and bisphenol A were 96.4 ng/L, 1010 ng/L, and 288 ng/L (Table 1), respectively, which are comparable to the detected concentrations of these ECs in water bodies. These LWQC can effectively reflect the exceedance of ECs in WWTP effluent. Additionally, the LWQC for carbamazepine [77] and ibuprofen [78] are consistent with those calculated by other researchers, while the LWQC for bisphenol-A [79,80] are one order of magnitude lower, which may be attributed to differences in the methodological approach employed.

The feasibility of the LWQC was further validated by integrating the reported data. As depicted in Figure 5, only the reported concentration of ibuprofen did not exceed its LWQC. However, for carbamazepine and ibuprofen, 14.3% and 45.5% of the reported data points, respectively, exceeded the corresponding LWQC. Therefore, this study concludes that the derived LWQC based on the SSD method can provide valuable reference data for controlling the discharge of ECs from WWTPs in China.

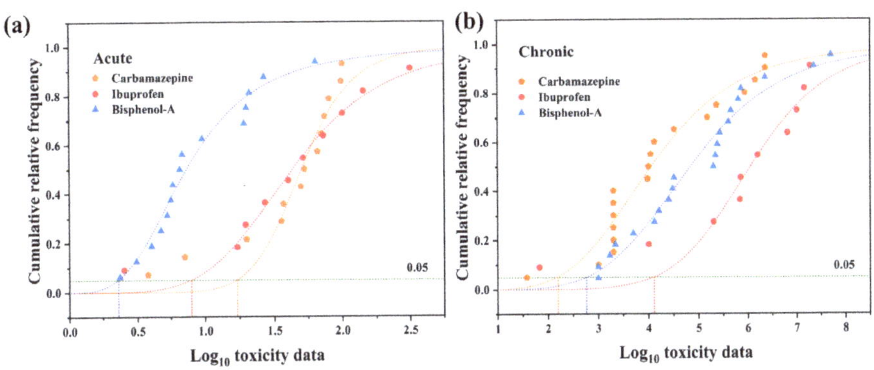


**Figure 4.** SSDs based on acute (**a**) and chronic (**b**) toxicity data for carbamazepine, ibuprofen, and bisphenol-A.

**Table 1.** Water quality criteria calculation results.

| Chemicals | Acute | | | | Chronic | | | |
|---|---|---|---|---|---|---|---|---|
| | Model | $SHC_5$ (mg/L) | SAF | SWQC (mg/L) | Model | $LHC_5$ (ng/L) | LAF | LWQC (ng/L) |
| Carbamazepine | Logistic distribution | 10.19 | 3 | 3.40 | Log-normal distribution | 192.8 | 2 | 96.4 |
| Ibuprofen | Logistic distribution | 5.59 | 3 | 1.86 | Logistic distribution | 3039.9 | 3 | 1010 |
| Bisphenol-A | Log-logistic distribution | 2.66 | 3 | 0.89 | Normal distribution | 575.3 | 2 | 288 |

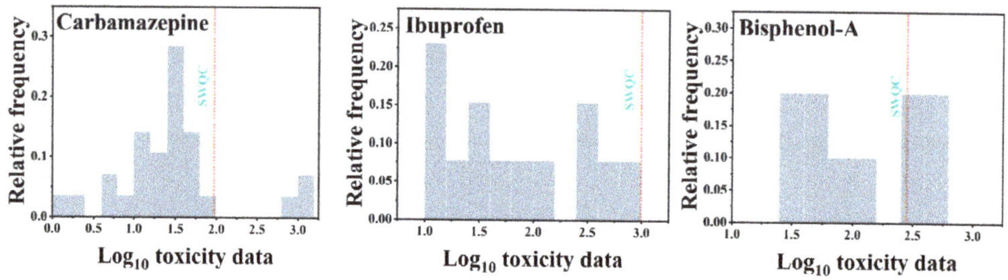


**Figure 5.** Logarithmic distribution of carbamazepine (**a**), ibuprofen (**b**), and bisphenol-A (**c**) detection data.

## 5. Future Prospects

Exposure data analysis indicates that the volume of data reported by WWTP effluent accounts for less than one-tenth of that reported for surface water [4]. Among the 34 provincial-level administrative regions in China, only 18 provinces reported the occurrence of ECs in WWTP effluent, with some provinces having very limited data. The reported data mainly focus on PHACs, while reports on EDCs are relatively scarce. The insufficient data significantly impair the accuracy of drawing a comprehensive and precise nationwide map of ECs in WWTP effluent, making it challenging to comprehensively and accurately understand the occurrence of ECs in WWTP effluent and the corresponding risks. Furthermore, the lack of toxicity data for some ECs prevents the calculation of their risk values, further exacerbating the limited understanding of EC risk levels in WWTP effluent. Therefore, future efforts should focus on generating more data to create a more objective and comprehensive fine-scale map of the occurrence and risk of ECs in WWTP effluent nationwide.

Regarding toxicity data, among the 18 high-risk ECs identified in this study, only three meet the basic requirements for deriving WQC using the SSD method, thus making it impossible to derive scientific and objective WQC for other high-risk ECs. In fact, we further screened the toxicity data of all PHACs with unacceptable risks based on the requirements of the SSD method. It was found that even for some PHACs with higher risk exceedance rates and detection frequencies, such as sulfamethoxazole, tetracycline, and erythromycin, although the toxicity data are adequate, WQC cannot be derived based on the SSD method. Therefore, it is necessary to derive WQC for more ECs based on more comprehensive and complete toxicity data to scientifically, accurately, objectively, and effectively control the discharge risks of ECs in WWTP effluent in the future.

## 6. Conclusions

In summary, we comprehensively investigated the occurrence and risks of PHACs and EDCs in WWTP effluent in China based on exposure data spanning from 2012 to 2022. Monitoring a total of 140 emerging contaminants (ECs), including 124 PHACs and 16 EDCs, revealed concentrations ranging from 0 to 706 μg/L. PHACs dominated the dataset, constituting 94.02% of the total. Through analyses encompassing overall exposure concentrations, regional risk assessment, risk exceedance rates, and detection frequencies, 18 ECs emerged as requiring close attention. These ECs were ofloxacin, norfloxacin, sulfamethoxazole, sulfapyridine, erythromycin, carbamazepine, roxithromycin, diclofenac, ciprofloxacin, caffeine, ibuprofen, metoprolol, bezafibrate, salicylic acid, bisphenol-A, estrone, nonylphenol, and octylphenol. Utilizing the SSD method, WQC were derived for three ECs meeting the derivation requirements: carbamazepine, ibuprofen, and bisphenol A, with respective LWQC of 96.4 ng/L, 1010 ng/L, and 288 ng/L. Except for ibuprofen, exposure data for carbamazepine and bisphenol A surpassed their LWQC to varying extents, underscoring the significance of these standards in regulating EC discharge in WWTP effluent. This study identifies major high-risk ECs and establishes LWQC based on nationwide data, offering a viable strategy for managing ECs discharge in WWTP effluent.

**Supplementary Materials:** The following supporting information can be downloaded at https://www.mdpi.com/article/10.3390/toxics12050309/s1: Table S1. Concentration levels of PHACs in the effluent of WWTPs in China [81–109]; Table S2. Concentration levels of EDCs in the effluent of WWTPs in China [110–123]; Table S3. Database of emerging contaminants detected in the effluent of WWTPs in China; Table S4. The selection criteria for AF in the derivation of PNEC; Table S5. PNEC of PHACs [124–181]; Table S6. PNEC of EDCs [174,182–193]; Table S7. Chronic toxicity data of carbamazepine on freshwater organisms [124,167,194–204]; Table S8. Acute toxicity data of carbamazepine on freshwater organisms [176,195,201,202,205–211]; Table S9. Chronic toxicity data of ibuprofen on freshwater organisms [167,168,212–217]; Table S10. Acute toxicity data of ibuprofen on freshwater organisms [155,176,209,211,214,217–219]; Table S11. Chronic toxicity data of bisphenol on freshwater organisms [191,220–236]; Table S12. Acute toxicity data of bisphenol on freshwater organisms [228,232,234,236–244]; Table S13. SSD fitting results of carbamazepine, ibuprofen, and bisphenol-A; Figure S1. Overview of the exposure database, including the ratio of number of detection ECs and the number of EC data by province; Figure S2. Boxplots for the calculated RQs for PHACs (a) and EDCs (b) detected in the effluent of WWTPs in China; Figure S3. The names and geographical locations of provinces in China.

**Author Contributions:** Conceptualization, Q.B. and Y.M.; methodology, W.Y.; software, W.Y.; validation, W.Y., Q.S. and Q.B.; formal analysis, W.Y.; investigation, W.Y.; resources, Q.B. and Y.M.; data curation, R.Z. and H.H.; writing—original draft preparation, W.Y.; writing—review and editing, Q.B., Y.M., L.Y. and J.T.; visualization, W.Y.; supervision, Q.B.; project administration, Q.B.; funding acquisition, Q.B. and Y.M. All authors have read and agreed to the published version of the manuscript.

**Funding:** This research was funded by the National Natural Science Foundation of China (grant No. 42277406; 21777188) and Jiangsu Province Ecology and Environment Protection Key Laboratory of Groundwater Monitoring, Surveillance, and Forewarning of Pollution (Project No. GWKL2204).

**Institutional Review Board Statement:** Not applicable.

**Informed Consent Statement:** Not applicable.

**Data Availability Statement:** Data are contained within the article and Supplementary Materials.

**Acknowledgments:** We thank the anonymous reviewers for their thorough reviews and constructive comments. All individuals have consented to the acknowledgment.

**Conflicts of Interest:** The authors declare that they have no known competing financial interests or personal relationships that could have appeared to influence the work reported in this paper.

## Abbreviations

Wastewater treatment plants, WWTPs; pharmaceuticals, PHACs; endocrine-disrupting chemicals, EDCs; long-term water quality criteria, LWQC; assessment factor, AF; species sensitivity distribution, SSD; not detected, N.D.; limit of detection, LOD; limit of quantitation, LOQ; toxicity percentile rank, TPR; US Environmental Protection Agency, US EPA; acute value for the same effect, AVE; chronic value for the same effect, CVE; China National Knowledge Infrastructure, CNKI; risk quotient, RQ; measured environmental concentration, MEC; predicted no-effect concentration, PNEC; hazardous concentration for 5% of species derived from toxicity data, $HC_5$; root mean square error, RSME; Anderson–Darling test, P(A-D); short-term water quality criteria, SWQC; hazardous concentration for 5% of species derived from acute toxicity data, $SHC_5$; short-term assessment factor, SAF; hazardous concentration for 5% of species derived from chronic toxicity data, $LHC_5$; long-term assessment factor, LAF.

## References


1. Ramírez-Malule, H.; Quiñones-Murillo, D.H.; Manotas-Duque, D. Emerging Contaminants as Global Environmental Hazards. A Bibliometr. Analysis. *Emerg. Contam.* **2020**, *6*, 179–193. [CrossRef]
2. Puri, M.; Gandhi, K.; Kumar, M.S. Emerging Environmental Contaminants: A Global Perspective on Policies and Regulations. *J. Environ. Manag.* **2023**, *332*, 117344. [CrossRef] [PubMed]
3. Khan, S.; Naushad, M.; Govarthanan, M.; Iqbal, J.; Alfadul, S.M. Emerging Contaminants of High Concern for the Environment: Current Trends and Future Research. *Environ. Res.* **2022**, *207*, 112609. [CrossRef]
4. Guo, J.; Tu, K.; Chou, L.; Zhang, Y.; Wei, S.; Zhang, X.; Yu, H.; Shi, W. Deep Mining of Reported Emerging Contaminants in China's Surface Water in the Past Decade: Exposure, Ecological Effects and Risk Assessment. *Water Res.* **2023**, *243*, 120318. [CrossRef] [PubMed]
5. Kasprzyk-Hordern, B.; Dinsdale, R.M.; Guwy, A.J. The Removal of Pharmaceuticals, Personal Care Products, Endocrine Disruptors and Illicit Drugs during Wastewater Treatment and Its Impact on the Quality of Receiving Waters. *Water Res.* **2009**, *43*, 363–380. [CrossRef] [PubMed]
6. Moldovan, Z.; Schmutzer, G.; Tusa, F.; Calin, R.; Alder, A.C. An Overview of Pharmaceuticals and Personal Care Products Contamination along the River Somes Watershed, Romania. *J. Environ. Monit.* **2007**, *9*, 986–993. [CrossRef] [PubMed]
7. Dai, G.; Wang, B.; Fu, C.; Dong, R.; Huang, J.; Deng, S.; Wang, Y.; Yu, G. Pharmaceuticals and Personal Care Products (PPCPs) in Urban and Suburban Rivers of Beijing, China: Occurrence, Source Apportionment and Potential Ecological Risk. *Environ. Sci.: Process. Impacts* **2016**, *18*, 445–455. [CrossRef] [PubMed]
8. Lin, X.; Xu, J.; Keller, A.A.; He, L.; Gu, Y.; Zheng, W.; Sun, D.; Lu, Z.; Huang, J.; Huang, X.; et al. Occurrence and Risk Assessment of Emerging Contaminants in a Water Reclamation and Ecological Reuse Project. *Sci. Total Environ.* **2020**, *744*, 140977. [CrossRef] [PubMed]
9. Bakopoulou, S.; Vasiloglou, V.; Kungolos, A. A Multicriteria Analysis Application for Evaluating the Possibility of Reusing Wastewater for Irrigation Purposes in a Greek Region. *Desalination Water Treat.* **2012**, *39*, 262–270. [CrossRef]
10. Lin, S.H.; Cheng, K.W. A New Sequencing Batch Reactor for Treatment of Municipal Sewage Wastewater for Agricultural Reuse. *Desalination* **2001**, *133*, 41–51. [CrossRef]
11. Bauer, S.; Linke, H.J.; Wagner, M. Optimizing Water-Reuse and Increasing Water-Saving Potentials by Linking Treated Industrial and Municipal Wastewater for a Sustainable Urban Development. *Water Sci. Technol.* **2020**, *81*, 1927–1940. [CrossRef] [PubMed]
12. Wei, X.; Hu, Y.; Zhu, Q.; Gao, J.; Liao, C.; Jiang, G. Co-Exposure and Health Risks of Several Typical Endocrine Disrupting Chemicals in General Population in Eastern China. *Environ. Res.* **2022**, *204*, 112366. [CrossRef] [PubMed]
13. Futran Fuhrman, V.; Tal, A.; Arnon, S. Why Endocrine Disrupting Chemicals (EDCs) Challenge Traditional Risk Assessment and How to Respond. *J. Hazard. Mater.* **2015**, *286*, 589–611. [CrossRef] [PubMed]
14. Posthuma, L.; Suter, G.W.; Traas, T.P. *Species Sensitivity Distributions in Ecotoxicology*; Taylor & Francis Ltd.: London, UK, 2001. [CrossRef]
15. Canadian Council of Minister of the Environment (CCME). *Canadian Environmental Quality Guidelines (CEQGs) Provide Science-Based Goals for the Quality of Aquatic and Terrestrial Ecosystems*; Canadian Council of Minister of the Environment: Manitoba, ON, Canada, 2018.
16. Niyogi, S.; Wood, C.M. Biotic Ligand Model, a Flexible Tool for Developing Site-Specific Water Quality Guidelines for Metals. *Environ. Sci. Technol.* **2004**, *38*, 6177–6192. [CrossRef] [PubMed]
17. USEPA. Guidelines for Deriving Numerical National Water Quality Criteria for the Protection of Aquatic Organisms and Their Uses. Available online: https://www.epa.gov/wqc/guidelines-deriving-numerical-national-water-quality-criteria-protection-aquatic-organisms-and (accessed on 22 March 2024).
18. Bu, Q.; Wang, B.; Huang, J.; Deng, S.; Yu, G. Pharmaceuticals and Personal Care Products in the Aquatic Environment in China: A Review. *J. Hazard. Mater.* **2013**, *262*, 189–211. [CrossRef] [PubMed]

19. Ministry of Ecology and Environment of People's Republic of China. *Technical Guideline for Deriving Water Quality Criteria for Freshwater Organisms*; Ministry of Ecology and Environment of People's Republic of China: Beijing, China, 2022.
20. National Ecological Environment Benchmark Expert Committee. *National Ecological Environment Benchmark Calculation Software Species Sensitivity Distribution Method*; Ministry of Ecology and Environment, People's Republic of China: Beijing, China, 2021.
21. Rodriguez-Mozaz, S.; Vaz-Moreira, I.; Varela Della Giustina, S.; Llorca, M.; Barceló, D.; Schubert, S.; Berendonk, T.U.; Michael-Kordatou, I.; Fatta-Kassinos, D.; Martinez, J.L.; et al. Antibiotic Residues in Final Effluents of European Wastewater Treatment Plants and Their Impact on the Aquatic Environment. *Environ. Int.* **2020**, *140*, 105733. [CrossRef] [PubMed]
22. Kwon, J.-W.; Rodriguez, J.M. Occurrence and Removal of Selected Pharmaceuticals and Personal Care Products in Three Wastewater-Treatment Plants. *Arch. Environ. Contam. Toxicol.* **2014**, *66*, 538–548. [CrossRef]
23. Balakrishna, K.; Rath, A.; Praveenkumarreddy, Y.; Guruge, K.S.; Subedi, B. A Review of the Occurrence of Pharmaceuticals and Personal Care Products in Indian Water Bodies. *Ecotoxicol. Environ. Saf.* **2017**, *137*, 113–120. [CrossRef] [PubMed]
24. Karthikeyan, K.G.; Meyer, M.T. Occurrence of Antibiotics in Wastewater Treatment Facilities in Wisconsin, USA. *Sci. Total Environ.* **2006**, *361*, 196–207. [CrossRef] [PubMed]
25. Crouse, B.A.; Ghoshdastidar, A.J.; Tong, A.Z. The Presence of Acidic and Neutral Drugs in Treated Sewage Effluents and Receiving Waters in the Cornwallis and Annapolis River Watersheds and the Mill CoveSewage Treatment Plant in Nova Scotia, Canada. *Environ. Res.* **2012**, *112*, 92–99. [CrossRef]
26. Zhao, W.; Guo, Y.; Lu, S.; Yan, P.; Sui, Q. Recent Advances in Pharmaceuticals and Personal Care Products in the Surface Water and Sediments in China. *Front. Environ. Sci. Eng.* **2016**, *10*, 2. [CrossRef]
27. Tran, N.H.; Hoang, L.; Nghiem, L.D.; Nguyen, N.M.H.; Ngo, H.H.; Guo, W.; Trinh, Q.T.; Mai, N.H.; Chen, H.; Nguyen, D.D.; et al. Occurrence and Risk Assessment of Multiple Classes of Antibiotics in Urban Canals and Lakes in Hanoi, Vietnam. *Sci. Total Environ.* **2019**, *692*, 157–174. [CrossRef]
28. Meng, F.; Sun, S.; Geng, J.; Ma, L.; Jiang, J.; Li, B.; Yabo, S.D.; Lu, L.; Fu, D.; Shen, J.; et al. Occurrence, Distribution, and Risk Assessment of Quinolone Antibiotics in Municipal Sewage Sludges throughout China. *J. Hazard. Mater.* **2023**, *453*, 131322. [CrossRef] [PubMed]
29. Zhang, L.; Zhu, Z.; Zhao, M.; He, J.; Zhang, X.; Hao, F.; Du, P. Occurrence, Removal, Emission and Environment Risk of 32 Antibiotics and Metabolites in Wastewater Treatment Plants in Wuhu, China. *Sci. Total Environ.* **2023**, *899*, 165681. [CrossRef] [PubMed]
30. Lu, S.; Lin, C.; Lei, K.; Xin, M.; Gu, X.; Lian, M.; Wang, B.; Liu, X.; Ouyang, W.; He, M. Profiling of the Spatiotemporal Distribution, Risks, and Prioritization of Antibiotics in the Waters of Laizhou Bay, Northern China. *J. Hazard. Mater.* **2022**, *424*, 127487. [CrossRef]
31. Sun, C.; Hu, E.; Liu, S.; Wen, L.; Yang, F.; Li, M. Spatial Distribution and Risk Assessment of Certain Antibiotics in 51 Urban Wastewater Treatment Plants in the Transition Zone between North and South China. *J. Hazard. Mater.* **2022**, *437*, 129307. [CrossRef] [PubMed]
32. Chaves, M.d.J.S.; Kulzer, J.; Lima, P. da R.P. de; Barbosa, S.C.; Primel, E.G. Updated Knowledge, Partitioning and Ecological Risk of Pharmaceuticals and Personal Care Products in Global Aquatic Environments. *Environ. Sci. Process. Impacts* **2022**, *24*, 1982–2008. [CrossRef]
33. Zheng, Q.; Zhang, R.; Wang, Y.; Pan, X.; Tang, J.; Zhang, G. Occurrence and Distribution of Antibiotics in the Beibu Gulf, China: Impacts of River Discharge and Aquaculture Activities. *Mar. Environ. Res.* **2012**, *78*, 26–33. [CrossRef]
34. Lin, T.; Yu, S.; Chen, W. Occurrence, Removal and Risk Assessment of Pharmaceutical and Personal Care Products (PPCPs) in an Advanced Drinking Water Treatment Plant (ADWTP) around Taihu Lake in China. *Chemosphere* **2016**, *152*, 1–9. [CrossRef]
35. Wu, C.; Huang, X.; Witter, J.D.; Spongberg, A.L.; Wang, K.; Wang, D.; Liu, J. Occurrence of Pharmaceuticals and Personal Care Products and Associated Environmental Risks in the Central and Lower Yangtze River, China. *Ecotoxicol. Environ. Saf.* **2014**, *106*, 19–26. [CrossRef]
36. Liu, J.; Lu, G.; Xie, Z.; Zhang, Z.; Li, S.; Yan, Z. Occurrence, Bioaccumulation and Risk Assessment of Lipophilic Pharmaceutically Active Compounds in the Downstream Rivers of Sewage Treatment Plants. *Sci. Total Environ.* **2015**, *511*, 54–62. [CrossRef] [PubMed]
37. Chen, H.; Liu, S.; Xu, X.-R.; Zhou, G.-J.; Liu, S.-S.; Yue, W.-Z.; Sun, K.-F.; Ying, G.-G. Antibiotics in the Coastal Environment of the Hailing Bay Region, South China Sea: Spatial Distribution, Source Analysis and Ecological Risks. *Mar. Pollut. Bull.* **2015**, *95*, 365–373. [CrossRef] [PubMed]
38. Santos, J.L.; Aparicio, I.; Alonso, E. Occurrence and Risk Assessment of Pharmaceutically Active Compounds in Wastewater Treatment Plants. A Case Study: Seville City (Spain). *Environ. Int.* **2007**, *33*, 596–601. [CrossRef] [PubMed]
39. Kondor, A.C.; Molnár, É.; Vancsik, A.; Filep, T.; Szeberényi, J.; Szabó, L.; Maász, G.; Pirger, Z.; Weiperth, A.; Ferincz, Á.; et al. Occurrence and Health Risk Assessment of Pharmaceutically Active Compounds in Riverbank Filtrated Drinking Water. *J. Water Process Eng.* **2021**, *41*, 102039. [CrossRef]
40. Wang, C.; Lu, Y.; Wang, C.; Xiu, C.; Cao, X.; Zhang, M.; Song, S. Distribution and Ecological Risks of Pharmaceuticals and Personal Care Products with Different Anthropogenic Stresses in a Coastal Watershed of China. *Chemosphere* **2022**, *303*, 135176. [CrossRef] [PubMed]
41. Ferrari, B.; Paxéus, N.; Giudice, R.L.; Pollio, A.; Garric, J. Ecotoxicological Impact of Pharmaceuticals Found in Treated Wastewaters: Study of Carbamazepine, Clofibric Acid, and Diclofenac. *Ecotoxicol. Environ. Saf.* **2003**, *55*, 359–370. [CrossRef] [PubMed]

42. Xu, N.; Shen, Y.; Jiang, L.; Jiang, B.; Li, Y.; Yuan, Q.; Zhang, Y. Occurrence and Ecological-Risk Levels of Antibiotic Pollution in the Coastal Waters of Eastern China. *Environ. Sci. Pollut. Res.* **2023**, *30*, 71371–71381. [CrossRef]
43. Choi, K.; Kim, Y.; Jung, J.; Kim, M.-H.; Kim, C.-S.; Kim, N.-H.; Park, J. Occurrences and Ecological Risks of Roxithromycin, Trimethoprim, and Chloramphenicol in the Han River, Korea. *Environ. Toxicol. Chem.* **2008**, *27*, 711–719. [CrossRef]
44. Yu, X.; Yu, F.; Li, Z.; Zhan, J. Occurrence, Distribution, and Ecological Risk Assessment of Pharmaceuticals and Personal Care Products in the Surface Water of the Middle and Lower Reaches of the Yellow River (Henan Section). *J. Hazard. Mater.* **2023**, *443*, 130369. [CrossRef]
45. Sun, S.; Chen, Y.; Lin, Y.; An, D. Occurrence, Spatial Distribution, and Seasonal Variation of Emerging Trace Organic Pollutants in Source Water for Shanghai, China. *Sci. Total Environ.* **2018**, *639*, 1–7. [CrossRef]
46. Acuña, V.; Ginebreda, A.; Mor, J.R.; Petrovic, M.; Sabater, S.; Sumpter, J.; Barceló, D. Balancing the Health Benefits and Environmental Risks of Pharmaceuticals: Diclofenac as an Example. *Environ. Int.* **2015**, *85*, 327–333. [CrossRef] [PubMed]
47. Gopal, C.M.; Bhat, K.; Ramaswamy, B.R.; Kumar, V.; Singhal, R.K.; Basu, H.; Udayashankar, H.N.; Vasantharaju, S.G.; Praveenkumarreddy, Y.; Shailesh; et al. Seasonal Occurrence and Risk Assessment of Pharmaceutical and Personal Care Products in Bengaluru Rivers and Lakes, India. *J. Environ. Chem. Eng.* **2021**, *9*, 105610. [CrossRef]
48. Tauxe-Wuersch, A.; De Alencastro, L.F.; Grandjean, D.; Tarradellas, J. Occurrence of Several Acidic Drugs in Sewage Treatment Plants in Switzerland and Risk Assessment. *Water Res.* **2005**, *39*, 1761–1772. [CrossRef] [PubMed]
49. Ashfaq, M.; Khan, K.N.; Rasool, S.; Mustafa, G.; Saif-Ur-Rehman, M.; Nazar, M.F.; Sun, Q.; Yu, C.-P. Occurrence and Ecological Risk Assessment of Fluoroquinolone Antibiotics in Hospital Waste of Lahore, Pakistan. *Environ. Toxicol. Pharmacol.* **2016**, *42*, 16–22. [CrossRef] [PubMed]
50. Mohd Nasir, F.A.; Praveena, S.M.; Aris, A.Z. Public Awareness Level and Occurrence of Pharmaceutical Residues in Drinking Water with Potential Health Risk: A Study from Kajang (Malaysia). *Ecotoxicol. Environ. Saf.* **2019**, *185*, 109681. [CrossRef]
51. de Souza, R.C.; Godoy, A.A.; Kummrow, F.; dos Santos, T.L.; Brandão, C.J.; Pinto, E. Occurrence of Caffeine, Fluoxetine, Bezafibrate and Levothyroxine in Surface Freshwater of São Paulo State (Brazil) and Risk Assessment for Aquatic Life Protection. *Environ. Sci. Pollut. Res.* **2021**, *28*, 20751–20761. [CrossRef] [PubMed]
52. Komori, K.; Suzuki, Y.; Minamiyama, M.; Harada, A. Occurrence of Selected Pharmaceuticals in River Water in Japan and Assessment of Their Environmental Risk. *Environ. Monit. Assess.* **2013**, *185*, 4529–4536. [CrossRef]
53. Singh, V.; Suthar, S. Occurrence, Seasonal Variations, and Ecological Risk of Pharmaceuticals and Personal Care Products in River Ganges at Two Holy Cities of India. *Chemosphere* **2021**, *268*, 129331. [CrossRef] [PubMed]
54. Li, S.; Wen, J.; He, B.; Wang, J.; Hu, X.; Liu, J. Occurrence of Caffeine in the Freshwater Environment: Implications for Ecopharmacovigilance. *Environ. Pollut.* **2020**, *263*, 114371. [CrossRef]
55. Ilechukwu, I.; Okonkwo, C.J.; Olusina, T.A.; Mpock, J.A.; Ilechukwu, C. Occurrence and Risk Assessment of Selected Pharmaceuticals in Water and Sediments of Usuma Dam, Abuja, Nigeria. *Int. J. Environ. Anal. Chem.* **2023**, *103*, 4398–4410. [CrossRef]
56. Qu, H.; Barrett, H.; Wang, B.; Han, J.; Wang, F.; Gong, W.; Wu, J.; Wang, W.; Yu, G. Co-Occurrence of Antiseptic Triclocarban and Chiral Anti-Inflammatory Ibuprofen in Environment: Association between Biological Effect in Sediment and Risk to Human Health. *J. Hazard. Mater.* **2021**, *407*, 124871. [CrossRef] [PubMed]
57. Lindqvist, N.; Tuhkanen, T.; Kronberg, L. Occurrence of Acidic Pharmaceuticals in Raw and Treated Sewages and in Receiving Waters. *Water Res.* **2005**, *39*, 2219–2228. [CrossRef] [PubMed]
58. Sörengård, M.; Campos-Pereira, H.; Ullberg, M.; Lai, F.Y.; Golovko, O.; Ahrens, L. Mass Loads, Source Apportionment, and Risk Estimation of Organic Micropollutants from Hospital and Municipal Wastewater in Recipient Catchments. *Chemosphere* **2019**, *234*, 931–941. [CrossRef] [PubMed]
59. Godoy, A.A.; Kummrow, F.; Pamplin, P.A.Z. Occurrence, Ecotoxicological Effects and Risk Assessment of Antihypertensive Pharmaceutical Residues in the Aquatic Environment—A Review. *Chemosphere* **2015**, *138*, 281–291. [CrossRef] [PubMed]
60. Zhang, K.; Zhao, Y.; Fent, K. Cardiovascular Drugs and Lipid Regulating Agents in Surface Waters at Global Scale: Occurrence, Ecotoxicity and Risk Assessment. *Sci. Total Environ.* **2020**, *729*, 138770. [CrossRef]
61. Mijangos, L.; Ziarrusta, H.; Ros, O.; Kortazar, L.; Fernández, L.A.; Olivares, M.; Zuloaga, O.; Prieto, A.; Etxebarria, N. Occurrence of Emerging Pollutants in Estuaries of the Basque Country: Analysis of Sources and Distribution, and Assessment of the Environmental Risk. *Water Res.* **2018**, *147*, 152–163. [CrossRef]
62. Peng, X.; Ou, W.; Wang, C.; Wang, Z.; Huang, Q.; Jin, J.; Tan, J. Occurrence and Ecological Potential of Pharmaceuticals and Personal Care Products in Groundwater and Reservoirs in the Vicinity of Municipal Landfills in China. *Sci. Total Environ.* **2014**, *490*, 889–898. [CrossRef]
63. Ashfaq, M.; Nawaz Khan, K.; Saif Ur Rehman, M.; Mustafa, G.; Faizan Nazar, M.; Sun, Q.; Iqbal, J.; Mulla, S.I.; Yu, C.-P. Ecological Risk Assessment of Pharmaceuticals in the Receiving Environment of Pharmaceutical Wastewater in Pakistan. *Ecotoxicol. Environ. Saf.* **2017**, *136*, 31–39. [CrossRef]
64. Isidori, M.; Lavorgna, M.; Nardelli, A.; Parrella, A.; Previtera, L.; Rubino, M. Ecotoxicity of Naproxen and Its Phototransformation Products. *Sci. Total Environ.* **2005**, *348*, 93–101. [CrossRef]
65. Korkmaz, N.E.; Savun-Hekimoğlu, B.; Aksu, A.; Burak, S.; Caglar, N.B. Occurrence, Sources and Environmental Risk Assessment of Pharmaceuticals in the Sea of Marmara, Turkey. *Sci. Total Environ.* **2022**, *819*, 152996. [CrossRef]

66. Cao, J.; Shi, J.; Han, R.; Li, Y.; Yang, Z. Seasonal Variations in the Occurrence and Distribution of Estrogens and Pharmaceuticals in the Zhangweinanyun River System. *Chin. Sci. Bull.* **2010**, *55*, 3138–3144. [CrossRef]
67. Wang, D.; Luo, Z.; Zhang, X.; Lin, L.; Du, M.; Du Laing, G.; Yan, C. Occurrence, Distribution and Risk Assessment of Estrogenic Compounds for Three Source Water Types in Ningbo City, China. *Environ. Earth Sci.* **2015**, *74*, 5961–5969. [CrossRef]
68. Yu, Y.; Wu, L.; Chang, A.C. Seasonal Variation of Endocrine Disrupting Compounds, Pharmaceuticals and Personal Care Products in Wastewater Treatment Plants. *Sci. Total Environ.* **2013**, *442*, 310–316. [CrossRef] [PubMed]
69. Du, B.; Fan, G.; Yu, W.; Yang, S.; Zhou, J.; Luo, J. Occurrence and Risk Assessment of Steroid Estrogens in Environmental Water Samples: A Five-Year Worldwide Perspective. *Environ. Pollut.* **2020**, *267*, 115405. [CrossRef] [PubMed]
70. Liu, J.; Zhang, L.; Lu, G.; Jiang, R.; Yan, Z.; Li, Y. Occurrence, Toxicity and Ecological Risk of Bisphenol A Analogues in Aquatic Environment—A Review. *Ecotoxicol. Environ. Saf.* **2021**, *208*, 111481. [CrossRef] [PubMed]
71. Wang, Y.; Wang, Q.; Hu, L.; Lu, G.; Li, Y. Occurrence of Estrogens in Water, Sediment and Biota and Their Ecological Risk in Northern Taihu Lake in China. *Environ. Geochem. Health* **2015**, *37*, 147–156. [CrossRef] [PubMed]
72. Hu, Y.; Zhu, Q.; Yan, X.; Liao, C.; Jiang, G. Occurrence, Fate and Risk Assessment of BPA and Its Substituents in Wastewater Treatment Plant: A Review. *Environ. Res.* **2019**, *178*, 108732. [CrossRef] [PubMed]
73. Liu, R.; Luo, X.; Shu, S.; Ding, J.; Zhang, G.; Wang, Z.; Zou, H.; Zhang, Y. Impact of Rainfall on the Occurrence, Spatiotemporal Distribution, and Partition Trend of Micropollutants in Taihu Lake, China: Bisphenol A and 4-Nonylphenol as Examples. *Ecotoxicol. Environ. Saf.* **2020**, *204*, 111064. [CrossRef] [PubMed]
74. Shen, J.; Sui, Y.; Feng, J.; Wang, X.; Li, X.; Jiang, S.; Zhang, Z.; Zi, J.; Sun, T.; Gao, Y.; et al. Environmental Occurrence and Ecological Risk of Bisphenol A in Erhai Lake Basin Away From Industrial Regions in China. *Pol. J. Environ. Stud.* **2020**, *30*, 841–850. [CrossRef]
75. Catenza, C.J.; Farooq, A.; Shubear, N.S.; Donkor, K.K. A Targeted Review on Fate, Occurrence, Risk and Health Implications of Bisphenol Analogues. *Chemosphere* **2021**, *268*, 129273. [CrossRef]
76. Zhong, M.; Wang, T.; Zhao, W.; Huang, J.; Wang, B.; Blaney, L.; Bu, Q.; Yu, G. Emerging Organic Contaminants in Chinese Surface Water: Identification of Priority Pollutants. *Engineering* **2022**, *11*, 111–125. [CrossRef]
77. Wu, J.; Shi, D.; Wang, S.; Yang, X.; Zhang, H.; Zhang, T.; Zheng, L.; Zhang, Y. Derivation of Water Quality Criteria for Carbamazepine and Ecological Risk Assessment in the Nansi Lake Basin. *Int. J. Environ. Res. Public Health* **2022**, *19*, 10875. [CrossRef]
78. Huang, Q.; Bu, Q.; Zhong, W.; Shi, K.; Cao, Z.; Yu, G. Derivation of Aquatic Predicted No-Effect Concentration (PNEC) for Ibuprofen and Sulfamethoxazole Based on Various Toxicity Endpoints and the Associated Risks. *Chemosphere* **2018**, *193*, 223–229. [CrossRef]
79. Wang, L.; Wang, Z.; Liu, J.; Ji, G.; Shi, L.; Xu, J.; Yang, J. Deriving the Freshwater Quality Criteria of BPA, BPF and BPAF for Protecting Aquatic Life. *Ecotoxicol. Environ. Saf.* **2018**, *164*, 713–721. [CrossRef]
80. Wang, Y.; Na, G.; Zong, H.; Ma, X.; Yang, X.; Mu, J.; Wang, L.; Lin, Z.; Zhang, Z.; Wang, J.; et al. Applying Adverse Outcome Pathways and Species Sensitivity–Weighted Distribution to Predicted-No-Effect Concentration Derivation and Quantitative Ecological Risk Assessment for Bisphenol A and 4-Nonylphenol in Aquatic Environments: A Case Study on Tianjin City, China. *Environ. Toxicol. Chem.* **2018**, *37*, 551–562. [CrossRef]
81. Zhou, Y.; Meng, J.; Zhang, M.; Chen, S.; He, B.; Zhao, H.; Li, Q.; Zhang, S.; Wang, T. Which Type Of pollutants Need to be Controlled with Priority in Wastewater Treatment Plants: Traditional or Emerging Pollutants? *Environ. Int.* **2019**, *131*, 104982. [CrossRef]
82. Yang, Y.-Y.; Liu, W.-R.; Liu, Y.-S.; Zhao, J.-L.; Zhang, Q.-Q.; Zhang, M.; Zhang, J.-N.; Jiang, Y.-X.; Zhang, L.-J.; Ying, G.-G. Suitability of Pharmaceuticals and Personal Care Products (PPCPs) and Artificial Sweeteners (ASs) as Wastewater Indicators in the Pearl River Delta, South China. *Sci. Total Environ.* **2017**, *590–591*, 611–619. [CrossRef]
83. Tang, Y.; Guo, L.-L.; Hong, C.-Y.; Bing, Y.-X.; Xu, Z.-C. Seasonal Occurrence, Removal and Risk Assessment of 10 Pharmaceuticals in 2 Sewage Treatment Plants of Guangdong, China. *Environ. Technol.* **2019**, *40*, 458–469. [CrossRef]
84. Wang, L.; Zhu, D.; Cao, Y.; Yu, X.; Hui, Y.; Li, W.; Wang, D. Seasonal changes and ecological risk assessment of pharmaceutical and personal care products in the effluents of wastewater treatment plants in Beijin. *Acta Sci. Circumstantiae* **2021**, *41*, 2922–2932. [CrossRef]
85. Shao, T.; Ben, W.; Su, D.; Zhang, H.; Hou, P.; Qiang, Z.; Zhang, Y. Quantitative determination of typical pharmaceuticals and their metabolites in municipal wastewater treatment plants. *Acta Sci. Circumstantiae* **2020**, *40*, 2136–2141. [CrossRef]
86. Yang, Z.; Li, J.; Zhang, S.; Xiang, F.; Tang, T.; Yang, Y. Pollution Level and Ecological Risk of Typical Antibiotics in Guiyang Wastewater Treatment Plants. *Environ. Sci.* **2019**, *40*, 3249–3256. [CrossRef]
87. Ye, P.; You, W.; Yang, B.; Chen, Y.; Wang, L.; Zhao, J.; Ying, G. Pollution Characteristics and Removal of Typical Pharmaceuticals in Hospital Wastewater and Municipal Wastewater Treatment Plants. *Environ. Sci.* **2021**, *42*, 2928–2936. [CrossRef]
88. Zhang, X.; Zhao, H.; Du, J.; Qu, Y.; Shen, C.; Tan, F.; Chen, J.; Quan, X. Occurrence, Removal, and Risk Assessment of Antibiotics in 12 Wastewater Treatment Plants from Dalian, China. *Environ. Sci. Pollut. Res.* **2017**, *24*, 16478–16487. [CrossRef]
89. Yao, L.; Chen, Z.-Y.; Dou, W.-Y.; Yao, Z.-K.; Duan, X.-C.; Chen, Z.-F.; Zhang, L.-J.; Nong, Y.-J.; Zhao, J.-L.; Ying, G.-G. Occurrence, Removal and Mass Loads of Antiviral Drugs in Seven Wastewater Treatment Plants with Various Treatment Processes. *Water Res.* **2021**, *207*, 117803. [CrossRef]

90. Ashfaq, M.; Li, Y.; Wang, Y.; Chen, W.; Wang, H.; Chen, X.; Wu, W.; Huang, Z.; Yu, C.-P.; Sun, Q. Occurrence, Fate, and Mass Balance of Different Classes of Pharmaceuticals and Personal Care Products in an Anaerobic-Anoxic-Oxic Wastewater Treatment Plant in Xiamen, China. *Water Res.* **2017**, *123*, 655–667. [CrossRef]
91. Duan, L.; Zhang, Y.; Wang, B.; Deng, S.; Huang, J.; Wang, Y.; Yu, G. Occurrence, Elimination, Enantiomeric Distribution and Intra-Day Variations of Chiral Pharmaceuticals in Major Wastewater Treatment Plants in Beijing, China. *Environ. Pollut.* **2018**, *239*, 473–482. [CrossRef]
92. Gao, L.; Shi, Y.; Li, W.; Niu, H.; Liu, J.; Cai, Y. Occurrence of Antibiotics in Eight Sewage Treatment Plants in Beijing, China. *Chemosphere* **2012**, *86*, 665–671. [CrossRef]
93. Wang, W.; Zhang, W.; Liang, H.; Gao, D. Occurrence and Fate of Typical Antibiotics in Wastewater Treatment Plants in Harbin, North-East China. *Front. Environ. Sci. Eng.* **2019**, *13*, 34. [CrossRef]
94. Gan, X.; Yan, Q.; Gao, X.; Zhang, Y.; Zi, C.; Peng, X.; Guo, J. Occurrence and Fate of Typical Antibiotics in a Wastewater Treatment Plant in Southwest China. *Environ. Sci.* **2014**, *35*, 1817–1823. [CrossRef]
95. Zhao, B.; Liu, Y.; Dong, W.; Li, M.; Lu, S.; Wang, W.; Fan, Y. Occurrence and Fate of Ten Sulfonamide Antibiotics in Typical Wastewater Treatment Plants in the City of Jinan of Northeastern China. *Desalination Water Treat.* **2020**, *206*, 340–348. [CrossRef]
96. Li, Q.; Liu, H.; Liu, J.; Jia, L.; Wang, Q. Occurrence and fate of sulfonamide and β-lactam antibiotics in wastewater treatment plants in Handan. *Environ. Chem.* **2018**, *37*, 1738–1745.
97. Wu, M.; Xiang, J.; Que, C.; Chen, F.; Xu, G. Occurrence and Fate of Psychiatric Pharmaceuticals in the Urban Water System of Shanghai, China. *Chemosphere* **2015**, *138*, 486–493. [CrossRef]
98. Hu, J.; Zhou, J.; Zhou, S.; Wu, P.; Tsang, Y.F. Occurrence and Fate of Antibiotics in a Wastewater Treatment Plant and Their Biological Effects on Receiving Waters in Guizhou. *Process Saf. Environ. Prot.* **2018**, *113*, 483–490. [CrossRef]
99. Du, J.; Fan, Y.; Qian, X. Occurrence and Behavior of Pharmaceuticals in Sewage Treatment Plants in Eastern China. *Front. Environ. Sci. Eng.* **2015**, *9*, 725–730. [CrossRef]
100. Wang, Y.; Li, Y.; Hu, A.; Rashid, A.; Ashfaq, M.; Wang, Y.; Wang, H.; Luo, H.; Yu, C.-P.; Sun, Q. Monitoring, Mass Balance and Fate of Pharmaceuticals and Personal Care Products in Seven Wastewater Treatment Plants in Xiamen City, China. *J. Hazard. Mater.* **2018**, *354*, 81–90. [CrossRef]
101. Lin, H.; Li, H.; Chen, L.; Li, L.; Yin, L.; Lee, H.; Yang, Z. Mass Loading and Emission of Thirty-Seven Pharmaceuticals in a Typical Municipal Wastewater Treatment Plant in Hunan Province, Southern China. *Ecotoxicol. Environ. Saf.* **2018**, *147*, 530–536. [CrossRef]
102. Ke, R.; Jiang, Y.; Huang, Q.; Chen, L. Investigative Screening of Pharmaceuticals in a Municipal Wastewater Treatment Plant in Shanghai. *Asian J. Ecotoxicol.* **2014**, *9*, 1146–1155.
103. Yin, Z.Y.; Min, L.; Jin, L.; Wang, F.; Sun, H. HPLC-MS/MS determination of eight PPCPs in sewage and sludge from three types of sewage treatment plants. *Environ. Chem.* **2018**, *37*, 1720–1727. [CrossRef]
104. Chen, L.; Fu, W.; Tan, Y.; Zhang, X. Emerging Organic Contaminants and Odorous Compounds in Secondary Effluent Wastewater: Identification and Advanced Treatment. *J. Hazard. Mater.* **2021**, *408*, 124817. [CrossRef]
105. Li, Y.; Niu, X.; Yao, C.; Yang, W.; Lu, G. Distribution, Removal, and Risk Assessment of Pharmaceuticals and Their Metabolites in Five Sewage Plants. *Int. J. Environ. Res. Public Health* **2019**, *16*, 4729. [CrossRef]
106. Leung, H.W.; Minh, T.B.; Murphy, M.B.; Lam, J.C.W.; So, M.K.; Martin, M.; Lam, P.K.S.; Richardson, B.J. Distribution, Fate and Risk Assessment of Antibiotics in Sewage Treatment Plants in Hong Kong, South China. *Environ. Int.* **2012**, *42*, 1–9. [CrossRef] [PubMed]
107. Chunhui, Z.; Liangliang, W.; Xiangyu, G.; Xudan, H. Antibiotics in WWTP Discharge into the Chaobai River, Beijing. *Arch. Environ. Prot.* **2016**, *42*, 48–57. [CrossRef]
108. Wang, K.; Zhuang, T.; Su, Z.; Chi, M.; Wang, H. Antibiotic Residues in Wastewaters from Sewage Treatment Plants and Pharmaceutical Industries: Occurrence, Removal and Environmental Impacts. *Sci. Total Environ.* **2021**, *788*, 147811. [CrossRef] [PubMed]
109. Li, W.-L.; Zhang, Z.-F.; Ma, W.-L.; Liu, L.-Y.; Song, W.-W.; Li, Y.-F. An Evaluation on the Intra-Day Dynamics, Seasonal Variations and Removal of Selected Pharmaceuticals and Personal Care Products from Urban Wastewater Treatment Plants. *Sci. Total Environ.* **2018**, *640–641*, 1139–1147. [CrossRef] [PubMed]
110. Xu, E.G.B.; Liu, S.; Ying, G.-G.; Zheng, G.J.S.; Lee, J.H.W.; Leung, K.M.Y. The Occurrence and Ecological Risks of Endocrine Disrupting Chemicals in Sewage Effluents from Three Different Sewage Treatment Plants, and in Natural Seawater from a Marine Reserve of Hong Kong. *Mar. Pollut. Bull.* **2014**, *85*, 352–362. [CrossRef] [PubMed]
111. Liang, H.; Gong, J.; Zhou, K.; Deng, L.; Chen, J.; Guo, L.; Jiang, M.; Lin, J.; Tang, H.; Liu, X. Removal Efficiencies and Risk Assessment of Endocrine-Disrupting Chemicals at Two Wastewater Treatment Plants in South China. *Ecotoxicol. Environ. Saf.* **2021**, *225*, 112758. [CrossRef] [PubMed]
112. Wang, K.; Huang, Z.; Wang, Y.; Li, Z.; Ye, X.; Yang, S.; Huang, Z. Pollution characteristics and ecological risk assessment of typical EDCs and PPCPs in a municipal sewage treatment plant in Huizhou. *Water Wastewater* **2017**, *53*, 152–154. [CrossRef]
113. Zhou, Y.; Zha, J.; Xu, Y.; Lei, B.; Wang, Z. Occurrences of Six Steroid Estrogens from Different Effluents in Beijing, China. *Environ. Monit. Assess.* **2012**, *184*, 1719–1729. [CrossRef]

114. Xu, G.; Ma, S.; Tang, L.; Sun, R.; Xiang, J.; Xu, B.; Bao, Y.; Wu, M. Occurrence, Fate, and Risk Assessment of Selected Endocrine Disrupting Chemicals in Wastewater Treatment Plants and Receiving River of Shanghai, China. *Environ. Sci. Pollut. Res.* **2016**, *23*, 25442–25450. [CrossRef]
115. Xu, Y.; Xu, N.; Llewellyn, N.R.; Tao, H. Occurrence and Removal of Free and Conjugated Estrogens in Wastewater and Sludge in Five Sewage Treatment Plants. *Environ. Sci. Process. Impacts* **2014**, *16*, 262–270. [CrossRef]
116. Ashfaq, M.; Li, Y.; Wang, Y.; Qin, D.; Rehman, M.S.U.; Rashid, A.; Yu, C.-P.; Sun, Q. Monitoring and Mass Balance Analysis of Endocrine Disrupting Compounds and Their Transformation Products in an Anaerobic-Anoxic-Oxic Wastewater Treatment System in Xiamen, China. *Chemosphere* **2018**, *204*, 170–177. [CrossRef] [PubMed]
117. He, Y.; Chen, W.; Zheng, X.; Wang, X.; Huang, X. Fate and Removal of Typical Pharmaceuticals and Personal Care Products by Three Different Treatment Processes. *Sci. Total Environ.* **2013**, *447*, 248–254. [CrossRef] [PubMed]
118. Sun, Q.; Wang, Y.; Li, Y.; Ashfaq, M.; Dai, L.; Xie, X.; Yu, C.-P. Fate and Mass Balance of Bisphenol Analogues in Wastewater Treatment Plants in Xiamen City, China. *Environ. Pollut.* **2017**, *225*, 542–549. [CrossRef] [PubMed]
119. Wu, L.; Lu, C.; Huang, N.; Zhong, M.; Teng, Y.; Tian, Y.; Ye, K.; Liang, L.; Hu, Z. Exploration of the Effect of Simultaneous Removal of EDCs in the Treatment Process of Different Types of Wastewater. *Water Sci. Technol.* **2022**, *87*, 436–453. [CrossRef] [PubMed]
120. Yang, M.; Wang, K.; Shen, Y.; Wu, M. Evaluation of Estrogenic Activity in Surface Water and Municipal Wastewater in Shanghai, China. *Bull. Environ. Contam. Toxicol.* **2011**, *87*, 215–219. [CrossRef] [PubMed]
121. Xu, W.; Yan, W.; Huang, W.; Miao, L.; Zhong, L. Endocrine-Disrupting Chemicals in the Pearl River Delta and Coastal Environment: Sources, Transfer, and Implications. *Environ. Geochem. Health* **2014**, *36*, 1095–1104. [CrossRef] [PubMed]
122. Jiang, R.; Liu, J.; Huang, B.; Wang, X.; Luan, T.; Yuan, K. Assessment of the Potential Ecological Risk of Residual Endocrine-Disrupting Chemicals from Wastewater Treatment Plants. *Sci. Total Environ.* **2020**, *714*, 136689. [CrossRef]
123. Qiang, Z.; Dong, H.; Zhu, B.; Qu, J.; Nie, Y. A Comparison of Various Rural Wastewater Treatment Processes for the Removal of Endocrine-Disrupting Chemicals (EDCs). *Chemosphere* **2013**, *92*, 986–992. [CrossRef]
124. Triebskorn, R.; Casper, H.; Scheil, V.; Schwaiger, J. Ultrastructural Effects of Pharmaceuticals (Carbamazepine, Clofibric Acid, Metoprolol, Diclofenac) in Rainbow Trout (*Oncorhynchus mykiss*) and Common Carp (*Cyprinus carpio*). *Anal. Bioanal. Chem.* **2007**, *387*, 1405–1416. [CrossRef]
125. Zhu, B.; Liu, L.; Gong, Y.-X.; Ling, F.; Wang, G.-X. Triazole-Induced Toxicity in Developing Rare Minnow (*Gobiocypris rarus*) Embryos. *Environ. Sci. Pollut. Res. Int.* **2014**, *21*, 13625–13635. [CrossRef]
126. Guo, J.; Bai, Y.; Chen, Z.; Mo, J.; Li, Q.; Sun, H.; Zhang, Q. Transcriptomic Analysis Suggests the Inhibition of DNA Damage Repair in Green Alga Raphidocelis Subcapitata Exposed to Roxithromycin. *Ecotoxicol. Environ. Saf.* **2020**, *201*, 110737. [CrossRef]
127. Huggett, D.B.; Brooks, B.W.; Peterson, B.; Foran, C.M.; Schlenk, D. Toxicity of Select Beta Adrenergic Receptor-Blocking Pharmaceuticals (B-Blockers) on Aquatic Organisms. *Arch. Environ. Contam. Toxicol.* **2002**, *43*, 229–235. [CrossRef] [PubMed]
128. Baran, W.; Sochacka, J.; Wardas, W. Toxicity and Biodegradability of Sulfonamides and Products of Their Photocatalytic Degradation in Aqueous Solutions. *Chemosphere* **2006**, *65*, 1295–1299. [CrossRef] [PubMed]
129. Kwon, B.; Kho, Y.; Kim, P.-G.; Ji, K. Thyroid Endocrine Disruption in Male Zebrafish Following Exposure to Binary Mixture of Bisphenol AF and Sulfamethoxazole. *Environ. Toxicol. Pharmacol.* **2016**, *48*, 168–174. [CrossRef] [PubMed]
130. Owen, S.F.; Huggett, D.B.; Hutchinson, T.H.; Hetheridge, M.J.; McCormack, P.; Kinter, L.B.; Ericson, J.F.; Constantine, L.A.; Sumpter, J.P. The Value of Repeating Studies and Multiple Controls: Replicated 28-day Growth Studies of Rainbow Trout Exposed to Clofibric Acid. *Enviro Toxic. Chem.* **2010**, *29*, 2831–2839. [CrossRef] [PubMed]
131. Lower, N. The Effects of Contaminants on Various Life-Cycle Stages of Atlantic Salmon (*Salmo salar* L.). Ph.D. Thesis, University of Portsmouth, Portsmouth, UK, 2008.
132. Ebringer, L.; Krajcovic, J.; Polónyi, J.; Lahitová, N.; Doupovcová, M.; Dobias, J. Tetracycline Reduces Fluoroquinolones-Induced Bleaching of Euglena Gracilis. *Mutat. Res.* **1996**, *340*, 141–149. [CrossRef]
133. Giusti, A.; Ducrot, V.; Joaquim-Justo, C.; Lagadic, L. Testosterone Levels and Fecundity in the Hermaphroditic Aquatic Snail Lymnaea Stagnalis Exposed to Testosterone and Endocrine Disruptors. *Environ. Toxicol. Chem.* **2013**, *32*, 1740–1745. [CrossRef] [PubMed]
134. Wilkinson, R.E.; Duncan, R.R. Seashore Paspalum (Paspalum Vaginatuh Swartz) Seminal Root Response to Calcium (45 $ca^{2+}$) Absorption Modifiers. *J. Plant Nutr.* **1994**, *17*, 1385–1392. [CrossRef]
135. Zhang, X.; Hecker, M.; Tompsett, A.R.; Park, J.-W.; Jones, P.D.; Newsted, J.; Au, D.; Kong, R.; Wu, R.S.S.; Giesy, J.P. Responses of the Medaka HPG Axis PCR Array and Reproduction to Prochloraz and Ketoconazole. *Environ. Sci. Technol.* **2008**, *42*, 6762–6769. [CrossRef]
136. Ayanda, O.; Oniye, S.; Auta, J.; Ajibola, V.; Bello, O. Responses of the African Catfish Clarias Gariepinus to Long-Term Exposure to Glyphosate- and Paraquat-Based Herbicides. *Afr. J. Aquat. Sci.* **2015**, *40*, 261–267. [CrossRef]
137. Crago, J.; Klaper, R. Place-Based Screening of Mixtures of Dominant Emerging Contaminants Measured in Lake Michigan Using Zebrafish Embryo Gene Expression Assay. *Chemosphere* **2018**, *193*, 1226–1234. [CrossRef] [PubMed]
138. Diaz, L. Pharmaceuticals in the Environment: Sources, Fate, Effects and Risks. *Waste Manag.* **2003**, *23*, 193. [CrossRef]
139. Zhu, Y.; Yang, D.; Duan, X.; Zhang, Y.; Chen, D.; Gong, Z.; Liu, C. Perfluorooctane Sulfonate Promotes Doxycycline-Induced Liver Tumor Progression in Male KrasV12 Transgenic Zebrafish. *Environ. Res.* **2021**, *196*, 110962. [CrossRef] [PubMed]
140. Cunha, E.; Machado, J. Parturition in Anodonta Cygnea Induced by Selective Serotonin Reuptake Inhibitors (SSRIs). *Can. J. Zool.* **2011**, *79*, 95–100. [CrossRef]

141. Divo, A.A.; Geary, T.G.; Jensen, J.B. Oxygen- and Time-Dependent Effects of Antibiotics and Selected Mitochondrial Inhibitors on Plasmodium Falciparum in Culture. *Antimicrob. Agents Chemother.* **1985**, *27*, 21–27. [CrossRef] [PubMed]
142. Passarelli, F.; Merante, A.; Pontieri, F.E.; Margotta, V.; Venturini, G.; Palladini, G. Opioid-Dopamine Interaction in Planaria: A Behavioral Study. *Comp. Biochem. Physiol. C Pharmacol. Toxicol. Endocrinol.* **1999**, *124*, 51–55. [CrossRef] [PubMed]
143. Mennillo, E.; Pretti, C.; Cappelli, F.; Luci, G.; Intorre, L.; Meucci, V.; Arukwe, A. Novel Organ-Specific Effects of Ketoprofen and Its Enantiomer, Dexketoprofen on Toxicological Response Transcripts and Their Functional Products in Salmon. *Aquat. Toxicol.* **2020**, *229*, 105677. [CrossRef] [PubMed]
144. Bereketoglu, C.; Pradhan, A.; Olsson, P.-E. Nonsteroidal Anti-Inflammatory Drugs (NSAIDs) Cause Male-Biased Sex Differentiation in Zebrafish. *Aquat. Toxicol.* **2020**, *223*, 105476. [CrossRef] [PubMed]
145. Lebreton, M.; Sire, S.; Carayon, J.-L.; Malgouyres, J.-M.; Vignet, C.; Géret, F.; Bonnafé, E. Low Concentrations of Oxazepam Induce Feeding and Molecular Changes in Radix Balthica Juveniles. *Aquat. Toxicol.* **2021**, *230*, 105694. [CrossRef]
146. Macphee, C.; Ruelle, R. Lethal Effects of 1888 Chemicals upon 4 Species of Fish from Western North America. Research technical Compeltion Report, Project A-013-IDA. Idaho Waters Digital Library, University of Idaho Library Digital Collections, 1969. Available online: https://www.lib.uidaho.edu/digital/iwdl/items/iwdl-196913.html (accessed on 15 June 2023).
147. Baldisserotto, B.; Brinn, R.P.; Brandão, F.R.; Gomes, L.C.; Abreu, J.S.; McComb, D.M.; Marcon, J.L. Ion Flux and Cortisol Responses of Cardinal Tetra, *Paracheirodon axelrodi* (Schultz, 1956), to Additives (Tetracycline, Tetracycline + Salt or Amquel®) Used during Transportation: Contributions to Amazonian Ornamental Fish Trade. *J. Appl. Ichthyol.* **2014**, *30*, 86–92. [CrossRef]
148. Gustafson, A.-L.; Stedman, D.B.; Ball, J.; Hillegass, J.M.; Flood, A.; Zhang, C.X.; Panzica-Kelly, J.; Cao, J.; Coburn, A.; Enright, B.P.; et al. Inter-Laboratory Assessment of a Harmonized Zebrafish Developmental Toxicology Assay—Progress Report on Phase I. *Reprod. Toxicol.* **2012**, *33*, 155–164. [CrossRef]
149. Yang, L.-H.; Ying, G.-G.; Su, H.-C.; Stauber, J.L.; Adams, M.S.; Binet, M.T. Growth-Inhibiting Effects of 12 Antibacterial Agents and Their Mixtures on the Freshwater Microalga Pseudokirchneriella Subcapitata. *Environ. Toxicol. Chem.* **2008**, *27*, 1201–1208. [CrossRef] [PubMed]
150. Bulut, C.; Kubilay, A.; Hanol Bektaş, Z.; Birden, B. Histopathological Effects of Formaldehyde (CH2O) on Rainbow Trout (*Oncorhynchus mykiss* Walbaum, 1792). *J. Limnol. Freshw. Fish. Res.* **2015**, *1*, 43.
151. Park, K.; Kwak, I.-S. Gene Expression of Ribosomal Protein mRNA in *Chironomus riparius*: Effects of Endocrine Disruptor Chemicals and Antibiotics. *Comp. Biochem. Physiol. Part C Toxicol. Pharmacol.* **2012**, *156*, 113–120. [CrossRef] [PubMed]
152. Zhang, X.; Gong, Z. Fluorescent Transgenic Zebrafish Tg(Nkx2.2a:mEGFP) Provides a Highly Sensitive Monitoring Tool for Neurotoxins. *PLoS ONE* **2013**, *8*, e55474. [CrossRef] [PubMed]
153. Rocco, L.; Frenzilli, G.; Fusco, D.; Peluso, C.; Stingo, V. Evaluation of Zebrafish DNA Integrity after Exposure to Pharmacological Agents Present in Aquatic Environments. *Ecotoxicol. Environ. Saf.* **2010**, *73*, 1530–1536. [CrossRef] [PubMed]
154. Eguchi, K.; Nagase, H.; Ozawa, M.; Endoh, Y.S.; Goto, K.; Hirata, K.; Miyamoto, K.; Yoshimura, H. Evaluation of Antimicrobial Agents for Veterinary Use in the Ecotoxicity Test Using Microalgae. *Chemosphere* **2004**, *57*, 1733–1738. [CrossRef] [PubMed]
155. Han, S.; Choi, K.; Kim, J.; Ji, K.; Kim, S.; Ahn, B.; Yun, J.; Choi, K.; Khim, J.S.; Zhang, X.; et al. Endocrine Disruption and Consequences of Chronic Exposure to Ibuprofen in Japanese Medaka (*Oryzias latipes*) and Freshwater Cladocerans *Daphnia magna* and Moina Macrocopa. *Aquat. Toxicol.* **2010**, *98*, 256–264. [CrossRef]
156. Tojo, J.; Santamariña, M.; Ubeira, F.; Leiro, J. Efficacy of Antiprotozoal Drugs against Gyrodactylosis in Rainbow Trout *(Oncorhynchus mykiss)*. *Bull. Eur. Assoc. Fish Pathol.* **1993**, *13*, 79–82.
157. Steinkey, D.; Lari, E.; Woodman, S.G.; Luong, K.H.; Wong, C.S.; Pyle, G.G. Effects of Gemfibrozil on the Growth, Reproduction, and Energy Stores of *Daphnia magna* in the Presence of Varying Food Concentrations. *Chemosphere* **2018**, *192*, 75–80. [CrossRef]
158. Oggier, D.M.; Weisbrod, C.J.; Stoller, A.M.; Zenker, A.K.; Fent, K. Effects of Diazepam on Gene Expression and Link to Physiological Effects in Different Life Stages in Zebrafish *Danio rerio*. *Environ. Sci. Technol.* **2010**, *44*, 7685–7691. [CrossRef] [PubMed]
159. Shi, H.; Sun, Z.; Liu, Z.; Xue, Y. Effects of Clotrimazole and Amiodarone on Early Development of Amphibian (*Xenopus tropicalis*). *Toxicol. Environ. Chem.* **2012**, *94*, 128–135. [CrossRef]
160. Nesbitt, R. Effects of Chronic Exposure to Ibuprofen and Naproxen on Florida Flagfish (*Jordanella floridae*) over One Complete Life-Cycle. Ph.D. Thesis, Ontario Tech University, Oshawa, ON, Canada, 2011.
161. Li, Y.; Ma, Y.; Yang, L.; Duan, S.; Zhou, F.; Chen, J.; Liu, Y.; Zhang, B. Effects of Azithromycin on Feeding Behavior and Nutrition Accumulation of *Daphnia magna* under the Different Exposure Pathways. *Ecotoxicol. Environ. Saf.* **2020**, *197*, 110573. [CrossRef]
162. Brain, R.A.; Johnson, D.J.; Richards, S.M.; Sanderson, H.; Sibley, P.K.; Solomon, K.R. Effects of 25 Pharmaceutical Compounds to Lemna Gibba Using a Seven-Day Static-Renewal Test. *Environ. Toxicol. Chem.* **2004**, *23*, 371–382. [CrossRef] [PubMed]
163. Nandurkar, H.P.; Zambare, S.P. Effect of Tetracycline and Chloramphenicol on Protein Contents in Different Tissues of Freshwater Bivalve, *Parreysia cylindrica* (Annandale & Prashad). *J. Am. Chem. Soc.* **2012**, *18*, 259–264. [CrossRef] [PubMed]
164. Reda, R.M.; Ibrahim, R.E.; Ahmed, E. N.G., El-Bouly, E.M. Effect of Oxytetracycline and Florfenicol as Growth Promoters on the Health Status of Cultured *Oreochromis niloticus*. *Egypt. J. Aquat. Res.* **2013**, *39*, 241–248. [CrossRef]
165. Han, G.H.; Hur, H.G.; Kim, S.D. Ecotoxicological Risk of Pharmaceuticals from Wastewater Treatment Plants in Korea: Occurrence and Toxicity to *Daphnia magna*. *Environ. Toxicol. Chem.* **2006**, *25*, 265–271. [CrossRef]
166. Białk-Bielińska, A.; Stolte, S.; Arning, J.; Uebers, U.; Böschen, A.; Stepnowski, P.; Matzke, M. Ecotoxicity Evaluation of Selected Sulfonamides. *Chemosphere* **2011**, *85*, 928–933. [CrossRef] [PubMed]

167. Overturf, M.D.; Overturf, C.L.; Baxter, D.; Hala, D.N.; Constantine, L.; Venables, B.; Huggett, D.B. Early Life-Stage Toxicity of Eight Pharmaceuticals to the Fathead Minnow, *Pimephales promelas*. *Arch. Environ. Contam. Toxicol.* **2012**, *62*, 455–464. [CrossRef]
168. Pascoe, D.; Karntanut, W.; Müller, C.T. Do Pharmaceuticals Affect Freshwater Invertebrates? A Study with the Cnidarian *Hydra vulgaris*. *Chemosphere* **2003**, *51*, 521–528. [CrossRef]
169. Parolini, M.; Quinn, B.; Binelli, A.; Provini, A. Cytotoxicity Assessment of Four Pharmaceutical Compounds on the Zebra Mussel (*Dreissena polymorpha*) Haemocytes, Gill and Digestive Gland Primary Cell Cultures. *Chemosphere* **2011**, *84*, 91–100. [CrossRef]
170. Schoenfuss, H.L.; Furlong, E.T.; Phillips, P.J.; Scott, T.-M.; Kolpin, D.W.; Cetkovic-Cvrlje, M.; Lesteberg, K.E.; Rearick, D.C. Complex Mixtures, Complex Responses: Assessing Pharmaceutical Mixtures Using Field and Laboratory Approaches. *Environ. Toxicol. Chem.* **2016**, *35*, 953–965. [CrossRef]
171. Zounková, R.; Klimešová, Z.; Nepejchalová, L.; Hilscherová, K.; Bláha, L. Complex Evaluation of Ecotoxicity and Genotoxicity of Antimicrobials Oxytetracycline and Flumequine Used in Aquaculture. *Environ. Toxicol. Chem.* **2011**, *30*, 1184–1189. [CrossRef]
172. Calleja, M.C.; Persoone, G.; Geladi, P. Comparative Acute Toxicity of the First 50 Multicentre Evaluation of In Vitro Cytotoxicity Chemicals to Aquatic Non-Vertebrates. Arch. *Environ. Contam. Toxicol.* **1994**, *26*, 69–78. [CrossRef]
173. Galus, M.; Kirischian, N.; Higgins, S.; Purdy, J.; Chow, J.; Rangaranjan, S.; Li, H.; Metcalfe, C.; Wilson, J.Y. Chronic, Low Concentration Exposure to Pharmaceuticals Impacts Multiple Organ Systems in Zebrafish. *Aquat. Toxicol.* **2013**, *132–133*, 200–211. [CrossRef]
174. Parolini, M.; Pedriali, A.; Binelli, A. Application of a Biomarker Response Index for Ranking the Toxicity of Five Pharmaceutical and Personal Care Products (PPCPs) to the Bivalve *Dreissena polymorpha*. *Arch. Environ. Contam. Toxicol.* **2013**, *64*, 439–447. [CrossRef]
175. Lister, A.L.; Van Der Kraak, G. An Investigation into the Role of Prostaglandins in Zebrafish Oocyte Maturation and Ovulation. *Gen. Comp. Endocrinol.* **2008**, *159*, 46–57. [CrossRef] [PubMed]
176. Quinn, B.; Gagné, F.; Blaise, C. An Investigation into the Acute and Chronic Toxicity of Eleven Pharmaceuticals (and Their Solvents) Found in Wastewater Effluent on the Cnidarian, Hydra Attenuata. *Sci. Total Environ.* **2008**, *389*, 306–314. [CrossRef] [PubMed]
177. Lützhøft, H.H.; Halling-Sørensen, B.; Jørgensen, S.E. Algal Toxicity of Antibacterial Agents Applied in Danish Fish Farming. *Arch. Environ. Contam. Toxicol.* **1999**, *36*, 1–6. [CrossRef]
178. Nunes, B.; Antunes, S.C.; Gomes, R.; Campos, J.C.; Braga, M.R.; Ramos, A.S.; Correia, A.T. Acute Effects of Tetracycline Exposure in the Freshwater Fish Gambusia Holbrooki: Antioxidant Effects, Neurotoxicity and Histological Alterations. *Arch. Environ. Contam. Toxicol.* **2015**, *68*, 371–381. [CrossRef]
179. Wollenberger, L.; Halling-Sørensen, B.; Kusk, K.O. Acute and Chronic Toxicity of Veterinary Antibiotics to *Daphnia magna*. *Chemosphere* **2000**, *40*, 723–730. [CrossRef] [PubMed]
180. Pruvot, B.; Quiroz, Y.; Voncken, A.; Jeanray, N.; Piot, A.; Martial, J.A.; Muller, M. A Panel of Biological Tests Reveals Developmental Effects of Pharmaceutical Pollutants on Late Stage Zebrafish Embryos. *Reprod. Toxicol.* **2012**, *34*, 568–583. [CrossRef] [PubMed]
181. Ando, T.; Nagase, H.; Eguchi, K.; Hirooka, T.; Nakamura, T.; Miyamoto, K.; Hirata, K. A Novel Method Using Cyanobacteria for Ecotoxicity Test of Veterinary Antimicrobial Agents. *Environ. Toxicol. Chem.* **2007**, *26*, 601–606. [CrossRef] [PubMed]
182. Selcer, K.W.; Verbanic, J.D. Vitellogenin of the Northern Leopard Frog (*Rana pipiens*): Development of an ELISA Assay and Evaluation of Induction after Immersion in Xenobiotic Estrogens. *Chemosphere* **2014**, *112*, 348–354. [CrossRef] [PubMed]
183. Liang, Y.-Q.; Xu, W.; Liang, X.; Jing, Z.; Pan, C.-G.; Tian, F. The Synthetic Progestin Norethindrone Causes Thyroid Endocrine Disruption in Adult Zebrafish. *Comp. Biochem. Physiol. Part C Toxicol. Pharmacol.* **2020**, *236*, 108819. [CrossRef] [PubMed]
184. Lee, W.; Kang, C.-W.; Su, C.-K.; Okubo, K.; Nagahama, Y. Screening Estrogenic Activity of Environmental Contaminants and Water Samples Using a Transgenic Medaka Embryo Bioassay. *Chemosphere* **2012**, *88*, 945–952. [CrossRef] [PubMed]
185. Thorpe, K.L.; Cummings, R.I.; Hutchinson, T.H.; Scholze, M.; Brighty, G.; Sumpter, J.P.; Tyler, C.R. Relative Potencies and Combination Effects of Steroidal Estrogens in Fish. *Environ. Sci. Technol.* **2003**, *37*, 1142–1149. [CrossRef] [PubMed]
186. Lei, B.; Kang, J.; Yu, Y.; Zha, J.; Li, W.; Wang, Z.; Wang, Y.; Wen, Y. Long-Term Exposure Investigating the Estrogenic Potency of Estriol in Japanese Medaka (*Oryzias latipes*). *Comp. Biochem. Physiol. Part C Toxicol. Pharmacol.* **2014**, *160*, 86–92. [CrossRef] [PubMed]
187. EG and G BIONOMICS. Initial Submission: The Chronic Toxicity of s-900d to the Water Flea (Daphnia Magna) with Cover Letter Dated 081492, USEPA: Health and Environmental Research on Line. 1992. Available online: https://hero.epa.gov/hero/index.cfm/reference/details/reference_id/1325554 (accessed on 15 June 2023).
188. Zhang, X.; Xiong, L.; Liu, Y.; Deng, C.; Mao, S. Histopathological and Estrogen Effect of Pentachlorophenol on the Rare Minnow (*Gobiocypris rarus*). *Fish. Physiol. Biochem.* **2014**, *40*, 805–816. [CrossRef]
189. Kazeto, Y.; Place, A.R.; Trant, J.M. Effects of Endocrine Disrupting Chemicals on the Expression of CYP19 Genes in Zebrafish (*Danio rerio*) Juveniles. *Aquat. Toxicol.* **2004**, *69*, 25–34. [CrossRef]
190. Villeneuve, D.L.; Garcia-Reyero, N.; Escalon, B.L.; Jensen, K.M.; Cavallin, J.E.; Makynen, E.A.; Durhan, E.J.; Kahl, M.D.; Thomas, L.M.; Perkins, E.J.; et al. Ecotoxicogenomics to Support Ecological Risk Assessment: A Case Study with Bisphenol A in Fish. *Environ. Sci. Technol.* **2012**, *46*, 51–59. [CrossRef]
191. Gagnaire, B.; Gagné, F.; André, C.; Blaise, C.; Abbaci, K.; Budzinski, H.; Dévier, M.-H.; Garric, J. Development of Biomarkers of Stress Related to Endocrine Disruption in Gastropods: Alkali-Labile Phosphates, Protein-Bound Lipids and Vitellogenin-like Proteins. *Aquat. Toxicol.* **2009**, *92*, 155–167. [CrossRef]

192. Bean, R.M.; Gibson, C.I.; Anderson, D.R. *Biocide By-Products in Aquatic Environments*; Pacific Northwest National Lab.: Richland, WA, USA, 1981. [CrossRef]
193. Croteau, M.C.; Davidson, M.; Duarte-Guterman, P.; Wade, M.; Popesku, J.T.; Wiens, S.; Lean, D.R.S.; Trudeau, V.L. Assessment of Thyroid System Disruption in *Rana pipiens* Tadpoles Chronically Exposed to UVB Radiation and 4-Tert-Octylphenol. *Aquat. Toxicol.* **2009**, *95*, 81–92. [CrossRef]
194. Jarvis, A.L.; Bernot, M.J.; Bernot, R.J. The Effects of the Psychiatric Drug Carbamazepine on Freshwater Invertebrate Communities and Ecosystem Dynamics. *Sci. Total Environ.* **2014**, *496*, 461–470. [CrossRef] [PubMed]
195. Melvin, S.D.; Cameron, M.C.; Lanctôt, C.M. Individual and Mixture Toxicity of Pharmaceuticals Naproxen, Carbamazepine, and Sulfamethoxazole to Australian Striped Marsh Frog Tadpoles (*Limnodynastes peronii*). *J. Toxicol. Environ. Health A* **2014**, *77*, 337–345. [CrossRef] [PubMed]
196. Li, Z.-H.; Velisek, J.; Zlabek, V.; Grabic, R.; Machova, J.; Kolarova, J.; Randak, T. Hepatic Antioxidant Status and Hematological Parameters in Rainbow Trout, *Oncorhynchus mykiss*, after Chronic Exposure to Carbamazepine. *Chem. Biol. Interact.* **2010**, *183*, 98–104. [CrossRef] [PubMed]
197. Fraz, S.; Lee, A.H.; Wilson, J.Y. Gemfibrozil and Carbamazepine Decrease Steroid Production in Zebrafish Testes (*Danio rerio*). *Aquat. Toxicol.* **2018**, *198*, 1–9. [CrossRef]
198. Dordio, A.V.; Belo, M.; Martins Teixeira, D.; Palace Carvalho, A.J.; Dias, C.M.B.; Picó, Y.; Pinto, A.P. Evaluation of Carbamazepine Uptake and Metabolization by Typha Spp., a Plant with Potential Use in Phytotreatment. *Bioresour. Technol.* **2011**, *102*, 7827–7834. [CrossRef] [PubMed]
199. Chen, H.; Zha, J.; Liang, X.; Li, J.; Wang, Z. Effects of the Human Antiepileptic Drug Carbamazepine on the Behavior, Biomarkers, and Heat Shock Proteins in the Asian Clam *Corbicula fluminea*. *Aquat. Toxicol.* **2014**, *155*, 1–8. [CrossRef]
200. Oetken, M.; Nentwig, G.; Löffler, D.; Ternes, T.; Oehlmann, J. Effects of Pharmaceuticals on Aquatic Invertebrates. Part I. The Antiepileptic Drug Carbamazepine. *Arch. Environ. Contam. Toxicol.* **2005**, *49*, 353–361. [CrossRef]
201. Zhang, W.; Zhang, M.; Lin, K.; Sun, W.; Xiong, B.; Guo, M.; Cui, X.; Fu, R. Eco-Toxicological Effect of Carbamazepine on *Scenedesmus obliquus* and *Chlorella pyrenoidosa*. *Environ. Toxicol. Pharmacol.* **2012**, *33*, 344–352. [CrossRef] [PubMed]
202. Lamichhane, K.; Garcia, S.N.; Huggett, D.B.; DeAngelis, D.L.; La Point, T.W. Chronic Effects of Carbamazepine on Life-History Strategies of *Ceriodaphnia dubia* in Three Successive Generations. *Arch. Environ. Contam. Toxicol.* **2013**, *64*, 427–438. [CrossRef] [PubMed]
203. Yan, S.; Chen, R.; Wang, M.; Zha, J. Carbamazepine at Environmentally Relevant Concentrations Caused DNA Damage and Apoptosis in the Liver of Chinese Rare Minnows (*Gobiocypris rarus*) by the Ras/Raf/ERK/P53 Signaling Pathway. *Environ. Pollut.* **2021**, *270*, 116245. [CrossRef]
204. Gust, M.; Gagné, F.; Berlioz-Barbier, A.; Besse, J.P.; Buronfosse, T.; Tournier, M.; Tutundjian, R.; Garric, J.; Cren-Olivé, C. Caged Mudsnail Potamopyrgus Antipodarum (Gray) as an Integrated Field Biomonitoring Tool: Exposure Assessment and Reprotoxic Effects of Water Column Contamination. *Water Res.* **2014**, *54*, 222–236. [CrossRef]
205. Weigt, S.; Huebler, N.; Strecker, R.; Braunbeck, T.; Broschard, T.H. Zebrafish (*Danio rerio*) Embryos as a Model for Testing Proteratogens. *Toxicology* **2011**, *281*, 25–36. [CrossRef]
206. DeLorenzo, M.E.; Fleming, J. Individual and Mixture Effects of Selected Pharmaceuticals and Personal Care Products on the Marine Phytoplankton Species *Dunaliella tertiolecta*. *Arch. Environ. Contam. Toxicol.* **2008**, *54*, 203–210. [CrossRef] [PubMed]
207. Jos, A.; Repetto, G.; Rios, J.C.; Hazen, M.J.; Molero, M.L.; del Peso, A.; Salguero, M.; Fernández-Freire, P.; Pérez-Martín, J.M.; Cameán, A. Ecotoxicological Evaluation of Carbamazepine Using Six Different Model Systems with Eighteen Endpoints. *Toxicol. Vitr.* **2003**, *17*, 525–532. [CrossRef] [PubMed]
208. Kim, Y.; Choi, K.; Jung, J.; Park, S.; Kim, P.-G.; Park, J. Aquatic Toxicity of Acetaminophen, Carbamazepine, Cimetidine, Diltiazem and Six Major Sulfonamides, and Their Potential Ecological Risks in Korea. *Environ. Int.* **2007**, *33*, 370–375. [CrossRef]
209. Cleuvers, M. Aquatic Ecotoxicity of Pharmaceuticals Including the Assessment of Combination Effects. *Toxicol. Lett.* **2003**, *142*, 185–194. [CrossRef]
210. Li, Z.-H.; Zlabek, V.; Velisek, J.; Grabic, R.; Machova, J.; Kolarova, J.; Li, P.; Randak, T. Acute Toxicity of Carbamazepine to Juvenile Rainbow Trout (*Oncorhynchus mykiss*): Effects on Antioxidant Responses, Hematological Parameters and Hepatic EROD. *Ecotoxicol. Environ. Saf.* **2011**, *74*, 319–327. [CrossRef]
211. Richards, S.M.; Cole, S.E. A Toxicity and Hazard Assessment of Fourteen Pharmaceuticals to *Xenopus laevis* Larvae. *Ecotoxicology* **2006**, *15*, 647–656. [CrossRef] [PubMed]
212. Kaza, M.; Nałącz-Jawecki, G.; Sawicki, J. The Toxicity of Selected Pharmaceuticals to the Aquatic Plant Lemna Minor. *Fresenius Environ. Bull.* **2007**, *16*, 524–531.
213. Dordio, A.; Ferro, R.; Teixeira, D.; Palace, A.J.; Pinto, A.P.; Dias, C.M.B. Study on the Use of Typha Spp. for the Phytotreatment of Water Contaminated with Ibuprofen. *Int. J. Environ. Anal. Chem.* **2011**, *91*, 654–667. [CrossRef]
214. Saravanan, M.; Devi, K.U.; Malarvizhi, A.; Ramesh, M. Effects of Ibuprofen on Hematological, Biochemical and Enzymological Parameters of Blood in an Indian Major Carp, *Cirrhinus mrigala*. *Environ. Toxicol. Pharmacol.* **2012**, *34*, 14–22. [CrossRef] [PubMed]
215. Ceballos-Laita, L.; Calvo, L.; Bes, M.; Fillat, M.; Peleato, M. Effects of Benzene and Several Pharmaceuticals on the Growth and Microcystin Production in Microcystis Aeruginosa PCC 7806. *Limnetica* **2015**, *34*, 237–246. [CrossRef]
216. Flippin, J.L.; Huggett, D.; Foran, C.M. Changes in the Timing of Reproduction Following Chronic Exposure to Ibuprofen in Japanese Medaka, *Oryzias latipes*. *Aquat. Toxicol.* **2007**, *81*, 73–78. [CrossRef] [PubMed]

217. Pounds, N.; Maclean, S.; Webley, M.; Pascoe, D.; Hutchinson, T. Acute and Chronic Effects of Ibuprofen in the Mollusc *Planorbis carinatus* (*Gastropoda: Planorbidae*). *Ecotoxicol. Environ. Saf.* **2008**, *70*, 47–52. [CrossRef]
218. Kim, J.-W.; Ishibashi, H.; Yamauchi, R.; Ichikawa, N.; Takao, Y.; Hirano, M.; Koga, M.; Arizono, K. Acute Toxicity of Pharmaceutical and Personal Care Products on Freshwater Crustacean (*Thamnocephalus platyurus*) and Fish (*Oryzias latipes*). *J. Toxicol. Sci.* **2009**, *34*, 227–232. [CrossRef]
219. Li, M.-H. Acute Toxicity of 30 Pharmaceutically Active Compounds to Freshwater Planarians, *Dugesia japonica*. *Toxicol. Environ. Chem.* **2013**, *95*, 1157–1170. [CrossRef]
220. Lv, X.; Zhou, Q.; Song, M.; Jiang, G.; Shao, J. Vitellogenic Responses of 17β-Estradiol and Bisphenol A in Male Chinese Loach (*Misgurnus anguillicaudatus*). *Environ. Toxicol. Pharmacol.* **2007**, *24*, 155–159. [CrossRef]
221. Qiu, W.; Shen, Y.; Pan, C.; Liu, S.; Wu, M.; Yang, M.; Wang, K.-J. The Potential Immune Modulatory Effect of Chronic Bisphenol A Exposure on Gene Regulation in Male Medaka (*Oryzias latipes*) Liver. *Ecotoxicol. Environ. Saf.* **2016**, *130*, 146–154. [CrossRef] [PubMed]
222. Haubruge, E.; Petit, F.; Gage, M.J. Reduced Sperm Counts in Guppies (*Poecilia reticulata*) Following Exposure to Low Levels of Tributyltin and Bisphenol A. *Proc. Biol. Sci.* **2000**, *267*, 2333–2337. [CrossRef] [PubMed]
223. Hatef, A.; Zare, A.; Alavi, S.M.H.; Habibi, H.R.; Linhart, O. Modulations in Androgen and Estrogen Mediating Genes and Testicular Response in Male Goldfish Exposed to Bisphenol A. *Environ. Toxicol. Chem.* **2012**, *31*, 2069–2077. [CrossRef] [PubMed]
224. Keiter, S.; Baumann, L.; Färber, H.; Holbech, H.; Skutlarek, D.; Engwall, M.; Braunbeck, T. Long-Term Effects of a Binary Mixture of Perfluorooctane Sulfonate (PFOS) and Bisphenol A (BPA) in Zebrafish (*Danio rerio*). *Aquat. Toxicol.* **2012**, *118–119*, 116–129. [CrossRef] [PubMed]
225. Mochida, K.; Fujii, K.; Kakuno, A.; Matsubara, T.; Ohkubo, N.; Adachi, S.; Yamauchi, K. Expression of Ubiquitin C-Terminal Hydrolase Is Regulated by Estradiol-17β in Testis and Brain of the Japanese Common Goby. *Fish. Physiol. Biochem.* **2003**, *28*, 435–436. [CrossRef]
226. Yang, F.-X.; Xu, Y.; Wen, S. Endocrine-Disrupting Effects of Nonylphenol, Bisphenol A, and p,p′-DDE on Rana Nigromaculata Tadpoles. *Bull. Environ. Contam. Toxicol.* **2005**, *75*, 1168–1175. [CrossRef]
227. Ha, M.-H.; Choi, J. Effects of Environmental Contaminants on Hemoglobin Gene Expression in *Daphnia magna*: A Potential Biomarker for Freshwater Quality Monitoring. *Arch. Environ. Contam. Toxicol.* **2009**, *57*, 330–337. [CrossRef] [PubMed]
228. Ladewig, V.; Jungmann, D.; Köhler, H.-R.; Licht, O.; Ludwichowski, K.-U.; Schirling, M.; Triebskorn, R.; Nagel, R. Effects of Bisphenol A on Gammarus Fossarum and Lumbriculus Variegatus in Artificial Indoor Streams. *Toxicol. Environ. Chem.* **2006**, *88*, 649–664. [CrossRef]
229. Lahnsteiner, F.; Berger, B.; Kletzl, M.; Weismann, T. Effect of Bisphenol A on Maturation and Quality of Semen and Eggs in the Brown Trout, *Salmo trutta f. fario*. *Aquat. Toxicol.* **2005**, *75*, 213–224. [CrossRef] [PubMed]
230. Staples, C.A.; Tilghman Hall, A.; Friederich, U.; Caspers, N.; Klecka, G.M. Early Life-Stage and Multigeneration Toxicity Study with Bisphenol A and Fathead Minnows (*Pimephales promelas*). *Ecotoxicol. Environ. Saf.* **2011**, *74*, 1548–1557. [CrossRef]
231. Van den Belt, K.; Verheyen, R.; Witters, H. Comparison of Vitellogenin Responses in Zebrafish and Rainbow Trout Following Exposure to Environmental Estrogens. *Ecotoxicol. Environ. Saf.* **2003**, *56*, 271–281. [CrossRef] [PubMed]
232. Jung, J.-W.; Kang, J.-S.; Choi, J.; Park, J.-W. Chronic Toxicity of Endocrine Disrupting Chemicals Used in Plastic Products in Korean Resident Species: Implications for Aquatic Ecological Risk Assessment. *Ecotoxicol. Environ. Saf.* **2020**, *192*, 110309. [CrossRef] [PubMed]
233. Zhu, L.; Yuan, C.; Wang, M.; Liu, Y.; Wang, Z.; Seif, M.M. Bisphenol A-Associated Alterations in DNA and Histone Methylation Affects Semen Quality in Rare Minnow *Gobiocypris rarus*. *Aquat. Toxicol.* **2020**, *226*, 105580. [CrossRef] [PubMed]
234. Plahuta, M.; Tišler, T.; Pintar, A.; Toman, M.J. Adverse Effects of Bisphenol A on Water Louse (*Asellus aquaticus*). *Ecotoxicol. Environ. Saf.* **2015**, *117*, 81–88. [CrossRef] [PubMed]
235. Mihaich, E.M.; Friederich, U.; Caspers, N.; Hall, A.T.; Klecka, G.M.; Dimond, S.S.; Staples, C.A.; Ortego, L.S.; Hentges, S.G. Acute and Chronic Toxicity Testing of Bisphenol A with Aquatic Invertebrates and Plants. *Ecotoxicol. Environ. Saf.* **2009**, *72*, 1392–1399. [CrossRef] [PubMed]
236. Zhang, W.; Xiong, B.; Sun, W.-F.; An, S.; Lin, K.-F.; Guo, M.-J.; Cui, X.-H. Acute and Chronic Toxic Effects of Bisphenol A on *Chlorella pyrenoidosa* and *Scenedesmus obliquus*. *Environ. Toxicol.* **2014**, *29*, 714–722. [CrossRef] [PubMed]
237. Ozmen, M.; Güngördü, A.; Erdemoglu, S.; Ozmen, N.; Asilturk, M. Toxicological Aspects of Photocatalytic Degradation of Selected Xenobiotics with Nano-Sized Mn-Doped $TiO_2$. *Aquat. Toxicol.* **2015**, *165*, 144–153. [CrossRef] [PubMed]
238. Watts, M.M.; Pascoe, D.; Carroll, K. Survival and Precopulatory Behaviour of *Gammarus pulex* (L.). Exposed to Two Xenoestrogens. *Water Res.* **2001**, *35*, 2347–2352. [CrossRef] [PubMed]
239. Wolkowicz, I.R.H.; Herkovits, J.; Pérez Coll, C.S. Stage-Dependent Toxicity of Bisphenol a on *Rhinella arenarum* (Anura, Bufonidae) Embryos and Larvae. *Environ. Toxicol.* **2014**, *29*, 146–154. [CrossRef]
240. Kashiwada, S.; Ishikawa, H.; Miyamoto, N.; Ohnishi, Y.; Magara, Y. Fish Test for Endocrine-Disruption and Estimation of Water Quality of Japanese Rivers. *Water Res.* **2002**, *36*, 2161–2166. [CrossRef]
241. Debenest, T.; Gagné, F.; Petit, A.-N.; André, C.; Kohli, M.; Blaise, C. Ecotoxicity of a Brominated Flame Retardant (Tetrabromobisphenol A) and Its Derivatives to Aquatic Organisms. *Comp. Biochem. Physiol. Part C Toxicol. Pharmacol.* **2010**, *152*, 407–412. [CrossRef] [PubMed]
242. Chan, W.K.; Chan, K.M. Disruption of the Hypothalamic-Pituitary-Thyroid Axis in Zebrafish Embryo-Larvae Following Waterborne Exposure to BDE-47, TBBPA and BPA. *Aquat. Toxicol.* **2012**, *108*, 106–111. [CrossRef] [PubMed]

243. Alexander, H.C.; Dill, D.C.; Smith, L.W.; Guiney, P.D.; Dorn, P. Bisphenol a: Acute Aquatic Toxicity. *Environ. Toxicol. Chem.* **1988**, *7*, 19–26. [CrossRef]
244. Jemec, A.; Tišler, T.; Erjavec, B.; Pintar, A. Antioxidant Responses and Whole-Organism Changes in *Daphnia magna* Acutely and Chronically Exposed to Endocrine Disruptor Bisphenol A. *Ecotoxicol. Environ. Saf.* **2012**, *86*, 213–218. [CrossRef] [PubMed]

*toxics*

MDPI

*Article*

# Dimethylcyclosiloxanes in Mobile Smart Terminal Devices: Concentrations, Distributions, Profiles, and Environmental Emissions

**Yuanna Xing [1,2], Yiming Ge [3], Shaoyou Lu [3], Tao Yang [1,2] and Xianzhi Peng [1,2,*]**

1 Guangzhou Institute of Geochemistry, Chinese Academy of Sciences, Guangzhou 510640, China; xingyuanna18@sina.com (Y.X.); yangtao21@mails.ucas.ac.cn (T.Y.)
2 University of Chinese Academy of Sciences, Beijing 100049, China
3 School of Public Health (Shenzhen), Shenzhen Campus of SunYat-sen University, Shenzhen 518107, China; geym5@mail2.sysu.edu.cn (Y.G.); lushy23@mail.sysu.edu.cn (S.L.)
* Correspondence: pengx@gig.ac.cn; Tel.: +86-136-4265-9081

**Abstract:** Dimethylcyclosiloxanes (DMCs) are utilized as vital monomers in the synthesis of organosilicon compounds, integral to the manufacture of mobile smart terminal devices. Toxicological studies have revealed potential endocrine-disrupting activity, reproductive toxicity, neurotoxicity, and other toxicities of the DMCs. This study investigated the concentrations and composition profiles of seven DMCs, including hexamethylcyclotrisiloxane (D3), octamethylcyclotetrasiloxane (D4), decamethylcyclopentasiloxane (D5), dodecamethylcyclohexasiloxane (D6), and tetradecamethylcycloheptasiloxane (D7), hexadecamethylcyclooctasiloxane (D8), and octadecamethylcyclononasiloxane (D9) in three types of mobile smart terminal device components (silicone rubber, adhesive, and plastics). Environmental emissions of DMCs from silicone rubber materials were also estimated to improve the recognition of their potential fate within the environment. D5–D9 were widely present in silicone rubber and adhesives with detection rates ranging from 91–95.5% and 50–100%, respectively, while D3 and D4 were more frequently detected in plastics, both showing a detection rate of 61.1%. Silicone rubber had the highest total DMCs (∑7DMCs) and a concentration of 802.2 mg/kg, which were dominated by D7, D8, and D9. DMCs detected in adhesives were dominated by D4, D5, and D6. The estimated emission of ∑DMCs released into the environment in China from silicone rubber used in mobile smart terminal devices exceeds 5000 tons per year. Further studies are needed on the presence of DMCs in various commodities and environmental media to assess their ecological and human health impacts, as well as the toxicological effects of D7–D9 for the appropriate regulation of these chemicals.

**Keywords:** dimethylcyclosiloxanes; mobile smart terminal devices; environmental emission; silicone polymers

## 1. Introduction

Dimethylcyclosiloxanes (DMCs) are a class of cyclic compounds with siloxanes as the main chain, characterized by low water solubility, high lipophilicity, and elevated vapor pressure [1,2]. Due to their thermal stability, lubricating properties, and diminished surface tension [3], these compounds play a pivotal role as fundamental raw materials and intermediates to synthesize silicone polymers. Notable derivatives include silicone oils, silicone resins, and silicone rubbers, contributing indispensably to a range of products, encompassing personal care products (PCPs), cosmetics, textiles, automobiles, construction, consumer electronics, baby products, and food [4–9]. According to the statistics released by the United States Environmental Protection Agency (EPA) in 2022, the total octamethylcyclotetrasiloxane (D4) production ranged from 750 million to 1 billion pounds during the year 2015. Based on a database from the European Chemicals Agency

(ECHA), D4, decamethylcyclopentasiloxane (D5), and dodecamethylcyclohexasiloxane (D6) are annually produced and imported into the European Economic Area in volumes of up to 1 million tons, 100,000 tons, and 10,000 tons, respectively, defined by EPA as high-production volume chemicals.

During industrial production, packaging, and product utilization, the release of DMCs into the environment is an inevitable consequence, facilitated by processes of volatilization and sewage discharge. Consequently, DMCs are widely present in various environmental components, such as air [2,10], waters [11,12], soil [13], sediments [14], and house dust [15,16], as well as organisms [17,18].

Organisms are exposed to DMCs primarily via inhalation, ingestion, or epidermal contact [19–21]. However, a review reported the accumulation and potential toxic effects of DMCs in aquatic organisms, owing to their high organic carbon partition coefficients as well as their marked lipophilicity and inherent resistance to biodegradation [22]. It is generally acknowledged that D4, beyond its characterization as an endocrine disruptor with potential estrogenic and antiandrogenic activity [23], could induce hepatic drug-metabolizing enzymes, liver enlargement, inflammatory responses, and hepatocyte death in rats [24,25]. Apart from this, both D4 and D5 are reproductively toxic to rats [26,27], with the former damaging the nervous system of mice offspring [28]. Notably, crayfish exposed to D6 exhibited altered antioxidant capacity and gene expression [29]. Toxicology studies also shed light on the possible link between DMCs and cancer. For instance, D4 and D5 act as dopamine agonists and inhibit prolactin secretion, thereby engendering prolonged stimulation of the endometrium and precipitating endometrial adenomas in rats [30,31]. DMCs may also induce breast cancer by facilitating damage to DNA [32]. A recent epidemiologic study established an association between volatile DMCs and non-alcoholic fatty liver disease in humans [33].

Considering the environmental persistence, bioconcentration, and potential health hazards of DMCs [34,35], researchers have paid mounting concern to these substances. In 2018, the ECHA emphasized the significance of DMCs by officially designating D4, D5, and D6 as Substances of Very High Concern (SVHC) with PBT (Persistent, Bioaccumulative, And Toxic) properties [36]. In June, 2023, ECHA published a restriction proposal to list D4, D5, and D6 under the Stockholm Convention on Persistent Organic Pollutants (POPs), proposing further restrictions to mitigate the risks to humans and the environment [37]. Although the deleterious effects of hexamethylcyclotrisiloxane (D3), D4, D5, and D6 have been the subjects of extensive research, with a consensus emerging on the imperative to limit their use to minimize their hazards, there is little research data on the toxicological effects of tetradecamethylcycloheptasiloxane (D7), hexadecamethylcyclooctasiloxane (D8), and octadecamethylcyclononasiloxane (D9), and their impacts on the ecological environment and human health.

Polysiloxanes synthesized with the participation of DMCs are the majority in use, accounting for more than 90% of the total use of organosilicon. In 2022, China's polysiloxane capacity and production reached 2.31 million tons and 1.91 million tons, respectively, occupying more than a half of the world [38]. Silicone rubber, silicone adhesives, and silicone copolymer-modified plastics are widely used silicone polymers. Silicone rubber facilitates the manufacture of complex-shaped products renowned for its exceptional waterproof performance, elasticity, and insulation [39,40], and is widely added to mobile smart terminal devices, especially materials that come into contact with human skin, such as remote device control buttons, cell phone cases, smartwatch bands, ear caps, or neck bands for Bluetooth headsets. Compared to ordinary polycarbonate (PC), silicone copolymer-modified PC boasts enhanced properties including flexibility, hydrolysis resistance, corrosion resistance, oxidation resistance, and flame retardancy, and are widely used as shells, covers, and structural components for mobile smart terminal devices [41,42]. The organic silicone adhesive, in addition to bonding and sealing, also has a moisture-proof, temperature, insulation, and shockproof role, accordingly enjoying widespread favor in the potting of various electronic devices [42–45]. In the intricate process of producing mobile smart terminal products with silicone polymers, DMCs can remain in the products due to incomplete reactions or eliminations, subsequently releasing to the environment during processes such as e-waste dismantling [46] and thermal treatment [47]. Particularly concerning is the substantial disposal of silicone rubber as household waste due to its low price and high frequency of replacement. In China, the annual demand for silicone rubber is as high as 2 million tons, with the predicted requirement for silicone adhesive in 2025 reaching 1350 tons [48]. Furthermore, China manages a diverse array of solid household refuse, encompassing metals, glass, organic matter, textiles, and cigarette butts [49]. However, most of the mobile smart terminal devices are not effectively handled [50]. Therefore, the environmental risk of DMCs caused by releasing from mobile smart terminal device components cannot be ignored.

DMCs have become a class of emerging contaminants due to their wide presence in human plasma and fat samples [51–53]. Previous studies have provided important insights into DMCs in PCPs, cosmetics, baby products, and food contact materials. Despite the extensive use of mobile smart terminal devices, a dearth of research exists regarding the content of DMCs within them and their potential environmental risks have been reported, notably concerning the presence of D7–D9. In addition, due to the volatility and long atmospheric half-life exceeding 10 days, the major environmental burdens of D3, D4, and D5 are present mainly in the air [54,55]. Consequently, the calculation of environmental emissions for DMCs with similar structures can provide clues for further understanding of the fate of DMCs in the environment.

In this context, this study investigated the occurrence and profiles of seven DMCs (D3–D9) in silicone rubber, silicone adhesive, and silicone copolymerized modified plastics currently produced and used in mobile smart terminal devices in China. The amount of $\sum$DMCs (the sum of D3–D9) released into the environment was also estimated.

## 2. Materials and Methods

### 2.1. Chemicals and Reagents

D3 (purity $\geq$ 99.9%) and D4 (purity $\geq$ 99.9%) were purchased from TMStandard. D5 (purity $\geq$ 98.4%), D6 (purity $\geq$ 97.4%), and D7 (purity $\geq$ 99.1%) were purchased from Anpel Resi Standard Technical Service. D8 (purity $\geq$ 98.2%), D9 (purity $\geq$ 96.0%), and acetone with chromatographic purity was purchased from Guangzhou Chemical Reagent Factory. An overview of the analyzed dimethylcyclosiloxanes is provided in Table 1.

**Table 1.** Information of targeted DMCs.

| Name | Abbreviation | CAS-No. | Boiling Points | Molar Weight |
|---|---|---|---|---|
| Hexamethylcyclotrisiloxane | D3 | 541-05-9 | 134 °C | 222.47 |
| Octamethylcyclotetrasiloxane | D4 | 556-67-2 | 175–176 °C | 296.62 |
| Decamethylcyclopentasiloxane | D5 | 541-02-6 | 205 °C | 370.72 |
| Dodecamethylcyclohexasiloxane | D6 | 540-97-6 | 245 °C | 444.93 |
| Tetradecamethylcycloheptasiloxane | D7 | 107-50-6 | 337 °C | 519.08 |
| Hexadecamethylcyclooctasiloxane | D8 | 556-68-3 | 290 °C | 593.23 |
| Octadecamethylcyclononasiloxane | D9 | 556-71-8 | 416 °C | 667.39 |

### 2.2. Sample Collection and Preparation

A total of 52 mobile smart terminal product material samples were purchased from official channels and online platforms in China, including silicone rubber samples ($n$ = 22), plastic samples ($n$ = 18), and adhesive samples ($n$ = 12). The silicone rubber samples were subcategorized into ear caps ($n$ = 9), watch bands ($n$ = 11), and case covers ($n$ = 2).

Adhesive samples were prepared in advance. Five grams of adhesive monomer was spread on a 5 cm $\times$ 50 cm tin foil and then placed for three days before sample extraction. Silicone rubber and plastic samples do not require special sample preparation before extraction.

The silicone rubber, plastic, and cured adhesive samples were cut into small pieces of 0.5 cm × 0.5 cm samples. One gram of the cut sample was placed in a glass tube and 10.0 mL of acetone was added. Then, the samples were ultrasonically extracted for 2 h. After resting to room temperature, the extract was transferred to a 15 mL tube and centrifuged at 20,000 rpm for 3 min. Then, the supernatant was passed through a 0.22 μm PTFE filter membrane (Jinteng Experimental Equipment, Tianjin, China.) and subjected to instrumental analysis.

### 2.3. Instrumental Analysis

DMCs were determined and quantified by a QP2020NX gas chromatography-mass spectrometer (GC-MS, Shimadzu Corporation, Japan) set to electron impact ionization (EI) mode. The target analytes were separated on an SH-VMS quartz capillary column (60 m × 0.32 mm × 1.8 μm). Helium (purity > 99.999%) was used as a carrier gas, with a flow rate of 1.0 mL/min. The injection volume was 1 μL with splitless mode and the inlet temperature was set to 230 °C. A programmed heating regimen was applied to the column temperature, ranging from 40 °C to 230 °C at a rate of 10 °C/min, and held for 15 min. The source temperature was 230 °C and the ionization voltage was 70 eV, with a transmission line temperature of 230 °C. Quantification was operated by selected ion monitoring with a solvent delay time of 4.5 min. Optimized instrumental parameters are presented in the Supporting Information (Supplementary Table S1).

### 2.4. Quality Assurance and Quality Control

Target analyte standards at different concentrations (1 mg/L, 10 mg/L, and 100 mg/L) were spiked into samples of silicone rubber, plastics, and adhesive to determine recoveries. The recoveries ranged from 82.3% to 107% with the relative standard deviations (RSDs) of 1.5% to 4.0%, collectively demonstrating the stability and reliability of the method. The standard curves of DMCs ranged from 0.1 mg/L to 5 mg/L, and the regression coefficients ($R^2$) were ≥0.999. The limits of quantitation (LOQ), defined as ten times of signal to noise, were all 1 mg/kg (Table S2).

### 2.5. Environmental Emissions of DMCs from Silicone Rubber

In this study, the estimated amount of DMCs released into the environment in China from silicone rubber used in mobile smart terminal devices was calculated according to the following equations:

$$M = M_a + M_b + M_c \tag{1}$$

where $M$ (kg) is the total weight of ∑DMCs (D3, D4, D5, D6, D7, D8, and D9) in all silicone rubber samples; $M_a$, $M_b$ and $M_c$ (kg) represent the total weight of ∑DMCs in the ear caps, watch bands, and case covers, respectively.

$$M_x = \frac{\sum_1^n (c \times m)}{n} \times Q_s \tag{2}$$

where $M_x$ (kg) represents $M_a$, $M_b$, or $M_c$; $c$ (mg/kg) is the concentration of ∑DMCs in the silicone rubber samples; $m$ (g) is the weight of the selected sample, with the weights of the ear caps in this study being three pairs; $n$ is the number of ear caps, watch bands and case covers analyzed in the present study, which is 9, 11, and 2, respectively; $Q_s$ (Pcs) is the marketed quantity of ear caps, watch bands, and case covers shipment quantities, which are expressed as the number of smart headphones (78.98 million), smart bracelet watches (56.88 million), and smartphones (351 million) shipped, respectively [56].

We performed the Kruskal–Wallis H test or Mann–Whitney U test for differences in the concentrations of DMCs in different samples since the determined DMCs concentrations were under non-normality. Differences were considered statistically significant if $p < 0.05$. Half of LOQ was replaced by the level below LOQ when calculating the average and median concentrations. Data processing was carried out using IBM SPSS 26.0.

## 3. Results and Discussion

### 3.1. Detection Rates and Concentrations of DMCs

A summary of detection rates and concentrations of seven DMCs (D3–D9) in the components of different mobile smart terminal devices were presented in Table 2. Overall, 86.5% of the samples analyzed ($n$ = 52) exhibited the presence of at least one detectable DMC. The median concentrations of DCMs varied considerably, ranging from 0.50 mg/kg (D3) to 11.0 mg/kg (D6).

**Table 2.** Concentrations of DMCs in silicone rubber, plastic, and adhesive samples (mg/kg).

| Samples | Statistics | D3 | D4 | D5 | D6 | D7 | D8 | D9 | D4+D5+D6 | D7+D8+D9 | ∑DMCs |
|---|---|---|---|---|---|---|---|---|---|---|---|
| Silicone rubber ($n$ = 22) | DR (%) | 9 | 59 | 95.5 | 95.5 | 95.5 | 91 | 91 | 95.5 | 95.5 | 95.5 |
| | Mean | 0.53 | 9.91 | 66.3 | 136.9 | 159.3 | 171.8 | 159.0 | 212.9 | 490.0 | 703.0 |
| | Median | 0.50 | 1.55 | 14.5 | 78.9 | 180.1 | 158.2 | 132.5 | 86.0 | 481.5 | 802.2 |
| | Range | <LOQ-1.1 | <LOQ-63 | <LOQ-335 | <LOQ-422 | <LOQ-479 | <LOQ-481 | <LOQ-438.9 | <LOQ-820 | <LOQ-1331 | <LOQ-1788 |
| Plastics ($n$ = 18) | DR (%) | 61.1 | 61.1 | 50 | 38.9 | 44.4 | 50 | 50 | 61.1 | 50 | 66.7 |
| | Mean | 22.1 | 19.0 | 9.36 | 5.32 | 4.46 | 4.53 | 3.69 | 33.1 | 12.2 | 66.9 |
| | Median | 2.95 | 2.15 | 0.50 | 0.50 | 0.50 | 0.75 | 1.65 | 2.40 | 2.40 | 8.25 |
| | Range | <LOQ-279 | <LOQ-181 | <LOQ-86 | <LOQ-42.5 | <LOQ-27.5 | <LOQ-33.6 | <LOQ-26.4 | <LOQ-309.5 | <LOQ-87.5 | <LOQ-631.6 |
| Adhesive ($n$ = 12) | DR (%) | 41.7 | 50 | 100 | 100 | 91.7 | 75 | 50 | 100 | 91.7 | 100 |
| | Mean | 2.11 | 5.33 | 28.9 | 43.8 | 38.2 | 41.9 | 48.4 | 78.5 | 129.0 | 208.6 |
| | Median | 0.50 | 0.95 | 2.35 | 7.35 | 2.10 | 1.20 | 0.90 | 12.1 | 5.0 | 21.5 |
| | Range | <LOQ-10.3 | <LOQ-27.7 | 1–119 | 1.8–160 | <LOQ-153 | <LOQ-169 | <LOQ-204 | 3.7–289.6 | <LOQ-515 | 6.3–807.9 |
| All ($n$ = 52) | DR (%) | 34.6 | 57.7 | 80.8 | 76.9 | 76.9 | 73.1 | 67.3 | 84.6 | 78.8 | 86.5 |
| | Mean | 8.52 | 12.0 | 38.0 | 79.9 | 77.8 | 83.9 | 79.7 | 119.7 | 241.3 | 368.7 |
| | Median | 0.50 | 1.55 | 6.60 | 11.0 | 7.35 | 8.10 | 7.55 | 33.3 | 23.2 | 70.8 |
| | Range | <LOQ-279 | <LOQ-181 | <LOQ-335 | <LOQ-422 | <LOQ-479 | <LOQ-481 | <LOQ-438.9 | <LOQ-820 | <LOQ-1331 | <LOQ-1788 |

∑DMCs: the sum concentration of D3, D4, D5, D6, D7, D8, and D9; DR: detection rates, $DR = \left(\frac{number\ of\ samples\ with\ DMCs\ detected}{total\ number\ of\ samples}\right) \times 100\%$.

Specifically, the DMCs were detected in 95.5% of silicone rubber samples, slightly below the 100% detection rate observed in silicone nipples from USA [9]. The detection rate of DMCs in adhesive samples was 100%, while that in plastic samples was 66.7%. The high detection rates of DMCs in this study reflected that the DMCs were prevalently applied to silicone polymers. However, individual DMCs exhibited notable variability among the three product samples. D5 was the most frequently detected DMC in all samples, consistent with the results of Canadian cosmetic products [7]. In contrast, D3 and D4 had lower detection rates, which were 34.6% and 57.7%, respectively. Moreover, the median concentrations of D3 and D4 were approximately one to two orders of magnitude lower than those of other DMCs, potentially attributed to their lower boiling points of 134 °C and 175–176 °C, respectively. Nonetheless, D3 and D4 were the most highly detected DMCs in plastics, which could be explained by the fact that DMCs with higher molecular weights, such as D6 and D7, are more commonly used in rubber products [57].

The median concentration of ∑DMCs in silicone rubber samples was 802.2 mg/kg, followed by 21.5 mg/kg in adhesive samples, and 8.25 mg/kg in plastic samples. There is a trend in silicone rubber samples where higher median concentrations of DMCs were associated with increased molecular weights of DMCs. However, D3 and D4 were predominant in plastic samples, and D6 dominated in adhesive samples. As depicted in Figure 1, among the three types of samples, the concentrations of individual DMCs were found to be the highest in silicone rubber, except for D4, which was the highest in plastic samples ($p < 0.05$). One plausible reason for this was that DMCs were typically incorporated in small amounts as additives during the manufacture of certain plastics [58], resulting in lower residue levels in plastics compared to silicone rubber and adhesive samples, where DCMs were added as raw materials.

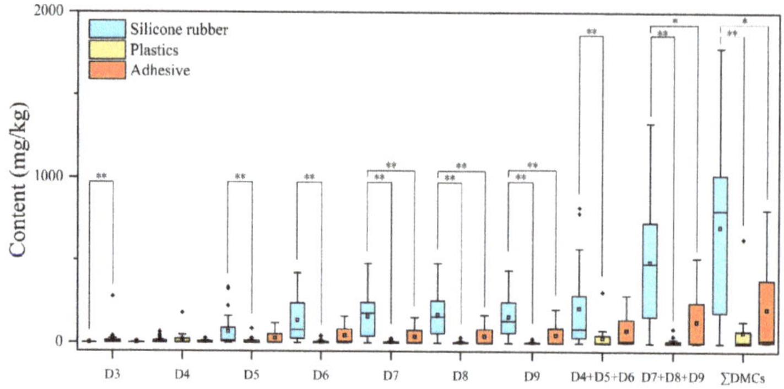


**Figure 1.** Concentrations (mg/kg) of DMCs in silicone rubber, plastic, and adhesive samples, respectively. The line in the box represents the median value; the bottom and the top of each box represent the 5th and 95th percentiles, respectively; * represents $p < 0.05$, and ** represents $p < 0.01$.

As shown in Table 3, the concentrations of D4 (<LOQ-63 mg/kg), D5 (<LOQ-335 mg/kg), and D6 (<LOQ-422 mg/kg) in the silicone rubbers in this study were comparable to those in the US silicone nipples [9], but were two to three orders of magnitude higher than those in Chinese soft rubber toys (median: 0.062–0.101 µg/g) and pacifiers (median: 0.107–1.46 µg/g) [58]. The concentrations of D4 (median: 2.15 mg/kg), D5 (median: 0.50 mg/kg), and D6 (median: 0.50 mg/kg) in the plastics in this study were also much higher than those in Chinese hard toys (median: $<0.7 \times 10^{-3}$ µg/g). Several studies have reported the concentrations of DMCs in other types of products, including baking molds (mean: 284–1325 mg/kg), Chinese personal care products (median: 0.06–1.70 µg/g), and European cosmetics and personal care products (median: 0.011–25.7 mg/g wet weight) [4,5,59]. In general, the amount of DMCs in silicone products made from different materials can vary greatly, even from the same material. But a common thread lies in the high vapor pressure of DMCs, leading to their continuous release into the environment during use. A higher baking temperature has been documented to increase the release of small molecular weight DMCs from baking molds [60]. The aforementioned mobile smart terminal products, such as personal computers, Bluetooth headsets, and cell phones, are normally used indoors, and the prolonged use of electronic products will heat up, which may be responsible for the increase in the volatilization of DMCs. While this conjecture requires further experimental validation, it raises concerns about the potential negative impacts on human health from long-term skin contact or inhalation of substantial amounts of DMCs.

**Table 3.** The levels of DMCs in different samples from various regions.

| Region | Samples | Sample Size | Statistics | D3 | D4 | D5 | D6 | D7 | D8 | D9 | Units | References |
|---|---|---|---|---|---|---|---|---|---|---|---|---|
| USA | Silicone nipples | 22 | DR (%)<br>Range | -<br>- | 100<br>0.6–49 | 100<br>0.6–269 | 100<br>0.3–108 | -<br>- | -<br>- | -<br>- | mg/kg | Zhang et al., 2012 [9] |
| Canada | Cosmetic products | 252 | DR (%) | 0.79 | 4.8 | 14 | 9.1 | - | - | - | | Wang et al., 2009 [7] |
| China | Soft rubber toys | 44 | DR (%)<br>Median | -<br>- | 100<br>0.062 | 91<br>0.086 | 100<br>0.101 | -<br>- | -<br>- | -<br>- | µg/g | Xu et al., 2017 [58] |
| | Pacifiers | 74 | DR (%)<br>Median | -<br>- | 100<br>0.107 | 100<br>0.924 | 100<br>1.46 | -<br>- | -<br>- | -<br>- | µg/g | |
| | Hard toys | 40 | DR (%)<br>Median | -<br>- | 10<br>$<0.6 \times 10^{-3}$ | 50<br>$<0.45 \times 10^{-3}$ | 40<br>$<0.7 \times 10^{-3}$ | -<br>- | -<br>- | -<br>- | µg/g | |
| Europe | Cosmetics and PCPs | 51 | Median | - | 0.011 | 25.7 | 0.64 | - | - | - | mg/g | Dudzina et al., 2014 [4] |
| Germany | Bakeware products | [illegible] | Mean | [illegible] | [illegible] | 284 | 753 | 1265 | 1127 | 1325 | mg/kg | Fromme et al., 2019 [5] |
| Shanghai, China | PCPs | 158 | Median | - | 0.29 | 1.70 | 0.66 | 0.06 | - | - | µg/g | Lu et al., 2011 [59] |

### *3.2. Composition Profiles of DMCs*

The compositional profiles of D3, D4+D5+D6, and D7+D8+D9 in silicone rubber, plastic, and adhesive samples are presented in Figure 2. Among the three products, the profiles of DMCs are quite different, indicating that the types and contents of DMCs required for different materials can vary dramatically. The compositional percentage in the adhesive was in the order of D4+D5+D6 (68.8%) > D7+D8+D9 (28.4%) > D3 (2.8%), while the proportions of components in the plastic were about the same. The sum concentration of D7, D8, and D9 in silicone rubber samples accounted for more than 80% of the ∑DMCs, which was five times higher than that of D4+D5+D6 (481.5 m g/kg vs. 86.0 mg/kg). This finding is similar to the results observed in bakeware, where the sum of the average concentrations of D7, D8, and D9 is approximately threefold higher than that of D4, D5, and D6 combined (3717 mg/kg vs. 1542 mg/kg) [5]. Among the analyzed DMCs, D3 exhibits the lowest boiling point of 134 °C, rendering it prone to volatilization during the production and use of the products, resulting in lower concentrations in the silicone rubber and adhesive samples. However, it remains to be verified whether the atmospheric concentration of D3 is the most minimal. Generally, D4, D5, and D6 are widely utilized as raw materials for the synthesis of silicone polymers across various fields such as children's products, cosmetics, electronic products, and food contact materials, and their conclusive evidence of human hazards has prompted regulatory concerns and restrictions. Thus, manufacturers have responded by reducing the amount of D4, D5, and D6 in products to comply with the production criteria. The well-documented persistence and high bioaccumulation of D4, D5, and D6 have long attracted worldwide attention. Canada is the first country to pay special consideration to the toxicity of DMCs, after which ECHA classified D4, D5, and D6 as SCHV substances and proposed to list them as POPs. The European Union also submitted a notification G/TBT/N/EU/989 to the World Trade Organization (WTO), which, in brief, restricted the application of these substances in textiles, leather, and rinse-off cosmetics in an attempt to minimize their emission [37]. On the other hand, D7, D8, and D9 are mainly intermediate by-products of the synthesis of silicone polymers, attracting less regulatory attention, and their use remains unrestricted. In this study, the compositional profiles of DMCs varied considerably across diverse samples, which, in addition to distinct restrictions, may be due to disparities in product manufacturing processes. However, until now, there is limited information on the toxicity of D7, D8, and D9. Indeed, D7, D8, and D9 have similar chemical structures to D4, D5, and D6, so they are speculated to have similar environmental and human hazardous effects. Considering the findings in this investigation, the proportion of D7, D8, and D9 within the ∑DMCs of the selected samples cannot be neglected, especially in silicone rubber samples. Hence, their presence in mobile smart terminal devices warrants careful consideration and further investigation.

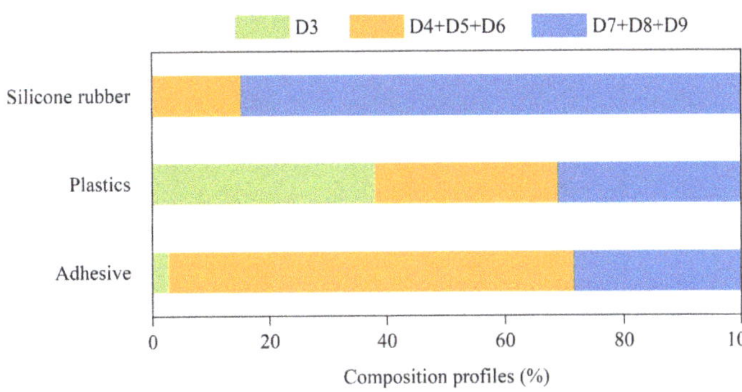


**Figure 2.** Composition profiles of DMCs in silicone rubber, plastics, and adhesive samples, respectively (%).

### 3.3. Environmental Emissions of DMCs from Silicone Rubber

The estimated amount of ∑DMCs released into the environment in China from silicone rubber used in mobile smart terminal devices was 5381.56 tons per year. The silicone rubber samples selected in this study comprised ear caps, watch bands, and case covers, which were replaced frequently and discarded as household waste due to their low price and a wide variety of styles. Substances like DMCs in these mobile smart terminal devices will eventually enter the environment through processes such as e-waste dismantling [46] and thermo-treatment [47]. Previous studies have reported half-lives exceeding 4 days for D6 and up to 10–30 days for D3–D5, facilitating their long-range environmental transfer [55,61]. Notably, airborne DMCs have an extremely high potential to form secondary organic aerosols [62], with inhalation representing the primary route of human exposure to interior airborne DMCs [63]. Furthermore, a study conducted in the Bohai Sea, China, demonstrated the biotrophic magnification of D4–D7 in aquatic food webs [64], thereby elevating the concentrations of DMCs accessible to humans via dietary intake. Collectively, DMCs released into the environment pose a threat to human health through multiple exposure routes. Unfortunately, it is known that 80% of the discarded mobile smart terminal devices are not properly recycled [50]. Thus, there is considerable merit in investigating and recognizing the environmental hazards of DMCs, not only in silicone rubber materials, but also in other device components. It is advisable to explore alternative materials such as metal or woven watchbands instead of silicone rubber ones, and sponge ear caps rather than silicone ear caps. Additionally, it is recommended for manufacturers of mobile smart terminal products to optimize the vulcanization process in silicone rubber production to diminish the content of DMCs.

Nevertheless, this study still has several limitations. First, while the chosen samples selected in our study, namely silicone rubber, plastics, and adhesives, were identified as the most representative and high-risk sources of DMCs, the sample sizes remain modest, with only 22, 18, and 12 specimens, respectively. Second, it is difficult to accurately calculate the environmental emissions of DMCs due to the varied types and proportions of adhesive and plastics used in different types of mobile smart terminal devices and further refinement of experimental methodologies is needed to ensure the validity and accuracy of the data. Third, there are also gaps in the toxicological studies and ecological hazards of D7, D8, and D9, which pose a significant obstacle to elucidating the mechanisms by which DMCs induce deleterious effects. However, this is the first study to provide valid information on the levels of DMCs in mobile smart terminal devices and a stable quantitative method. Future studies could encompass large-scale product categories and expanded sample sizes, as well as incorporate degradation byproducts of DMCs, such as those resulting from UV irradiation or oxidation, to comprehensively scrutinize the environmental and human health implications of DMCs. Toxicological studies of D7-D9 and its congeners could facilitate the development of mathematical models, elucidating the metabolism of DMCs in humans, enhancing our microscopic comprehension of the characteristics of DMCs. Furthermore, the quantification of DMCs in commodities, food, environmental media, and human body fluids (e.g., urine, blood, and breast milk) will allow an effective estimation of DMCs releases and their risk to humans, which can inform the establishment of more precise regulatory thresholds for the application of DMCs.

## 4. Conclusions

The DMCs were widely detected in over 95% of silicone materials used in mobile smart terminal devices. D7, D8, and D9 were dominant in silicone rubbers, while D4, D5, and D6 were predominant in adhesives, collectively constituting over 80% and 60% of ∑DMCs concentrations, respectively. The estimated annual release of ∑DMCs into the environment in China from silicone rubber used in mobile smart terminal devices exceeds 5000 tons. Further studies are needed on the toxicological effects of D7–D9 and the possible degradation byproducts of DMCs. Focus should also be placed on the fate of DMCs in the environment, their internal and external exposure to humans, and risk assessments to clarify their threat to human health and the regulation of these chemicals.

**Supplementary Materials:** The following supporting information can be downloaded at: https://www.mdpi.com/article/10.3390/toxics12040287/s1, Table S1: Retention times, qualitative ions, quantitative ions and ion-pair ratios of DMCs; Table S2: The standard curve, regression coefficients ($R^2$), and limit of quantitation of each analyte.

**Author Contributions:** Methodology, T.Y.; validation, T.Y. and Y.X.; formal analysis, Y.X. and Y.G.; investigation, Y.X.; writing—original draft, Y.X. and Y.G.; supervision, S.L. and X.P. All authors have read and agreed to the published version of the manuscript.

**Funding:** This research was funded by the National Natural Science Foundation of China (No. 42277424).

**Institutional Review Board Statement:** Not applicable.

**Informed Consent Statement:** Not applicable.

**Data Availability Statement:** Data are available from the corresponding author by request.

**Acknowledgments:** Acknowledgments are due to those who contributed to this research.

**Conflicts of Interest:** The authors declare that they have no known competing financial interest or personal relationships that could have appeared to influence the work reported in this paper.

## References

1. Horii, Y.; Minomo, K.; Lam, J.C.W.; Yamashita, N. Spatial Distribution and Accumulation Profiles of Volatile Methylsiloxanes in Tokyo Bay, Japan: Mass Loadings and Historical Trends. *Sci. Total Environ.* **2022**, *806*, 150821. [CrossRef] [PubMed]
2. Molinier, B.; Arata, C.; Katz, E.; Lunderberg, D.; Liu, Y.; Misztal, P.; Nazaroff, W.; Goldstein, A. Volatile Methyl Siloxanes and Other Organosilicon Compounds in Residential Air. *Environ. Sci. Technol.* **2022**, *56*, 15427–15436. [CrossRef] [PubMed]
3. Gerhards, R.; Seston, R.M.; Kozerski, G.E.; McNett, D.A.; Boehmer, T.; Durham, J.A.; Xu, S. Basic Considerations to Minimize Bias in Collection and Analysis of Volatile Methyl Siloxanes in Environmental Samples. *Sci. Total Environ.* **2022**, *851*, 158275. [CrossRef] [PubMed]
4. Dudzina, T.; von Goetz, N.; Bogdal, C.; Biesterbos, J.; Hungerbühler, K. Concentrations of Cyclic Volatile Methylsiloxanes in European Cosmetics and Personal Care Products: Prerequisite for Human and Environmental Exposure Assessment. *Environ. Int.* **2014**, *62*, 86–94. [CrossRef] [PubMed]
5. Fromme, H.; Witte, M.; Fembacher, L.; Gruber, L.; Hagl, T.; Smolic, S.; Fiedler, D.; Sysoltseva, M.; Schober, W. Siloxane in Baking Moulds, Emission to Indoor Air and Migration to Food during Baking with an Electric Oven. *Environ. Int.* **2019**, *126*, 145–152. [CrossRef] [PubMed]
6. Varaprath, S.; Stutts, D.H.; Kozerski, G.E. A Primer on the Analytical Aspects of Silicones at Trace Levels-Challenges and Artifacts—A Review. *Silicon Chem.* **2006**, *3*, 79–102. [CrossRef]
7. Wang, R.; Moody, R.; Koniecki, D.; Zhu, J. Low Molecular Weight Cyclic Volatile Methylsiloxanes in Cosmetic Products Sold in Canada: Implication for Dermal Exposure. *Environ. Int.* **2009**, *35*, 900–904. [CrossRef] [PubMed]
8. Zammarano, M.; Cazzetta, V.; Nazaré, S.; Shields, J.R.; Kim, Y.S.; Hoffman, K.M.; Maffezzoli, A.; Davis, R. Smoldering and Flame Resistant Textiles via Conformal Barrier Formation. *Adv. Mater. Interfaces* **2016**, *3*, 1600617. [CrossRef] [PubMed]
9. Zhang, K.; Wong, J.W.; Begley, T.H.; Hayward, D.G.; Limm, W. Determination of Siloxanes in Silicone Products and Potential Migration to Milk, Formula and Liquid Simulants. *Food Addit. Contam. Part. A Chem. Anal. Control Expo. Risk Assess.* **2012**, *29*, 1311–1321. [CrossRef]
10. Horii, Y.; Ohtsuka, N.; Minomo, K.; Takemine, S.; Motegi, M.; Hara, M. Distribution Characteristics of Methylsiloxanes in Atmospheric Environment of Saitama, Japan: Diurnal and Seasonal Variations and Emission Source Apportionment. *Sci. Total Environ.* **2021**, *754*, 142399. [CrossRef] [PubMed]
11. Guo, J.; Zhou, Y.; Zhang, B.; Zhang, J. Distribution and Evaluation of the Fate of Cyclic Volatile Methyl Siloxanes in the Largest Lake of Southwest China. *Sci. Total Environ.* **2019**, *657*, 87–95. [CrossRef] [PubMed]
12. Nu Nguyen, H.M.; Khieu, H.T.; Ta, N.A.; Le, H.Q.; Nguyen, T.Q.; Do, T.Q.; Hoang, A.Q.; Kannan, K.; Tran, T.M. Distribution of Cyclic Volatile Methylsiloxanes in Drinking Water, Tap Water, Surface Water, and Wastewater in Hanoi, Vietnam. *Environ. Pollut.* **2021**, *285*, 117260. [CrossRef] [PubMed]
13. Zhang, Y.; Shen, M.; Tian, Y.; Zeng, G. Cyclic Volatile Methylsiloxanes in Sediment, Soil, and Surface Water from Dongting Lake, China. *J. Soil. Sediment.* **2018**, *18*, 2063–2071. [CrossRef]
14. Wang, D.-G.; Alaee, M.; Steer, H.; Tait, T.; Williams, Z.; Brimble, S.; Svoboda, L.; Barresi, E.; DeJong, M.; Schachtschneider, J.; et al. Determination of Cyclic Volatile Methylsiloxanes in Water, Sediment, Soil, Biota, and Biosolid Using Large-Volume Injection-Gas Chromatography-Mass Spectrometry. *Chemosphere* **2013**, *93*, 741–748. [CrossRef] [PubMed]

15. Tran, T.; Abualnaja, K.; Asimakopoulos, A.; Covaci, A.; Gevao, B.; Johnson-Restrepo, B.; Kumosani, T.; Malarvannan, G.; Minh, T.; Moon, H.; et al. A Survey of Cyclic and Linear Siloxanes in Indoor Dust and Their Implications for Human Exposures in Twelve Countries. *Environ. Int.* **2015**, *78*, 39–44. [CrossRef] [PubMed]
16. Zhu, Y.; Tang, Z.; He, Y.; Wang, F.; Lyu, Y. Occurrence of Methylsiloxanes in Indoor Store Dust in China and Potential Human Exposure. *Environ. Res.* **2023**, *218*, 114969. [CrossRef] [PubMed]
17. Wang, D.; de Solla, S.; Lebeuf, M.; Bisbicos, T.; Barrett, G.; Alaee, M. Determination of Linear and Cyclic Volatile Methylsiloxanes in Blood of Turtles, Cormorants, and Seals from Canada. *Sci. Total Environ.* **2017**, *574*, 1254–1260. [CrossRef] [PubMed]
18. Zhi, L.; Xu, L.; He, X.; Zhang, C.; Cai, Y. Distribution of Methylsiloxanes in Benthic Mollusks from the Chinese Bohai Sea. *J. Environ. Sci.* **2019**, *76*, 199–207. [CrossRef] [PubMed]
19. Guo, J.; Zhou, Y.; Wang, Y.; Zhang, B.; Zhang, J. Assessment of Internal Exposure to Methylsiloxanes in Children and Associated Non-Dietary Exposure Risk. *Environ. Int.* **2021**, *154*, 106672. [CrossRef] [PubMed]
20. Krenczkowska, D.; Mojsiewicz-Pienkowska, K.; Wielgomas, B.; Bazar, D.; Jankowski, Z. Ex Vivo Human Skin Is Not a Barrier for Cyclic Siloxanes (Cyclic Silicones): Evidence of Diffusion, Bioaccumulation, and Risk of Dermal Absorption Using a New Validated GC-FID Procedure. *Pharmaceutics* **2020**, *12*, 586. [CrossRef] [PubMed]
21. Niu, H.; Su, X.; Li, Q.; Zhao, J.; Hou, M.; Dong, S.; Yan, X.; Sun, J.; Feng, J. Dimethylsiloxanes in Dust from Nine Indoor Microenvironments of Henan Province: Occurrence and Human Exposure Assessment. *Sci. Total Environ.* **2023**, *903*, 166546. [CrossRef] [PubMed]
22. Sun, H.; Li, D.; Xu, L.; Qiu, C.; Wang, S.; Liu, N.; Sun, L. Research Progress on the Distribution, Behavior and Effects of Cyclic Volatile Methylsiloxanes in Organisms. *J. Soil. Sediment.* **2022**, *41*, 193–204. [CrossRef]
23. Kumari, K.; Singh, A.; Marathe, D. Cyclic Volatile Methyl Siloxanes (D4, D5, and D6) as the Emerging Pollutants in Environment: Environmental Distribution, Fate, and Toxicological Assessments. *Environ. Sci. Pollut. Res.* **2023**. [CrossRef] [PubMed]
24. Franzen, A.; Greene, T.; Van Landingham, C.; Gentry, R. Toxicology of Octamethylcyclotetrasiloxane (D4). *Toxicol. Lett.* **2017**, *279*, 2–22. [CrossRef] [PubMed]
25. Zhang, J.; Falany, J.; Xie, X.; Falany, C. Induction of Rat Hepatic Drug Metabolizing Enzymes by Dimethylcyclosiloxanes. *Chem. Biol. Interact.* **2000**, *124*, 133–147. [CrossRef] [PubMed]
26. Lee, J.; Kim, K.; Park, S.-M.; Kwon, J.-S.; Jeung, E.-B. Effects of Decamethylcyclopentasiloxane on Reproductive Systems in Female Rats. *Toxics* **2023**, *11*, 302. [CrossRef] [PubMed]
27. Siddiqui, W.H.; Stump, D.G.; Plotzke, K.P.; Holson, J.F.; Meeks, R.G. A Two-Generation Reproductive Toxicity Study of Octamethylcyclotetrasiloxane (D4) in Rats Exposed by Whole-Body Vapor Inhalation. *Reprod. Toxicol.* **2007**, *23*, 202–215. [CrossRef] [PubMed]
28. Tran, D.N.; Park, S.-M.; Jung, E.-M.; Jeung, E.-B. Prenatal Octamethylcyclotetrasiloxane Exposure Impaired Proliferation of Neuronal Progenitor, Leading to Motor, Cognition, Social and Behavioral Functions. *Int. J. Mol. Sci.* **2021**, *22*, 12949. [CrossRef]
29. Hossain, M.M.; Yuan, Y.; Huang, H.; Wang, Z.; Baki, M.A.; Dai, Z.; Rizwan, M.; Xiong, S.; Cao, M.; Tu, S. Exposure to Dodecamethylcyclohexasiloxane (D6) Affects the Antioxidant Response and Gene Expression of *Procambarus clarkii*. *Sustainability* **2021**, *13*, 3495. [CrossRef]
30. Jean, P.A.; Plotzke, K.P. Chronic Toxicity and Oncogenicity of Octamethylcyclotetrasiloxane (D4) in the Fischer 344 Rat. *Toxicol. Lett.* **2017**, *279*, 75–97. [CrossRef]
31. Young, L.J.; Morfeld, P. Statistical Considerations for a Chronic Bioassay Study: Exposure to Decamethylcyclopentasiloxane (D5) and Incidence of Uterine Endometrial Adenocarcinomas in a 2-Year Inhalation Study with Fischer Rats. *Regul. Toxicol. Pharmacol.* **2016**, *74*, S14–S24. [CrossRef] [PubMed]
32. Farasani, A.; Darbre, P.D. Exposure to Cyclic Volatile Methylsiloxanes (cVMS) Causes Anchorage-Independent Growth and Reduction of BRCA1 in Non-Transformed Human Breast Epithelial Cells. *J. Appl. Toxicol.* **2017**, *37*, 454–461. [CrossRef] [PubMed]
33. Zhang, B.; Zhou, Y.; Guo, J. Association of Volatile Methylsiloxanes Exposure with Non-Alcoholic Fatty Liver Disease among Chinese Adults. *Environ. Pollut.* **2023**, *334*, 122128. [CrossRef]
34. Cantu, M.; Gobas, F. Bioaccumulation of Dodecamethylcyclohexasiloxane (D6) in Fish. *Chemosphere* **2021**, *281*, 130948. [CrossRef] [PubMed]
35. Wang, D.-G.; Norwood, W.; Alaee, M.; Byer, J.D.; Brimble, S. Review of Recent Advances in Research on the Toxicity, Detection, Occurrence and Fate of Cyclic Volatile Methyl Siloxanes in the Environment. *Chemosphere* **2013**, *93*, 711–725. [CrossRef] [PubMed]
36. ECHA. Candidate List of Substances of Very High Concern for Authorisation. Available online: https://echa.europa.eu/candidate-list-table (accessed on 22 January 2024).
37. ECHA. List of Substances Proposed as POPs. Available online: https://echa.europa.eu/list-of-substances-proposed-as-pops/-/dislist/details/0b0236e184f17c3e (accessed on 12 April 2024).
38. ACMI. *2023 China Silicone Series Product Market Report Was Released Grandly*; Advanced Chemical Materials Institution: Beijing, China, 2023. Available online: https://acmi.org.cn/marketresearch/info.aspx?itemid=508&eqid=8dd6ecea000024680000000464537057 (accessed on 17 December 2023).
39. Arshad; Nekahi, A.; McMeekin, S.G.; Farzaneh, M. Measurement of Surface Resistance of Silicone Rubber Sheets under Polluted and Dry Band Conditions. *Electr. Eng.* **2018**, *100*, 1729–1738. [CrossRef]
40. Li, G.; Gong, J.M.; Tan, J.Z.; Zhu, D.S.; Jia, W.H.; Lu, X.J. Acidic-Thermal Ageing Effect on Compression Stress Relaxation of Silicone Rubber. *Strength. Mater.* **2019**, *51*, 660–666. [CrossRef]

41. Mollah, M.S.I.; Kwon, Y.-D.; Islam, M.M.; Seo, D.-W.; Jang, H.-H.; Lim, Y.-D.; Lee, D.-K.; Kim, W.-G. Synthesis and Characterization of Polycarbonates Containing Terminal and Chain Interior Siloxane. *Polym. Bull.* **2012**, *68*, 1551–1564. [CrossRef]
42. Yuan, D.; Cai, X. Synthesis of a Silicon-Containing Flame Retardant and Its Synergistic Effect with Potassium-4-(Phenylsulfonyl) Benzenesulfonate (KSS) in Polycarbonate (PC). *Chin. J. Polym. Sci.* **2013**, *31*, 1352–1358. [CrossRef]
43. Antosik, A.K.; Makuch, E.; Gziut, K. Influence of Modified Attapulgite on Silicone Pressure-Sensitive Adhesives Properties. *J. Polym. Res.* **2022**, *29*, 135. [CrossRef]
44. Antosik, A.K.; Weisbrodt, M.; Mozelewska, K.; Czech, Z.; Piątek-Hnat, M. Impact of Environmental Conditions on Silicone Pressure-Sensitive Adhesives. *Polym. Bull.* **2020**, *77*, 6625–6639. [CrossRef]
45. Storozhenko, P.A.; Minas'yan, R.M.; Polivanov, A.N.; Nikitushkin, I.V.; Minas'yan, O.I. New Thermally Conductive Silicone Adhesive Sealants. *Polym. Sci. Ser. D* **2017**, *10*, 221–224. [CrossRef]
46. He, X.; Xu, L.; Zhang, C.; Cai, Y. Pollution Characteristics of Methyl Siloxanes in Soil from an Electronic Waste(e-Waste) Dismantling Area in Taizhou, China. *J. Soil. Sediment.* **2016**, *35*, 2287–2294. [CrossRef]
47. Xu, L.; Huang, Z.; Zhang, Q.; Xiang, X.; Zhang, S.; Cai, Y. Methylsiloxanes and Their Brominated Products in One E-Waste Recycling Area in China: Emission, Environmental Distribution, and Elimination. *Environ. Sci. Technol.* **2020**, *54*, 4267–4274. [CrossRef] [PubMed]
48. ACMI. *Adhesives Market Report*; ACMI: Beijing, China, 2022. Available online: https://acmi.org.cn/Achievement/info.aspx?itemid=482 (accessed on 23 January 2024).
49. Amghar, N.; Moreno, V.; Sánchez-Jiménez, P.E.; Perejón, A.; Pérez-Maqueda, L.A. Ca-Based Materials Derived from Calcined Cigarette Butts for $CO_2$ Capture and Thermochemical Energy Storage. *J. Environ. Sci.* **2024**, *140*, 230–241. [CrossRef] [PubMed]
50. Forti, V.; Bald, C.P.; Kuehr, R.; Bel, G. The Global E-Waste Monitor 2020. E-Waste Monitor. Available online: https://ewastemonitor.info/gem-2020/ (accessed on 26 January 2024).
51. Fromme, H.; Cequier, E.; Kim, J.; Hanssen, L.; Hilger, B.; Thomsen, C.; Chang, Y.; Völkel, W. Persistent and Emerging Pollutants in the Blood of German Adults: Occurrence of Dechloranes, Polychlorinated Naphthalenes, and Siloxanes. *Environ. Int.* **2015**, *85*, 292–298. [CrossRef] [PubMed]
52. Hanssen, L.; Warner, N.; Braathen, T.; Odland, J.; Lund, E.; Nieboer, E.; Sandanger, T. Plasma Concentrations of Cyclic Volatile Methylsiloxanes (cVMS) in Pregnant and Postmenopausal Norwegian Women and Self-Reported Use of Personal Care Products (PCPs). *Environ. Int.* **2013**, *51*, 82–87. [CrossRef]
53. Xu, L.; Shi, Y.; Liu, N.; Cai, Y. Methyl Siloxanes in Environmental Matrices and Human Plasma/Fat from Both General Industries and Residential Areas in China. *Sci. Total Environ.* **2015**, *505*, 454–463. [CrossRef] [PubMed]
54. ECHA. Background Document to the Opinion on the Annex XV Dossier Proposing Restrictions on Octamethylcyclotetrasiloxane (D4) and Decamethylcyclopentasiloxane (D5), 2016. Available online: https://echa.europa.eu/documents/10162/fefaa3a2-ffc0-4b74-4ec8-3c869d4adae7 (accessed on 29 March 2024).
55. Xiao, R.; Zammit, I.; Wei, Z.; Hu, W.-P.; MacLeod, M.; Spinney, R. Kinetics and Mechanism of the Oxidation of Cyclic Methylsiloxanes by Hydroxyl Radical in the Gas Phase: An Experimental and Theoretical Study. *Environ. Sci. Technol.* **2015**, *49*, 13322–13330. [CrossRef] [PubMed]
56. IDC. IDC Media Center. IDC: The Premier Global Market Intelligence Company. Available online: https://www.idc.com/about/press (accessed on 26 January 2024).
57. Horii, Y.; Kannan, K. Survey of Organosilicone Compounds, Including Cyclic and Linear Siloxanes, in Personal-Care and Household Products. *Arch. Environ. Contam. Toxicol.* **2008**, *55*, 701–710. [CrossRef] [PubMed]
58. Xu, L.; Zhi, L.; Cai, Y. Methylsiloxanes in Children Silicone-Containing Products from China: Profiles, Leaching, and Children Exposure. *Environ. Int.* **2017**, *101*, 165–172. [CrossRef] [PubMed]
59. Lu, Y.; Yuan, T.; Wang, W.; Kannan, K. Concentrations and Assessment of Exposure to Siloxanes and Synthetic Musks in Personal Care Products from China. *Environ. Pollut.* **2011**, *159*, 3522–3528. [CrossRef] [PubMed]
60. Helling, R.; Mieth, A.; Altmann, S.; Simat, T.J. Determination of the Overall Migration from Silicone Baking Moulds into Simulants and Food Using 1H-NMR Techniques. *Food Addit. Contam. Part. A Chem. Anal. Control Expo. Risk Assess.* **2009**, *26*, 395–407. [CrossRef] [PubMed]
61. Kim, J.; Xu, S. Quantitative Structure-reactivity Relationships of Hydroxyl Radical Rate Constants for Linear and Cyclic Volatile Methylsiloxanes. *Environ. Toxicol. Chem.* **2017**, *36*, 3240–3245. [CrossRef] [PubMed]
62. Xu, J.; Harrison, R.M.; Song, C.; Hou, S.; Wei, L.; Fu, P.; Li, H.; Li, W.; Shi, Z. PM2.5-Bound Silicon-Containing Secondary Organic Aerosols (Si-SOA) in Beijing Ambient Air. *Chemosphere* **2022**, *288 Pt 1*, 132377. [CrossRef] [PubMed]
63. Tran, T.M.; Hoang, A.Q.; Le, S.T.; Minh, T.B.; Kannan, K. A Review of Contamination Status, Emission Sources, and Human Exposure to Volatile Methyl Siloxanes (VMSs) in Indoor Environments. *Sci. Total Environ.* **2019**, *691*, 584–594. [CrossRef]
64. Cui, S.; Fu, Q.; An, L.; Yu, T.; Zhang, F.; Gao, S.; Liu, D.; Jia, H. Trophic Transfer of Cyclic Methyl Siloxanes in the Marine Food Web in the Bohai Sea, China. *Ecotoxicol. Environ. Saf.* **2019**, *178*, 86–93. [CrossRef] [PubMed]

toxics

MDPI

*Article*

# Bisphenols in Aquatic Products from South China: Implications for Human Exposure

**Yinhai Chen *, Xiurong Chen, Wenchi Lin, Jinghong Chen, Yuejun Zhu and Zhanghong Guo**

Center for Disease Control and Prevention of Shantou, Shantou 515041, China; 13829450098@163.com (X.C.); zhguo2889@163.com (Z.G.)
* Correspondence: stcdcagsea@163.com; Tel.: +86-137-1991-5229

**Abstract:** In this study, 245 representative samples of aquatic products were selected from local markets in Shenzhen by stochastic sampling. The samples comprised eight species and fell into three aquatic product categories: fish, crustaceans, and bivalves. A total of eight BPs were determined by liquid chromatography coupled with mass spectrometry, namely, bisphenol A (BPA), bisphenol AF (BPAF), bisphenol AP (BPAP), bisphenol B (BPB), bisphenol S (BPS), bisphenol P (BPP), bisphenol Z (BPZ), and bisphenol F (BPF). All BPs were detected in aquatic products, except for BPAF, indicating pervasive contamination by BPs in aquatic products. BPS demonstrated the highest detection rate both before and after enzymatic hydrolysis, whereas BPAP exhibited the lowest detection rate before enzymatic hydrolysis and BPB displayed the lowest detection rate after enzymatic hydrolysis. The concentration difference before and after enzymatic hydrolysis proved to be statistically significant. Moreover, 49–96% of BPs in aquatic products were found in the combined state, underscoring the essentiality of conducting detections on aquatic product samples following enzymatic hydrolysis. While the health risks associated with ingesting BPs residues through aquatic product consumption were found to be minimal for residents at risk of exposure, the results suggest the necessity for more stringent regulations governing the consumption of aquatic products.

**Keywords:** bisphenols; aquatic products; levels; risk assessment

## 1. Introduction

Bisphenols (BPs) are recognized as an important class of synthetic chemicals that are widely used as raw materials to make epoxy resins and polycarbonate plastics. These materials are used in the manufacturing of a number of products, such as food and drink containers, thermal paper receipts, and baby products [1–3]. Polycarbonate plastic production involves synthesizing bisphenol A (BPA). Polycarbonate plastics are popular among manufacturers worldwide due to their economic viability, color adaptability, thermal stability, and corrosion resistance. This popularity is reflected in the annual production of approximately 5 million tons of BPA [4]. It has been reported that BPA has been found in a wide range of environmental components, including surface water, drinking water, soil, and sediment [5–8]. Humans are exposed to BPA via ingestion, inhalation, and dermal absorption through different sources. BPA is classified as an endocrine-disrupting chemical (EDC) due to its estrogenic and antiandrogenic properties, raising concerns about its potential impact on human health. Moreover, BPA is linked to various adverse health effects, including obesity, cardiovascular disease, cognitive decline, diabetes, and reproductive defects [9–13].

Due to restrictions on the use of BPA in various countries, alternative compounds called BPA analogs have been developed as substitutes. Examples of these analogs include Bisphenol AF (BPAF), bisphenol AP (BPAP), bisphenol B (BPB), bisphenol S (BPS), bisphenol P (BPP), bisphenol Z (BPZ), and bisphenol F (BPF). However, studies on the toxicity of these analogs have shown that they have antiandrogenic and estrogenic properties that are comparable to, if not more potent than, those of BPA [14,15]. As an illustration, a study conducted on young female rats showed that BPS could affect genes linked to the dopamine and serotonin systems, as well as 5α-reductase in the prefrontal cortex [16]. According to Ullah et al., adult rats exposed to BPB experience oxidative stress in the testes, which leads to a decrease in daily sperm viability and production [17]. Additionally, maternal exposure to BPF has been found to increase anxiety and depressive behaviors in the offspring of mice [18]. Moreover, BPAF induces inhibition of testosterone production by altering genes and proteins in the testosterone biosynthesis pathway in adult male rats [19]. As a result, the extensive usage and environmental persistence of these compounds pose a significant concern for public health due to their potential impact on biological systems.

Aquatic products, including fish, shrimp, shellfish, etc., are distinguished by their delightful taste and substantial nutritional value. Given these attributes, they have evolved into essential sources of crucial nutrients such as vitamins, protein, and trace elements [20]. Aquatic products are a major source of animal protein for residents living in coastal areas. China's demand for aquatic product consumption has been rising recently [21]. Between 2008 and 2019, China's annual consumption of aquatic products averaged 53 million tons, as reported. Nevertheless, these products may harbor heavy metals or other emerging organic pollutants (EOPs) like BPs, parabens, and polychlorinated biphenyls (PCBs). The entry of these pollutants into aquatic products can occur through diverse pathways, encompassing water, sediment, aquaculture, transportation, processing, and storage [21,22]. These contaminants, which are highly concentrated in aquatic products, progressively build up through the food chain and food web and have a negative impact on human health. Numerous countries, including China, have established stringent limit standards for substances universally recognized as harmful to human health. However, no country has instituted definitive guidelines for certain emerging pollutants, like BPs. Consequently, accurately measuring the content of BPs in aquatic products becomes imperative for relevant regulatory authorities to establish safety dosage limits for associated food items.

Situated as a coastal city in South China, Shenzhen hosts several aquaculture bases. The economic development of Shenzhen has inevitably led to environmental pollution, which is exacerbated by human activities that may contribute to contamination. This contamination poses potential health impacts for residents who consume aquatic products in the region. To fill the current data gap, 245 representative aquatic product samples in Shenzhen were collected stochastically from local markets. The samples included eight species and three aquatic product categories: fish, crustaceans, and bivalves. High-performance liquid chromatography–tandem mass spectrometry (HPLC-MS/MS) was used to detect eight BPs (BPA, BPAF, BPAP, BPB, BPS, BPP, BPZ, and BPF) from aquatic products. In this paper, we aim to achieve the following objectives: (1) quantify BP levels in aquatic products; (2) investigate the factors affecting BP levels; and (3) estimate the human health risks associated with aquatic product consumption.

## 2. Materials and Methods

### *2.1. Reagents and Materials*

BPA, BPAF, BPAP, BPB, BPS, BPP, BPF, BPZ, $^{13}C_{12}$-BPA, $^{13}C_{12}$-BPS, $^{13}C_{12}$-BPAF, $^{13}C_{12}$-BPB, and BPF-$d_{10}$ were purchased from Toronto Research Chemicals (Toronto, ON, Canada). Methanol, acetonitrile, n-hexane, and β-glucuronidase (aqueous solution > 100,000 units/mL, sulfatase activity < 20,000 units/mL) were purchased from Shanghai Anpu Experimental Technology (Shanghai, China). Acetic acid and ammonium acetate were purchased from Shanghai Jingchun Biochemical Technology (Shanghai, China). Solid-phase extraction (SPE) cartridges (Oasis PRiME HLB, 200 mg/6 mL) were purchased from Waters (Framingham, MA, USA). Ultra-pure water was provided by the Millipore water purification system (Billerica, MA, USA).

### 2.2. Sampling and Sample Preparation

To mitigate any potential sampling bias, a total of 245 representative aquatic product samples were stochastically collected from local markets in Shenzhen. The samples represented three aquatic product categories, namely, fish (including six species—*Oreochromis mossambicus, Lateolabrax maculatus, Carassius auratus, Ctenopharyngodon idella, Aristichys nobilis* and *Trachinotus ovatus*, $n = 171$), crustaceans (one species of *Metapenaeus ensis*, $n = 56$), and bivalves (one species of *Paphia undulata*, $n = 18$). After the edible parts were homogenized, the samples were stored in airtight bags at −20 °C until further analysis.

For the enzymatic group samples, 2.0 g samples were weighed, then 50 μL mixed internal standard was added. The pH was adjusted with 3 mL 0.1 mol/L ammonium acetate buffer and an additional 20 μL β-glucuronidase. Samples were shaken at 37 °C for 12–16 h, avoiding light. Then, 9 mL acetonitrile was added for extraction, followed by shaking at 2500 rpm (the relative centrifugal force, RCF: 560× *g*) for 5 min, and 10 min ultrasonication. The samples were centrifuged at 9000 rpm (RCF: 7258× *g*) for 5 min, and the process was repeated twice. After supernatant removal and addition of 2 mL n-hexane, the samples were shaken for 45 s and centrifuged at 9000 rpm (RCF: 7258× *g*) for 5 min, then the lower phase was obtained and passed through the solid-phase extraction (SPE) column. Next, 2.5 mL of the extract was blown nearly dry with nitrogen and redissolved in 60% methanol solution. Finally, it was filtered through a 0.22 μm membrane for instrumental analysis.

The procedure for samples without enzymatic digestion was similar to those with enzymatic digestion. A volume of 50 μL mixed internal standard was added to a 50 mL pointed bottom centrifuge tube containing 2.0 g of aquatic product samples, and then 10 mL acetonitrile was added for extraction. The following steps were the same as those described for the enzymatic group samples.

### 2.3. Instrumental Analysis

Target chemicals were analyzed using a 20A HPLC system (Shimadzu, Tokyo, Japan) coupled with a Q-Trap 5500 tandem mass spectrometer (MS/MS; Applied Biosystems, Foster City, CA, USA). Chromatographic separation of the analytes was performed on an Atlantis C18 column (1.7 μm, 3.0 × 100 mm, Waters, Dublin, Ireland), accompanied by mobile phases of water and methanol. The gradient elution conditions and the mass spectrometric information of the target chemicals are shown in Tables S1 and S2, respectively. The injection volume was 5 μL and the column temperature was set at 40 °C. The flow rate was set at 0.3 mL/min. The mass spectrometer was performed in negative electron spray ionization (ESI) mode. The source temperature was set as 550 °C and the ionization voltage was −4500 V.

### 2.4. Quality Assurance and Quality Control (QA/QC)

To show potential contamination during the experiment, a blank sample was analyzed every 25 samples. No target compounds were detected in the blanks. The linear range of all target chemicals was 0.1–50 μg/L with the correlation coefficient ($R^2$) being above 0.99. The limits of detection (LOD), defined as the signal-to-noise (S/N) ratio = 3, ranged from 0.002 to 0.09 ng/g and the limits of quantification (LOQ), defined as the S/N ratio = 10, ranged from 0.007 to 0.03 ng/g. The recoveries were between 84.8% and 111%. The reaction stability of the instrument was confirmed by 100 μg/L mixed internal standard solution. The LODs, LOQs, and recoveries of the applied analytical methods are described in detail in Table S3.

### 2.5. Health Risk Assessments

The estimated daily intake (EDI) of BPs due to consumption of aquatic products was calculated according to the following Equation (1):

$$\text{EDI} = \frac{C_a \times CR}{BW} \tag{1}$$

where $C_a$ (ng/g ww) is the concentration of various BPs in aquatic products. *CR* (g/day) is the daily consumption of aquatic products, and according to the China Fishery Statistical Yearbook 2020, a typical Chinese resident consumed aquatic products at a rate of 76.16 g/day in 2019. *BW* (kg) is the average body weight of the population, which is on average 63.9 kg for males and 54.0 kg for females according to the *Report on the Nutrition and Chronic Disease Status of Chinese Residents 2020*.

The hazard quotient (HQ) was used to preliminarily assess the health risk of exposure to BPs through a single route. The value can be obtained through the following Equation (2):

$$HQ = \frac{EDI}{ADI} \quad (2)$$

where EDI (ng/kg bw/day) is the estimated daily intake of each BP calculated previously; ADI (ng/kg bw/day) is the allowable daily intake; and 50 μg/kg bw/day was assigned according to European Food Safety Authority (EFSA) and U.S. Environmental Protection Agency (EPA).

### 2.6. Data Analysis

Statistical analysis was performed using Excel 2016, IBM SPSS 26.0, and Python 3.7. Descriptive statistics, including mean, median, maximum, and minimum values, were employed to characterize the concentration of BPs in aquatic products. Pearson's correlation coefficient (bilateral) test or Spearman's rank correlation test were used to analyze the correlation between various substances. Inter-group differences between enzymatic and non-enzymatic hydrolysis were examined using the Mann–Whitney U test. Statistical significance was established at $p < 0.05$.

## 3. Results and Discussion

### 3.1. Levels of Bisphenols in Aquatic Products

Table 1 displays the detection rates and concentration levels of BPs in aquatic products sold in Shenzhen. The concentration and detection rates of BPs before enzymatic hydrolysis were 0.08–94.1 ng/g ww and 6.12–78.8%, respectively. Following enzymatic digestion, the samples showed different detection rates of BPs, from 33.9% to 97.1%, and the concentrations of BPs, from 0.01 ng/g ww to 162 ng/g ww. These results indicated that the aquatic products were found to be generally contaminated with BPs, and BPS had the highest detection rate both before and after enzymatic digestion. The highest detection rate for BPS (50.8%) was also observed in the muscle tissue of fish from distinct aquatic environments [23]. BPAF was essentially undetectable, regardless of enzymatic digestion. The BPA and BPF concentrations after enzymatic digestion in this experiment ranged from 0.01 to 19.9 ng/g ww and from 0.20 to 138 ng/g ww, respectively. The levels of BPA in marine and freshwater fish from Hong Kong (0.31–19.25 ng/g ww) were comparable to the results after enzymatic digestion in this study. However, the BPF levels (2.30–26.37 ng/g ww) were considerably lower compared to those observed post enzymatic digestion [24]. The mean concentration of BPA in mollusks from the Bohai Sea recorded at 1.03 ng/g ww, and the mean concentration of BPF at 0.747 ng/g ww, were both lower than the outcomes of the current study [25]. In contrast, the BPA levels found in adult shad from the Ganges River, India, ranging between 172 and 288 ng/g ww, significantly exceeded the results of the present study [26].

**Table 1.** Detection rate and content level of BPs in aquatic products (ng/g ww).

| Sample | Process | Parameters | BPA | BPS | BPF | BPP | BPZ | BPAP | BPB |
|---|---|---|---|---|---|---|---|---|---|
| *Metapenaeus ensis* | Non-enzymolysis | DR (%) | 42.9 | 82.1 | 42.9 | 23.2 | 16.1 | 5.60 | 25.0 |
| | | Median | 0.36 | 0.80 | 2.54 | 0.63 | 0.47 | 0.06 | 1.15 |
| | | Mean | 1.29 | 1.78 | 3.54 | 0.64 | 0.60 | 0.05 | 1.91 |
| | | Minimum | 0.01 | 0.02 | 1.05 | 0.04 | 0.04 | 0.01 | 0.09 |
| | | Maximum | 9.84 | 12.7 | 15.8 | 1.73 | 1.73 | 0.07 | 11.9 |
| | Enzymolysis | DR (%) | 69.6 | 98.2 | 51.8 | 53.7 | 82.1 | 69.6 | 23.2 |
| | | Median | 2.16 | 1.76 | 7.32 | 3.37 | 1.54 | 1.28 | 0.36 |
| | | Mean | 2.49 | 4.34 | 16.7 | 7.30 | 2.04 | 1.11 | 2.34 |
| | | Minimum | 0.31 | 0.03 | 0.26 | 0.01 | 0.01 | 0.01 | 0.19 |
| | | Maximum | 8.25 | 26.5 | 93.1 | 37.5 | 8.79 | 3.12 | 18.2 |
| *Aristichys nobilis* | Non-enzymolysis | DR (%) | 33.3 | 94.4 | 27.8 | 27.8 | 11.1 | 5.56 | 22.2 |
| | | Median | 0.24 | 2.39 | 2.20 | 0.68 | 1.24 | 0.04 | 6.48 |
| | | Mean | 0.73 | 3.13 | 2.44 | 3.77 | 1.24 | 0.04 | 6.76 |
| | | Minimum | 0.05 | 0.13 | 0.49 | 0.24 | 0.44 | 0.04 | 0.36 |
| | | Maximum | 3.39 | 9.66 | 4.46 | 15.2 | 2.03 | 0.04 | 13.7 |
| | Enzymolysis | DR (%) | 83.3 | 100 | 38.9 | 50.0 | 66.7 | 72.2 | 38.9 |
| | | Median | 1.33 | 4.36 | 42.5 | 5.81 | 2.16 | 1.35 | 10.2 |
| | | Mean | 1.85 | 4.58 | 33.2 | 8.27 | 2.05 | 1.04 | 7.23 |
| | | Minimum | 0.05 | 0.48 | 3.37 | 1.56 | 0.33 | 0.03 | 0.18 |
| | | Maximum | 7.47 | 10.5 | 66.8 | 16.9 | 3.08 | 1.63 | 11.9 |
| *Oreochromis mossambicus* | Non-enzymolysis | DR (%) | 34.3 | 68.7 | 31.4 | 14.3 | 11.4 | 2.90 | 37.1 |
| | | Median | 0.19 | 4.79 | 1.88 | 0.47 | 0.47 | 0.03 | 0.95 |
| | | Mean | 1.24 | 4.80 | 2.60 | 0.83 | 0.99 | 0.03 | 5.56 |
| | | Minimum | 0.02 | 0.72 | 0.25 | 0.18 | 0.03 | 0.03 | 0.00 |
| | | Maximum | 8.23 | 9.35 | 7.24 | 2.29 | 3.01 | 0.03 | 33.7 |
| | Enzymolysis | DR (%) | 65.7 | 97.1 | 40.0 | 22.8 | 51.4 | 51.4 | 22.8 |
| | | Median | 0.97 | 5.20 | 26.9 | 0.96 | 1.28 | 1.31 | 2.71 |
| | | Mean | 1.19 | 11.0 | 30.2 | 1.29 | 1.22 | 1.00 | 3.08 |
| | | Minimum | 0.04 | 1.38 | 0.25 | 0.56 | 0.08 | 0.01 | 0.47 |
| | | Maximum | 2.99 | 161 | 74.3 | 2.91 | 2.62 | 1.43 | 7.80 |
| *Paphia undulata* | Non-enzymolysis | DR (%) | 44.4 | 72.2 | 33.3 | 27.8 | 16.7 | 5.56 | 33.3 |
| | | Median | 0.44 | 4.42 | 2.43 | 0.86 | 0.60 | 0.03 | 0.42 |
| | | Mean | 0.52 | 4.55 | 2.15 | 2.10 | 0.83 | 0.03 | 0.67 |
| | | Minimum | 0.14 | 1.89 | 0.36 | 0.15 | 0.01 | 0.03 | 0.09 |
| | | Maximum | 1.37 | 8.60 | 4.29 | 7.17 | 1.87 | 0.03 | 1.64 |
| | Enzymolysis | DR (%) | 88.9 | 94.4 | 50.0 | 72.2 | 72.2 | 72.2 | 27.8 |
| | | Median | 1.70 | 4.30 | 8.45 | 6.98 | 1.31 | 1.38 | 8.67 |
| | | Mean | 1.85 | 6.66 | 15.0 | 6.31 | 1.09 | 1.00 | 7.22 |
| | | Minimum | 0.07 | 1.28 | 0.42 | 0.54 | 0.33 | 0.10 | 0.30 |
| | | Maximum | 6.01 | 29.9 | 49.1 | 12.3 | 2.07 | 1.62 | 9.59 |
| *Trachinotus ovatus* | Non-enzymolysis | DR (%) | 18.7 | 93.7 | 68.7 | 43.7 | 18.7 | 6.25 | 31.2 |
| | | Median | 0.17 | 3.09 | 3.38 | 0.41 | 0.43 | 0.08 | 0.99 |
| | | Mean | 0.22 | 3.47 | 3.06 | 0.73 | 1.14 | 0.08 | 0.97 |
| | | Minimum | 0.04 | 0.34 | 0.41 | 0.08 | 0.03 | 0.08 | 0.04 |
| | | Maximum | 0.44 | 8.48 | 6.30 | 2.97 | 2.97 | 0.08 | 1.81 |
| | Enzymolysis | DR (%) | 75.0 | 100 | 43.7 | 62.5 | 68.7 | 62.5 | 31.2 |
| | | Median | 1.06 | 4.41 | 4.49 | 1.20 | 1.34 | 1.25 | 3.39 |
| | | Mean | 1.48 | 6.19 | 15.8 | 3.98 | 1.33 | 0.94 | 8.09 |
| | | Minimum | 0.25 | 2.59 | 0.70 | 0.13 | 0.03 | 0.01 | 0.43 |
| | | Maximum | 4.81 | 19.4 | 78.6 | 20.9 | 2.96 | 1.60 | 19.2 |

**Table 1.** *Cont.*

| Sample | Process | Parameters | BPA | BPS | BPF | BPP | BPZ | BPAP | BPB |
|---|---|---|---|---|---|---|---|---|---|
| *Lateolabrax maculatus* | Non-enzymolysis | DR (%) | 35.3 | 82.3 | 50.0 | 14.7 | 14.7 | 2.94 | 23.5 |
| | | Median | 0.33 | 2.61 | 2.48 | 0.88 | 0.85 | 0.04 | 0.45 |
| | | Mean | 1.73 | 4.35 | 2.78 | 0.75 | 0.67 | 0.04 | 1.40 |
| | | Minimum | 0.02 | 0.01 | 0.06 | 0.38 | 0.27 | 0.04 | 0.10 |
| | | Maximum | 9.30 | 25.0 | 8.45 | 0.99 | 0.96 | 0.04 | 5.55 |
| | Enzymolysis | DR (%) | 52.9 | 97.0 | 32.3 | 41.2 | 44.1 | 50.0 | 38.2 |
| | | Median | 1.32 | 6.21 | 4.03 | 1.95 | 1.84 | 1.31 | 5.16 |
| | | Mean | 1.85 | 15.1 | 6.89 | 2.47 | 1.77 | 0.93 | 4.59 |
| | | Minimum | 0.00 | 0.63 | 0.71 | 0.35 | 0.11 | 0.01 | 0.14 |
| | | Maximum | 5.19 | 122 | 30.7 | 6.45 | 3.42 | 1.85 | 12.4 |
| *Ctenopharyngodon idella* | Non-enzymolysis | DR (%) | 33.3 | 69.4 | 55.6 | 19.4 | 16.7 | 8.33 | 47.2 |
| | | Median | 0.19 | 2.56 | 3.23 | 0.19 | 0.22 | 0.03 | 1.65 |
| | | Mean | 0.40 | 6.30 | 4.09 | 0.41 | 0.40 | 0.03 | 3.19 |
| | | Minimum | 0.02 | 0.06 | 0.40 | 0.01 | 0.01 | 0.02 | 0.02 |
| | | Maximum | 2.27 | 94.1 | 21.0 | 1.32 | 1.32 | 0.03 | 16.8 |
| | Enzymolysis | DR (%) | 72.2 | 94.4 | 30.6 | 47.2 | 61.1 | 72.2 | 41.7 |
| | | Median | 1.67 | 4.20 | 25.2 | 1.21 | 1.99 | 1.26 | 1.63 |
| | | Mean | 1.99 | 8.48 | 34.7 | 1.84 | 1.86 | 1.07 | 3.52 |
| | | Minimum | 0.11 | 0.36 | 1.44 | 0.04 | 0.51 | 0.01 | 0.18 |
| | | Maximum | 7.44 | 74.5 | 138 | 5.39 | 3.52 | 2.09 | 15.3 |
| *Carassius auratus* | Non-enzymolysis | DR (%) | 50.0 | 78.1 | 43.7 | 31.2 | 28.1 | 12.5 | 50.0 |
| | | Median | 0.20 | 2.68 | 0.91 | 0.46 | 0.44 | 0.04 | 1.09 |
| | | Mean | 0.70 | 4.87 | 1.55 | 0.91 | 1.16 | 0.04 | 1.42 |
| | | Minimum | 0.05 | 0.13 | 0.15 | 0.06 | 0.07 | 0.02 | 0.01 |
| | | Maximum | 3.32 | 25.9 | 4.73 | 3.10 | 4.23 | 0.07 | 6.79 |
| | Enzymolysis | DR (%) | 78.3 | 96.8 | 53.3 | 59.4 | 75.0 | 71.8 | 53.1 |
| | | Median | 1.56 | 6.86 | 25.8 | 1.65 | 1.84 | 1.27 | 4.97 |
| | | Mean | 2.81 | 11.1 | 41.1 | 2.83 | 2.12 | 1.10 | 8.69 |
| | | Minimum | 0.07 | 1.58 | 0.21 | 0.13 | 0.08 | 0.01 | 0.28 |
| | | Maximum | 19.9 | 40.7 | 131 | 14.6 | 7.23 | 1.68 | 34.3 |
| All samples | Non-enzymolysis | DR (%) | 37.9 | 78.8 | 44.1 | 23.3 | 16.7 | 6.10 | 33.9 |
| | | Median | 0.27 | 2.69 | 2.41 | 0.55 | 0.47 | 0.03 | 1.03 |
| | | Mean | 0.98 | 3.95 | 2.99 | 1.10 | 0.83 | 0.04 | 2.70 |
| | | Minimum | 0.01 | 0.01 | 0.06 | 0.001 | 0.01 | 0.01 | 0.01 |
| | | Maximum | 9.84 | 94.1 | 20.9 | 15.2 | 4.23 | 0.08 | 33.7 |
| | Enzymolysis | DR (%) | 71.1 | 97.1 | 42.8 | 48.9 | 65.7 | 64.9 | 33.9 |
| | | Median | 1.51 | 4.33 | 11.9 | 2.14 | 1.61 | 1.28 | 2.76 |
| | | Mean | 2.04 | 8.56 | 24.1 | 4.54 | 1.79 | 1.05 | 5.33 |
| | | Minimum | 0.01 | 0.03 | 0.20 | 0.01 | 0.01 | 0.01 | 0.14 |
| | | Maximum | 19.9 | 162 | 138 | 37.5 | 8.79 | 3.12 | 34.3 |

DR—detection rate.

Table S4 presents the outcomes of the comparative analysis of enzymatically and non-enzymatically dissolved BPs. Except for BPB ($p > 0.05$), the concentrations of all BPs in enzymatically and non-enzymatically dissolved states exhibited statistically significant differences ($p < 0.05$). The mean concentrations of BPs increased by a factor of 1.97 to 26.25 from pre-treatment levels. This observation is likely attributed to the action of β-glucuronidase/sulfate lyase [27], which hydrolyzes bound BPs in the organism, thereby transitioning it to the free state, resulting in elevated concentrations as observed in the results. The bound state content of BPs in aquatic products ranged from 49.0% to 96.0%, as depicted in Figure S1. Elevated concentrations and detection rates after enzymatic digestion indicated a higher proportion of bound BPs in aquatic products. Various studies have utilized comparable hydrolysis techniques to explore the BPs content in aquatic products. Wong et al. (2017) identified BPA concentrations in the free state ranging from 0.83 to 19.25 ng/g across 20 fish species, a range consistent with the results of the current study [24]. In a study by Wang et al. (2020), the investigation into eight distinct BP types in carp revealed proportions in the free state ranging from 21% (BPS) to 48% (BPAP) [28]. Conversely, the present study indicated that the proportions of BPA, BPS, and BPB in the fish were higher compared to their investigation. The metabolic pathways of BPs in organisms exhibit considerable variation across different sources of exposure [29]. Furthermore, the specific type of aquatic product and variations in conditions such as temperature, as well as distinctions between marine and freshwater environments, may further influence the metabolism of these substances in organisms.

### *3.2. Compositional Profiles and Potential Sources*

Figure 1 illustrates the composition profiles of detected BPs in aquatic products both without and with enzymatic digestion. In the non-enzymatic samples, the highest concentration was attributed to BPS, constituting 31.4% of the total free-form BPs, followed by BPF (23.7%), BPB (21.4%), BPP (8.7%), BPA (7.8%), BPZ (6.6%), and BPAP (0.3%), respectively. As for the enzyme-treated aquatic samples, BPF emerged as the predominant contributor in both free and bound states, accounting for 50.9% of the total BPs, followed by BPS (18%), BPB (11.2%), BPP (9.6%), BPA (4.3%), BPZ (3.8%), and BPAP (2.2%), respectively. It is noteworthy to mention that a considerable proportion of the composition in both enzymatically and non-enzymatically digested samples is composed of the combination of BPF and BPS, as opposed to the conventional assumption that the highest levels are linked to BPA, which has been reported in the majority of studies [30]. Recent studies have revealed that BPA is gradually being replaced by BPS and BPF in various manufacturing processes, leading to their high prevalence in aquatic environments. This trend is reflected in the gradual decline of BPA concentration, with its analogs, particularly BPS and BPF, occupying an increasingly substantial portion of the aquatic environment [31]. Some studies even found that the concentration of BPS can be equal to that of BPA [32], and higher concentrations of BPF in surface waters may be anticipated in the near future [33]. The content of BPs in organisms is not only related to exposure but may also be affected by metabolism. In subsequent studies, more attention should be paid to the aquatic environment and organisms.

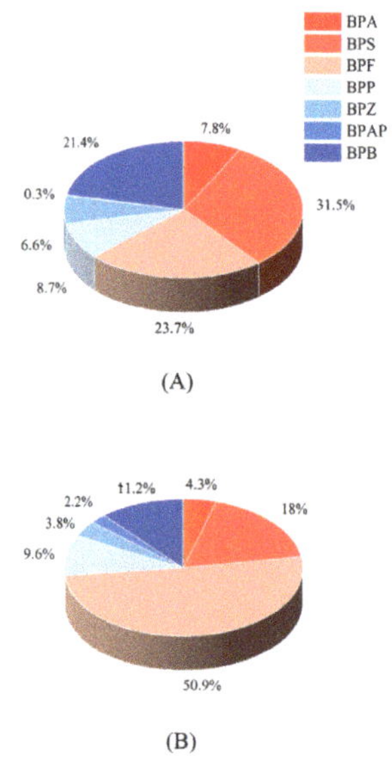


**Figure 1.** Compositional profiles of BPs in non-enzymatic samples (**A**) and enzymatic (**B**) aquatic products.

For a more in-depth examination of the sources of BPs, pairs of the identified BPs in enzymatically digested samples were analyzed using Spearman's correlation ($p < 0.05$), as illustrated in Figure 2, with the correlation coefficients ranging from 0.001 to 0.698. The results of the analysis reveal a statistically significant positive association between BPB and BPA, BPF, BPP, BPZ, and BPAP. However, no significant correlation was observed between BPS and other BPs, implying a divergence in their sources. Moreover, a strong positive correlation was identified between BPA and BPAP, BPZ, BPP, and BPF. Notably, the relatively high correlation coefficients for BPF-BPAP, BPP-BPZ, and BPP-BPAP suggest the possibility of common sources, such as sewage treatment plants, municipal solid waste, and landfill leachate [34]. The addition of different analogs to goods in place of BPA might have contributed to the aforementioned findings. Therefore, further research is needed to identify independent sources of BPA analogs.

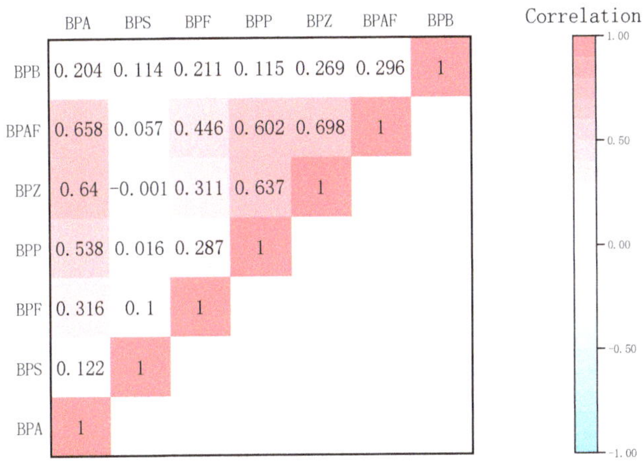


**Figure 2.** Spearman's correlation between pairs of BPs in aquatic products.

### *3.3. Differences in Bisphenols Levels across Types of Aquatic Product*

The distribution of BPs in different aquatic products is shown in Figure 3. In different aquatic products, the concentration ranking from high to low before enzymatic hydrolysis is as follows: *Aristichys nobilis*, *Oreochromis mossambicus*, *Ctenopharyngodon idella*, *Lateolabrax maculatus*, *Paphia undulata*, *Carassius auratus*, *Metapenaeus ensis*, and *Trachinotus ovatus*. After enzymatic hydrolysis, the concentration ranking changes to *Carassius auratus*, *Aristichys nobilis*, *Ctenopharyngodon idella*, *Oreochromis mossambicus*, *Paphia undulata*, *Trachinotus ovatus*, *Metapenaeus ensis*, and *Lateolabrax maculatus*. The analysis indicates a consistently high concentration of BPs in samples of *Aristichys nobilis* and *Ctenopharyngodon idella*. Before enzymatic hydrolysis, BPS exhibits the highest concentration in samples except for *Oreochromis mossambicus*, *Aristichys nobilis*, and *Metapenaeus ensis*, where a high concentration of BPB or BPF is observed. After enzymatic hydrolysis, BPF emerges as the compound with the highest proportion of BPs in samples, except for *Lateolabrax maculatus*, which is still rich in BPS.

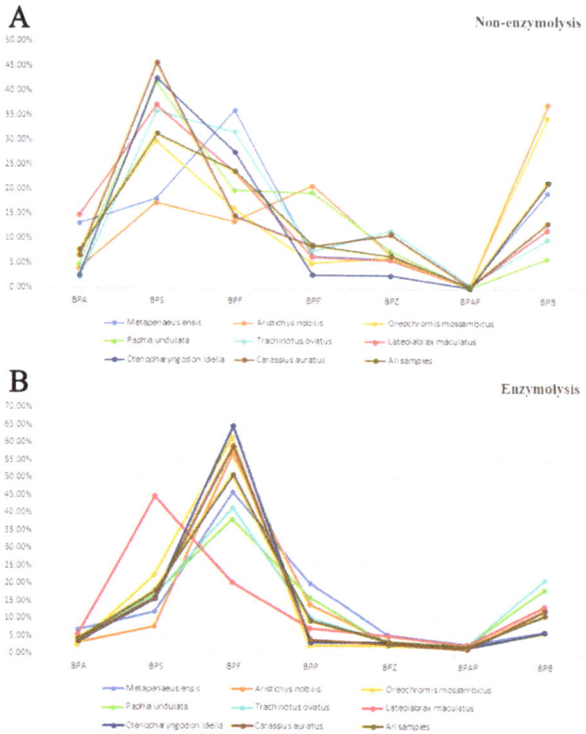


**Figure 3.** Distribution of BPs in different aquatic products. (**A**) Distribution of BPs in different aquatic products without enzymolysis; (**B**) distribution of BPs in different aquatic products with enzymolysis.

In contrast to our findings, previous studies have only detected BPA in fish muscles, with an average concentration of 16.2 ng/g. BPB was detectable in fish livers, with an average concentration of 7.3 ng/g. However, BPF was not detected in any of the samples [35]. Aligning more closely with our study, Rios-Fuster et al. observed that BPA exhibited the highest concentration in fish, followed by BPS and BPF [36]. The concentration of BPA in shrimps shows correspondence between our study (*Metapenaeus ensis*: 2.49 ng/g) and another study located in Pearl River Estuary (*Penaeuschin ensis*: 0.16–4.36 ng/g) [37]. The concentration of BPA (1.63 vs. 1.85 ng/g) and BPB (6.62 vs. 7.22 ng/g) in clams, as reported by Antía Lestido-Cardama, aligns closely with our study. However, the concentration of BPF (14.97 vs. 1.52 ng/g) in their study is approximately ten times lower than observed in our investigation [38]. In a separate study, it was noted that the average concentration of BPA in Rangia cuneata, a type of bivalve, was 174 ng/g, which is approximately 100 times higher than the concentration observed in *Paphia undulata* in the current study [39]. It is evident that the concentration and distribution of BPs in sea products exhibit significant variations across different areas.

### 3.4. Exposure Risk Assessment

The EDIs of $\sum_7$BPs (the sum concentration of all the studied BPs) in different aquatic products are shown in Table 2. The 95% quantiles of EDI were in the range of 63.76–157.4 ng/kg bw/day in males and 75.45–186.3 ng/kg bw/day in females. In the consumption of identical types of aquatic products, except for *Trachinotus ovatus*, females exhibited greater EDI compared to males. The highest EDI, with a 95% quantile, is associated with the consumption of *Carassius auratus*, followed by *Oreochromis mossambicus*, *Aristichys nobilis*, *Ctenopharyngodon idella*, *Paphia undulata*, *Metapenaeus ensis*, *Lateolabrax maculatus*, and *Trachinotus ovatus*, which indicated that the intake of BPs varies across different aquatic products.

**Table 2.** EDI (ng/kg bw/day) and HQ value of $\sum_7$BPs in edible aquatic products of different genders.

| Species | Risk Assessment | Male | Female |
|---|---|---|---|
| *Metapenaeus ensis* | 50% quantile of EDI | 13.34 | 18.83 |
| | 95% quantile of EDI | 78.81 | 93.26 |
| | HQ | 0.0015 | 0.0018 |
| *Aristichys nobilis* | 50% quantile of EDI | 24.95 | 25.03 |
| | 95% quantile of EDI | 97.12 | 114.9 |
| | HQ | 0.0019 | 0.0022 |
| *Oreochromis mossambicus* | 50% quantile of EDI | 9.25 | 23.79 |
| | 95% quantile of EDI | 104.9 | 124.2 |
| | HQ | 0.0020 | 0.0024 |
| *Paphia undulata* | 50% quantile of EDI | 16.85 | 220.2 |
| | 95% quantile of EDI | 86.35 | 102.1 |
| | HQ | 0.0017 | 0.0020 |
| *Trachinotus ovatus* | 50% quantile of EDI | 18.47 | 16.86 |
| | 95% quantile of EDI | 63.76 | 75.45 |
| | HQ | 0.0012 | 0.0015 |
| *Lateolabrax maculatus* | 50% quantile of EDI | 14.97 | 21.12 |
| | 95% quantile of EDI | 78.62 | 93.03 |
| | HQ | 0.0015 | 0.0018 |
| *Ctenopharyngodon idella* | 50% quantile of EDI | 11.43 | 19.73 |
| | 95% quantile of EDI | 88.74 | 105.0 |
| | HQ | 0.0017 | 0.0021 |
| *Carassius auratus* | 50% quantile of EDI | 16.35 | 16.86 |
| | 95% quantile of EDI | 157.4 | 186.3 |
| | HQ | 0.0031 | 0.0037 |
| All samples | 50% quantile of EDI | 11.73 | 16.86 |
| | 95% quantile of EDI | 104.6 | 123.7 |
| | HQ | 0.0020 | 0.0024 |
| | The Hazard Index | 0.0146 | 0.0175 |

EDI: estimated daily intake; HQ: hazard quotient; the Hazard Index (summing HQs).

To further evaluate human exposure risks via aquatic product consumption, the HQ value was calculated using 95% quantile of the EDI value of aquatic products consumed by residents (Table 2). The HQ ranged from 0.0012 to 0.0031 for males and from 0.015 to 0.037 for females. Residents who consume *Carassius auratus* run a higher risk of BPs exposure compared to those who consume other aquatic products, as evidenced by the significantly higher HQ value of this species ($p < 0.05$). Females have a higher risk of BPs exposure than males, as indicated by the higher HQ value. The present study employed a relatively high EDI value and assumed a 100% absorption rate of BPs in the human body. It is noteworthy that the potential reduction in BPs content during food preparation was overlooked, leading to possible overestimation of human exposure to BPs. Despite these limitations, the calculated HQ values remained well below one, signifying that there is no significant health risk associated with the consumption of aquatic products by the residents.

## 4. Conclusions

In this study, a comprehensive understanding of the detrimental effects of BPs on organisms, as well as their direct or indirect impacts on humans, can be obtained by systematically measuring and analyzing their levels in aquatic products. The analysis of 245 aquatic products sold in Shenzhen revealed detectable levels of BPs, and the associated risks for residents by gender were assessed. All BPs were present in aquatic products, except for BPAF, indicating widespread contamination. BPS exhibited the highest detection rate, while BPAP had the lowest rate before enzymatic hydrolysis, and BPB had the lowest rate after enzymatic hydrolysis. The concentration difference before and after enzymatic hydrolysis was statistically significant. Notably, 49–96% of BPs in aquatic products existed in a combined state, underscoring the importance of enzymatic hydrolysis in sample detection. There was a significant correlation between different BPs, with the exception of BPS, suggesting possible shared sources. The risks associated with male and female residents' consumption of aquatic products containing BPs residues were negligible. However, the results highlight the need for stricter regulations regarding the intake of aquatic products. To reduce the human health risks associated with BPs, several suggestions were proposed, such as a systematic investigation of BPs' fate in environmental media, the formulation of discharge standards for BPs pollution in industrial wastewater, and implementation of BP remediation projects.

**Supplementary Materials:** The following supporting information can be downloaded at: https://www.mdpi.com/article/10.3390/toxics12020154/s1, Figure S1: Average concentration distribution of free and bound BPs in aquatic products; Table S1: Mobile phase and gradient elution program; Table S2: Mass spectrometric information of the target compounds; Table S3: Linear equations and related quality control indicators for each analyte; Table S4: Test the difference of concentration levels of BPs in aquatic products with or without enzymatic hydrolysis.

**Author Contributions:** Methodology, Y.C.; validation, X.C. and W.L.; formal analysis, W.L.; investigation, X.C., J.C., Y.Z. and Z.G.; writing—original draft, Y.C.; supervision, Y.C. All authors have read and agreed to the published version of the manuscript.

**Funding:** This research received no external funding.

**Institutional Review Board Statement:** Not applicable.

**Informed Consent Statement:** Not applicable.

**Data Availability Statement:** Data are available from the corresponding author by request.

**Acknowledgments:** Acknowledgments are due to those who contributed to this research.

**Conflicts of Interest:** The authors declare that they have no known competing financial interests or personal relationships that could have appeared to influence the work reported in this paper.

## References

1. Khalili Sadrabad, E.; Hashemi, S.A.; Nadjarzadeh, A.; Askari, E.; Akrami Mohajeri, F.; Ramroudi, F. Bisphenol A release from food and beverage containers—A review. *Food Sci. Nutr.* **2023**, *11*, 3718–3728. [CrossRef] [PubMed]
2. Banaderakhshan, R.; Kemp, P.; Breul, L.; Steinbichl, P.; Hartmann, C.; Fürhacker, M. Bisphenol A and its alternatives in Austrian thermal paper receipts, and the migration from reusable plastic drinking bottles into water and artificial saliva using UHPLC-MS/MS. *Chemosphere* **2022**, *286 Pt 3*, 131842. [CrossRef]
3. Cao, X.L.; Corriveau, J.; Popovic, S.; Clement, G.; Beraldin, F.; Dufresne, G. Bisphenol a in baby food products in glass jars with metal lids from Canadian markets. *J. Agric. Food Chem.* **2009**, *57*, 5345–5351. [CrossRef] [PubMed]
4. Yamazaki, E.; Yamashita, N.; Taniyasu, S.; Lam, J.; Lam, P.K.; Moon, H.B.; Jeong, Y.; Kannan, P.; Achyuthan, H.; Munuswamy, N.; et al. Bisphenol A and other bisphenol analogues including BPS and BPF in surface water samples from Japan, China, Korea and India. *Ecotoxicol. Environ. Saf.* **2015**, *122*, 565–572. [CrossRef]
5. Yan, Z.; Liu, Y.; Yan, K.; Wu, S.; Han, Z.; Guo, R.; Chen, M.; Yang, Q.; Zhang, S.; Chen, J. Bisphenol analogues in surface water and sediment from the shallow Chinese freshwater lakes: Occurrence, distribution, source apportionment, and ecological and human health risk. *Chemosphere* **2017**, *184*, 318–328. [CrossRef] [PubMed]
6. Santhi, V.A.; Sakai, N.; Ahmad, E.D.; Mustafa, A.M. Occurrence of bisphenol A in surface water, drinking water and plasma from Malaysia with exposure assessment from consumption of drinking water. *Sci. Total Environ.* **2012**, *427–428*, 332–338. [CrossRef] [PubMed]
7. Xu, Y.; Hu, A.; Li, Y.; He, Y.; Xu, J.; Lu, Z. Determination and occurrence of bisphenol A and thirteen structural analogs in soil. *Chemosphere* **2021**, *277*, 130232. [CrossRef]
8. Jin, H.; Zhu, L. Occurrence and partitioning of bisphenol analogues in water and sediment from Liaohe River Basin and Taihu Lake, China. *Water Res.* **2016**, *103*, 343–351. [CrossRef]
9. Zhang, Y.; Dong, T.; Hu, W.; Wang, X.; Xu, B.; Lin, Z.; Hofer, T.; Stefanoff, P.; Chen, Y.; Wang, X.; et al. Association between exposure to a mixture of phenols, pesticides, and phthalates and obesity: Comparison of three statistical models. *Environ. Int.* **2019**, *123*, 325–336. [CrossRef]

10. Moon, S.; Yu, S.H.; Lee, C.B.; Park, Y.J.; Yoo, H.J.; Kim, D.S. Effects of bisphenol A on cardiovascular disease: An epidemiological study using National Health and Nutrition Examination Survey 2003–2016 and meta-analysis. *Sci. Total Environ.* **2021**, *763*, 142941. [CrossRef]
11. Kahn, L.G.; Philippat, C.; Nakayama, S.F.; Slama, R.; Trasande, L. Endocrine-disrupting chemicals: Implications for human health. The lancet. *Diabetes Endocrinol.* **2020**, *8*, 703–718. [CrossRef] [PubMed]
12. Sowlat, M.H.; Lotfi, S.; Yunesian, M.; Ahmadkhaniha, R.; Rastkari, N. The association between bisphenol A exposure and type-2 diabetes: A world systematic review. *Environ. Sci. Pollut. Res. Int.* **2016**, *23*, 21125–21140. [CrossRef] [PubMed]
13. Murata, M.; Kang, J.H. Bisphenol A (BPA) and cell signaling pathways. *Biotechnol. Adv.* **2018**, *36*, 311–327. [CrossRef] [PubMed]
14. Chen, D.; Kannan, K.; Tan, H.; Zheng, Z.; Feng, Y.L.; Wu, Y.; Widelka, M. Bisphenol Analogues Other Than BPA: Environmental Occurrence, Human Exposure, and Toxicity-A Review. *Environ. Sci. Technol.* **2016**, *50*, 5438–5453. [CrossRef]
15. Moreman, J.; Lee, O.; Trznadel, M.; David, A.; Kudoh, T.; Tyler, C.R. Acute Toxicity, Teratogenic, and Estrogenic Effects of Bisphenol A and Its Alternative Replacements Bisphenol S, Bisphenol F, and Bisphenol AF in Zebrafish Embryo-Larvae. *Environ. Sci. Technol.* **2017**, *51*, 12796–12805. [CrossRef] [PubMed]
16. Castro, B.; Sánchez, P.; Torres, J.M.; Ortega, E. Bisphenol A, bisphenol F and bisphenol S affect differently 5α-reductase expression and dopamine-serotonin systems in the prefrontal cortex of juvenile female rats. *Environ. Res.* **2015**, *142*, 281–287. [CrossRef]
17. Ullah, A.; Pirzada, M.; Jahan, S.; Ullah, H.; Turi, N.; Ullah, W.; Siddiqui, M.F.; Zakria, M.; Lodhi, K.Z.; Khan, M.M. Impact of low-dose chronic exposure to bisphenol A and its analogue bisphenol B, bisphenol F and bisphenol S on hypothalamo-pituitary-testicular activities in adult rats: A focus on the possible hormonal mode of action. *Food Chem. Toxicol.* **2018**, *121*, 24–36. [CrossRef]
18. Ohtani, N.; Iwano, H.; Suda, K.; Tsuji, E.; Tanemura, K.; Inoue, H.; Yokota, H. Adverse effects of maternal exposure to bisphenol F on the anxiety- and depression-like behavior of offspring. *J. Vet. Med. Sci.* **2017**, *79*, 432–439. [CrossRef]
19. Feng, Y.; Yin, J.; Jiao, Z.; Shi, J.; Li, M.; Shao, B. Bisphenol AF may cause testosterone reduction by directly affecting testis function in adult male rats. *Toxicol. Lett.* **2012**, *211*, 201–209. [CrossRef]
20. Wang, X.; Chen, Y.; Xiao, X.; Li, G. Recent advances in sample preparation technologies for analysis of harmful substances in aquatic products. *Se Pu Chin. J. Chromatogr.* **2021**, *39*, 34–45. [CrossRef]
21. Wang, H.; Mao, W.F.; Jiang, D.G.; Liu, S.J.; Zhang, L. Cumulative Risk Assessment of Exposure to Heavy Metals through Aquatic Products in China. *Biomed. Environ. Sci.* **2021**, *34*, 606–615.
22. Ngoubeyou, P.S.K.; Wolkersdorfer, C.; Ndibewu, P.P.; Augustyn, W. Toxicity of polychlorinated biphenyls in aquatic environments—A review. *Aquat. Toxicol.* **2022**, *251*, 106284. [CrossRef]
23. Liu, J.; Guo, J.; Cai, Y.; Ren, J.; Lu, G.; Li, Y.; Ji, Y. Multimedia distribution and ecological risk of bisphenol analogues in the urban rivers and their bioaccumulation in wild fish with different dietary habits. *Process Saf. Environ. Prot.* **2022**, *164*, 309–318. [CrossRef]
24. Wong, Y.M.; Li, R.; Lee, C.K.F.; Wan, H.T.; Wong, C.K.C. The measurement of bisphenol A and its analogues, perfluorinated compounds in twenty species of freshwater and marine fishes, a time-trend comparison and human health based assessment. *Mar. Pollut. Bull.* **2017**, *124*, 743–752. [CrossRef]
25. Liao, C.; Kannan, K. Species-specific accumulation and temporal trends of bisphenols and benzophenones in mollusks from the Chinese Bohai Sea during 2006–2015. *Sci. Total Environ.* **2019**, *653*, 168–175. [CrossRef]
26. Kundu, S.; Biswas, A.; Ray, A.; Roy, S.; Das Gupta, S.; Ramteke, M.H.; Kumar, V.; Das, B.K. Bisphenol A contamination in Hilsa shad and assessment of potential health hazard: A pioneering investigation in the national river Ganga, India. *J. Hazard. Mater.* **2024**, *461*, 132532. [CrossRef]
27. Hanioka, N.; Naito, T.; Narimatsu, S. Human UDP-glucuronosyltransferase isoforms involved in bisphenol A glucuronidation. *Chemosphere* **2008**, *74*, 33–36. [CrossRef] [PubMed]
28. Wang, Q.; Chen, M.; Qiang, L.; Wu, W.; Yang, J.; Zhu, L. Toxicokinetics and bioaccumulation characteristics of bisphenol analogues in common carp (*Cyprinus carpio*). *Ecotoxicol. Environ. Saf.* **2020**, *191*, 110183. [CrossRef] [PubMed]
29. Liu, J.; Martin, J.W. Prolonged Exposure to Bisphenol A from Single Dermal Contact Events. *Environ. Sci. Technol.* **2017**, *51*, 9940–9949. [CrossRef]
30. Zhao, N.; Hu, H.; Zhao, M.; Liu, W.; Jin, H. Occurrence of Free-Form and Conjugated Bisphenol Analogues in Marine Organisms. *Environ. Sci. Technol.* **2021**, *55*, 4914–4922. [CrossRef] [PubMed]
31. Liu, J.; Zhang, L.; Lu, G.; Jiang, R.; Yan, Z.; Li, Y. Occurrence, toxicity and ecological risk of Bisphenol A analogues in aquatic environment—A review. *Ecotoxicol. Environ. Saf.* **2021**, *208*, 111481. [CrossRef] [PubMed]
32. Wu, L.H.; Zhang, X.M.; Wang, F.; Gao, C.J.; Chen, D.; Palumbo, J.R.; Guo, Y.; Zeng, E.Y. Occurrence of bisphenol S in the environment and implications for human exposure: A short review. *Sci. Total Environ.* **2018**, *615*, 87–98. [CrossRef] [PubMed]
33. Tišler, T.; Krel, A.; Gerželj, U.; Erjavec, B.; Dolenc, M.S.; Pintar, A. Hazard identification and risk characterization of bisphenols A, F and AF to aquatic organisms. *Environ. Pollut.* **2016**, *212*, 472–479. [CrossRef] [PubMed]
34. Wang, H.; Tang, Z.; Liu, Z.H.; Zeng, F.; Zhang, J.; Dang, Z. Occurrence, spatial distribution, and main source identification of ten bisphenol analogues in the dry season of the Pearl River, South China. *Environ. Sci. Pollut. Res. Int.* **2022**, *29*, 27352–27365. [CrossRef] [PubMed]

35. Barboza, L.G.A.; Cunha, S.C.; Monteiro, C.; Fernandes, J.O.; Guilhermino, L. Bisphenol A and its analogs in muscle and liver of fish from the North East Atlantic Ocean in relation to microplastic contamination. Exposure and risk to human consumers. *J. Hazard. Mater.* **2020**, *393*, 122419. [CrossRef] [PubMed]
36. Rios-Fuster, B.; Alomar, C.; Paniagua González, G.; Garcinuño Martínez, R.M.; Soliz Rojas, D.L.; Fernández Hernando, P.; Deudero, S. Assessing microplastic ingestion and occurrence of bisphenols and phthalates in bivalves, fish and holothurians from a Mediterranean marine protected area. *Environ. Res.* **2022**, *214 Pt 3*, 114034. [CrossRef]
37. Diao, P.; Chen, Q.; Wang, R.; Sun, D.; Cai, Z.; Wu, H.; Duan, S. Phenolic endocrine-disrupting compounds in the Pearl River Estuary: Occurrence, bioaccumulation and risk assessment. *Sci. Total Environ.* **2017**, *584–585*, 1100–1107. [CrossRef]
38. Lestido-Cardama, A.; Petrarca, M.; Monteiro, C.; Ferreira, R.; Marmelo, I.; Maulvault, A.L.; Anacleto, P.; Marques, A.; Fernandes, J.O.; Cunha, S.C. Seasonal occurrence and risk assessment of endocrine-disrupting compounds in Tagus estuary biota (NE Atlantic Ocean coast). *J. Hazard. Mater.* **2023**, *444 Pt A*, 130387. [CrossRef]
39. Graca, B.; Rychter, A.; Staniszewska, M.; Smolarz, K.; Sokołowski, A.; Bodziach, K. Bioaccumulation of phenolic endocrine disruptors in the clam *Rangia cuneata*: Storage in shells and influence of size and sex. *Environ. Res.* **2021**, *197*, 111181. [CrossRef]

toxics

MDPI

*Article*

# Adsorption Behavior and Kinetics of 1,4-Dioxane by Carbon Aerogel

**Tianyu Lu [1,†], Huihui Huang [1,†], Guifen Lv [1,*], Fei Li [2,3], Ren-jie Song [1,*] and Yuting Cai [1]**

1 Key Laboratory of Jiangxi Province for Persistent Pollutants Control and Resources Recycle, Nanchang Hangkong University, Nanchang 330063, China; lty756425125@163.com (T.L.); 2102085700033@stu.nchu.edu.cn (H.H.); yutingcai@163.com (Y.C.)
2 Beijing Construction Engineering Group Environmental Remediation Co., Ltd., Beijing 100015, China; lifei@bceer.com
3 National Engineering Laboratory for Site Remediation Technologies, Beijing 100015, China
* Correspondence: summerlvguifen@163.com (G.L.); srj0731@hnu.edu.cn (R.-j.S.)
† These authors contributed equally to this work.

**Abstract:** 1,4-dioxane is a potential carcinogen in water and is difficult to deal with due to its robust cycloether bond and complete miscibility with water. To remove 1,4-dioxane in an economically viable and environmentally friendly way, a series of carbon aerogels were synthesized as adsorbents for 1,4-dioxane. The experiment results showed that adsorption performances were closely related to the preparation conditions of carbon aerogels, such as the molar ratio, heating rate, pyrolysis temperature and residence time, which were carefully controlled. Scanning electron microscope analysis revealed the presence of a three-dimensional porous network structure in carbon aerogels. Brunauer–Emmett–Teller analysis results demonstrated an increase in specific surface area (673.89 $m^2/g$) and total pore volume after carbonization, with an increase in mesoporous porosity and a decrease in microporosity. When considering each variable individually, the highest specific surface area of prepared carbon aerogels was achieved at a pyrolysis temperature of 800 °C, a holding time of 1 h, and a heating rate of 2 °C/min. Under optimal experimental conditions, the adsorption removal of 1,4-dioxane by carbon aerogels exceeded 95%, following quasi-second-order kinetics and Langmuir isothermal adsorption isotherms, indicating that monolayer adsorption on the surface of carbon aerogels occurred. The maximum adsorption capacity obtained was 67.28 mg/g at a temperature of 318 K, which was attributed to the presence of a large proportion of mesopores and abundant micropores simultaneously in carbon aerogels. Furthermore, with the interference of chlorinated solvents such as trichloroethylene (TCE), the removal efficiency of 1,4-dioxane had no obvious inhibition effect. Regeneration experiments showed that after five continuous cycles, the carbon aerogels still kept a comparable adsorption capacity, which illustrates its potential application in 1,4-dioxane-polluted water purification.

**Keywords:** carbon aerogels; sol–gel method; adsorption; 1,4-dioxane; trichloroethylene

## 1. Introduction

1,4-dioxane (1,4-D) has been identified as an impurity in numerous personal care products and cosmetics, as well as an undesired by-product in various industrial processes including antifreeze production, surfactant manufacturing, and paint formulation. The specific physical properties of 1,4-dioxane are shown in Table 1. The historical utilization, improper storage practices, and continuous discharge of 1,4-D have led to extensive contamination in drinking water sources, surface waters, and groundwater [1]. Once released into the environment, 1,4-D persists and exhibits high mobility due to its robust cyclic ether linkage and complete miscibility with water [2–4]. Toxicological studies have indicated that exposure to 1,4-D can cause health harms such as nasal cavity carcinomas and liver and gall bladder carcinomas [5]. Therefore, it has been classified as a Group 2B probable human carcinogen by the International Agency for Research on Cancer (IARC) [6]. Data show that approximately 21% of the drinking water supplies in the United States are contaminated with 1,4-D [7]. Counties including Japan, South Korea, and Canada have proposed a guideline for the concentration of 1,4-D in drinking water of 50 μg $L^{-1}$. Though 1,4-D is not federally regulated in the US, several states have established notification levels and guidelines. The acceptable concentrations of 1,4-D are low, typically in the range of (sub-) parts per billion (ppb), e.g., New York State set up a level of 1 μg $L^{-1}$, and several other states have established water criteria with the concentration ranging from 0.3 to 7.2 μg $L^{-1}$ [6]. The prevalence, persistency, and toxicity of this emerging pollutant have prompted increasing attention, highlighting the necessity for cost-effective treatments to mitigate its exposure risk [8].

**Table 1.** Physical properties of 1,4-D.

| Property | Value | Structure |
| --- | --- | --- |
| Molecular weight | 88.11 g/mol | 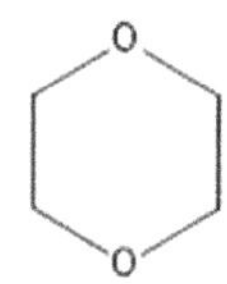 |
| Water solubility | Miscible | |
| Density (25 °C) | 1.0329 g/mL | |
| Vapor pressure | 4 kPa at 20 °C | |
| Octanol–water partition coefficient ($LogK_{ow}$) | −0.27 | |
| Organic carbon partition coefficient ($Log\ K_{oc}$) | 1.23 | |
| Henry's law constant at 25 °C | $4.8 \times 10^{-6}$ atm $m^3$ $mol^{-1}$ | |

Furthermore, its low Henry's law constant, low Kow and low Koc limit the effective remediation of 1,4-D via conventional technologies. For instance, air stripping is not efficient in treating 1,4-D due to its low volatility and high hydrophilicity [9]. Advanced oxidation processes (AOPs) [10] are effective alternatives to treat 1,4-D, including electrochemical methods, photo-catalysis [11], Fenton reaction methods [12], etc. The AOPs exhibit high removal efficiency and a short reaction time. However, these methods necessitate substantial energy consumption and incur significant costs, often resulting in secondary environmental pollution. Additionally, the degradation products generated by AOPs typically consist of low-molecular-weight organic compounds that require further treatment. In summary, there is a demand for economically viable and environmentally friendly approaches to 1,4-D treatment.

Adsorption has been a conventional and successful method for the removal of contaminants due to its simplicity, cost-effectiveness, and ease of operation. As shown in Table 2, some porous materials have been studied for the adsorption of 1,4-D. Among these adsorbents, some showed low adsorption capacity towards 1,4-D or entail a complex preparation process and have high costs. Pollutants such as persistent organics [13,14], dyes [15] and inorganic metals [16,17] can be adsorbed efficiently via granular activated carbons (GAC). Unlike other contaminants, 1,4-D cannot be removed efficiently by GAC because of its special characteristics including low solubility, high water solubility and other characteristics [18,19]. Nevertheless, studies have demonstrated the effective removal of 1,4-D by Ambersorb 560 (A560) [20], a carbonaceous adsorbent synthesized by the Dow Chemical Company, Midland, USA. A560 possesses a high surface area and porosity that facilitate specific adsorption of 1,4-D. Additionally, it can be regenerated using low-pressure steam. Nonetheless, A560 is relatively expensive due to its complex manufacturing process. Therefore, the development of alternative adsorbents is desirable to meet the increasing demand for the sustainable and economical management of dioxane-impacted water worldwide.

Aerogels are porous materials with many unique characteristics such as low bulk density, large surface area, low thermal conductivity, and porosity > 95%. Aerogels can be made of $SiO_2$, graphene, $MoS_2$ and so on. Pekala et al. [21] first reported the synthesis of porous resorcinol (R) and formaldehyde (F) aerogels, and after undergoing carbonization in an inert atmosphere, the RF aerogel underwent a transformation into a carbon aerogel (CA) [22]. CAs exhibit the characteristics of being porous, lightweight, and possessing a large specific surface area [23]. They have the ability to adsorb various pollutants, including inorganic metal ions [24–26] and organic pollutants [27]. For instance, CAs have shown excellent adsorption capacity on pesticides [28], 4-nitrophenol [29] and dyes [30]. In particular, CAs have shown a promising adsorption performance on some pollutants that were difficult to be adsorbed on GAC, especially for cyclic substances such as benzene rings or antibiotics with large molecular weights [31].

**Table 2.** Comparison of adsorption properties of other materials for 1,4-dioxane.

| Adsorbent | Initial Solute Concentration | Dosage | $Q_{max}$ (mg-1,4-D g-Adsorbent$^{-1}$) | Cycles (Recovery Rate) | Equilibrium Time | Reference |
|---|---|---|---|---|---|---|
| Norit 1240 GAC | 100 mg/L | 1 g | ~38 | _ | 72 h | [19] |
| Ambersorb 560 | 80 mg/L | _ | ~40 | _ | _ | [32] |
| Titanium Silicate-1 | 298 K, 500 mg/L | 100 mg | 85.17 | 3 (~100%) | 2 min | [33] |
| Sawdust GAC | 296 K | _ | 0.41 | _ | 16–24 h | [18] |
| ZSM-5 zeolite | 303 K, 100 mg/L | _ | 22.44 to 107.36 | _ | 1–8 h | [34] |
| CA125-800-1-2° | 308 K, 20 mg/L | 20 mg | 67.28 | 5 (~100%) | 1 h | [This work] |

However, little is known regarding the effectiveness of CAs for 1,4-D adsorption and the associated underlying mechanisms. In this study, through the sol–gel method, resorcinol and formaldehyde are utilized as precursors, while cetyltrimethylammonium bromide (CTAB) is introduced as a surfactant and catalyst. The continuous cross-linking and room temperature drying of resorcinol and formaldehyde result in the formation of a phenolic resin-based aerogel during the reaction process. Carbonization at high temperatures under an inert atmosphere is employed to prepare CAs. This study not only investigates the impact of different carbonization procedures on the adsorption performance and structural characteristics of CAs, comparing their adsorption rates with A560 and activated carbon for 1,4-D removal, but also analyzes the kinetic and thermodynamic models governing CAs' adsorption process. Additionally, this research explores the underlying mechanism behind 1,4-D removal by CAs through comprehensive surface characterization. In addition, as stable and renewable adsorbents, CAs elicit potential values and unparalleled interest in the fields of water treatment and remediation due to their efficient and economic benefits.

## 2. Materials and Methods

### *2.1. Chemicals and Materials*

Resorcinol ($C_6H_6O_2$, 99%) and formaldehyde ($CH_2O$, 36~38%) were provided by TCL Chemical Products Co., Ltd. (Guangzhou, China), and cetyl trimethyl ammonium bromide (CTAB) was purchased from Adamax Reagent Co., Ltd. (Shanghai, China). 1,4-dioxane ($C_4H_8O_2$,99%) was provided by Anneji Chemical Technology Co., Ltd. (Shanghai China), and deionized water (resistivity < 18.2 MΩ, Millipore system) was used. Ambersorb 560 (A560) was purchased from the Dow Chemical Company, and GAC (200 mesh) was purchased from Thain Chemical Technology Co., Ltd. (Shanghai, China).

### *2.2. CAs Preparation*

The preparation of CAs was based on the sol-gel polymerization method [35]. Briefly, a predetermined amount of resorcinol (R), 37% formaldehyde solution, cetyltrimethylammonium bromide (CTAB) and deionized water were added into a glass beaker, while stirring well at room temperature for 30 min, and then transferred into a glass vial. The vial was sealed and placed into an oil bath pot at 85 °C for 5 days, and the obtained organic aerogel was cured at room temperature for 2 days, followed by 60 °C for 24 h and then 105 °C for 3 h in an oven. Finally, the dried organic aerogel was heated to the point of carbonization with a set heating rate and residence time in a tube furnace. The molar ratios of R/CTAB were from 50 to 500, the carbonization temperatures ranged between 700 °C and 800 °C, the heating rates were 2 °C/min and 5 °C/min, respectively, and the residence time was in the range of 1 h to 2 h, respectively. They were denoted as CAa-b-c-d, where a represented the molar ratio R/CTAB, b represented the carbonization temperature, c represented the residence time, and d represented the heating rates. The specific preparation process is shown in Figure 1.

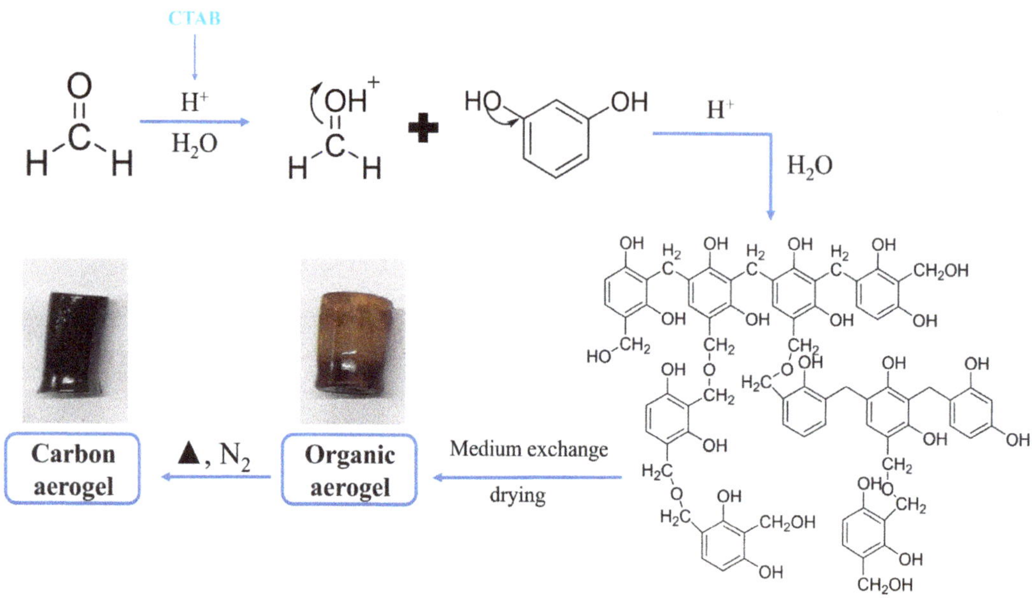


**Figure 1.** Preparation process of CAs.

### 2.3. Characterization

The specific surface area and pore size of the CAs were measured using a specific surface area analyzer (BELSORP-max, MicrotracBEL Japan, Inc., Osaka, Japan). The surface morphology was investigated via scanning electron microscopy (SEM, S-3400N, Hitachi, Japan). Powder X-ray diffraction (XRD) pattern was performed on a D8-Advance-A25 diffractometer with Cu K$\alpha$ radiation ($\lambda$ = 1.5418 Å) at 40 kV and 200 mA and a step size of 0.02° in the 5°–90° 2$\theta$ range at room temperature (25 °C). The pH of the point zero charge was measured via a potentiometric titrimeter (ZDJ-4A, Leici, Shanghai Instrument Electrical Scientific Instrument Co., Ltd., Shanghai, China).

### 2.4. Adsorption Experiments

Adsorbents screening experiments were evaluated in batch tests prepared in 25 mL glass flasks containing 60 mg of adsorbents (CAs, A560, GAC) and 10 mL 1,4-D with initial 20 mg/L concentrations, and the temperature was measured at 298 K. The mixture was shaken in a thermostatic shaker at 25 °C and 150 rpm for 12 h at room temperature to achieve equilibrium. The supernatant was filtered and analyzed for the concentrations of 1,4-D determined via gas chromatography with flame ionization detection (GC-FID). All experiments were conducted in 3 replicates.

For kinetics experiments, the dosage of CAs was administered at 20 mg and with varying initial 1,4-D concentrations ranging from 20 to 160 mg/L. Moreover, 0.5 mL of supernatant was removed using a syringe and filtered before storage for analysis. Two kinetic models, pseudo-first-order and pseudo-second-order equations which included all steps of the adsorption such as external film diffusion, adsorption, and internal particle diffusion, were used to analyze the adsorption processes. The formula is shown in Equations (1) and (2).

$$ln(Q_e - Q_t) = lnQ_e - k_1 t \quad (1)$$

$$\frac{t}{Q_t} = \frac{1}{k_2 {Q_e}^2} + \frac{1}{Q_e} t \quad (2)$$

For adsorption isotherms, batch experiments were prepared similarly as described above, but with varying initial 1,4-D concentrations ranging from 20 to 480 mg/L, while the experimental temperature varied between 298 K and 308 K. The maximum sorption capability ($Q_m$) and other parameters were computed by fitting with classic isotherm models (e.g., Langmuir and Freundlich) [36].

The Langmuir model is shown in Equation (3):

$$\frac{C_e}{Q_e} = \frac{1}{Q_m K_l} + \frac{C_e}{Q_m} \quad (3)$$

The Freundlich model is shown in Equation (4):

$$logQ_e = logK_f + \frac{1}{n} logC_e \quad (4)$$

where $Q_e$ (mg/g) and $C_e$ (mg/L) were the amounts of adsorbed 1,4-D per unit mass of adsorbent at equilibrium. $Q_m$ (mg/g) was the maximum amount of 1,4-D per unit mass of the adsorbent to form a complete monolayer on the surface. $K_l$ (L/mg) was a constant related to the affinity of the binding sites on the adsorbent. $K_f$ (L/mg) and 1/n were the Freundlich model constants, indicating the capacity and intensity of adsorption, respectively.

Adsorption thermodynamics enables the investigation of various factors, including temperature, on the adsorption process. In terms of temperature's impact, a set of thermodynamic parameters can be derived to substantiate its influence on 1,4-D adsorption via CAs. These parameters primarily encompass Gibbs free energy ($\Delta G$), adsorption enthalpy change ($\Delta H$), and adsorption entropy change ($\Delta S$), which can be computed using Equations (5) and (6):

$$\Delta G = -RTlnK \quad (5)$$

$$lnK = \frac{\Delta S}{R} - \frac{\Delta H}{RT} \quad (6)$$

In the given formula, R represents the universal gas constant with a value of 8.314 (J/mol/K). T denotes the absolute Kelvin temperature (K) of the adsorption solution, while K signifies the dimensionless thermodynamic equilibrium constant. The $K_L$ (L/mg) value from Langmuir's thermodynamic curve is multiplied by $10^6$ to obtain its corresponding numerical value. Herein, consider 1/T the *x*-axis and lnK the *y*-axis. Furthermore, $\Delta H$ and $\Delta S$ can be determined by calculating the slope and intercept of an equation that relates lnK to 1/T.

### 2.5. Analytical Methods

1,4-D concentration was detected via GC-FID (Agilent 7890A, Thermo, Waltham, MA, USA) coupled with a Supelco SLB™-5 ms fused silica capillary column (30 m length × 0.2 mm ID × 0.25 µm film). The direct injection volume of the filtered aqueous sample was 1 µL. Nitrogen was used as the carrier gas with a constant flow rate of 6.0 mL/min. The inlet temperature was set at 250 °C, and the samples were split at the ratio of 2:1 by the split flow of 12 mL/min. The oven temperature started from 110 °C for 1 min, then ramped up to 180 °C at the rate of 15 °C/min, and was held for 6 min. The detector temperature was maintained at 250 °C.

### 2.6. Regeneration Experiment

1,4-D-laden CA was regenerated via extraction with ethanol and then by heating the used CAs in an oven at 110 °C for 12 h. Adsorption experiments were repeated by using the same CA in consecutive cycles. After each desorption cycle, the CA was reused in a new adsorption process. The adsorption–desorption test of the adsorbent was repeated for 5 consecutive cycles. Data for the adsorption and regeneration experiments were calculated as the average value of the three replicates.

## 3. Results

### 3.1. The Investigation of Adsorption Properties

A series of experiments for CA preparation were performed to optimize the 1,4-D removals using CAs. The CAs' synthetic conditions included the R/CTAB ratio, pyrolysis temperature, residence time and heating rate. As Figure 2 showed, the concentration of 1,4-D is 20 mg/L with an addition amount of CAs at 60 mg and an adsorption volume of 10 mL at room temperature (298 K). The preparation conditions affected the 1,4-D removal greatly. A560 and GAC were used as controls, and the removal amounts were 87.89% and 33.47%, respectively. All CAs showed much higher removal than GAC. The R/CTAB ratio varied from 50 to 200, and the highest 1,4-D removal was obtained at R/CTAB = 125, which was 95.08%. When the pyrolysis temperature increased from 700 °C to 800 °C, the 1,4-D removal increased first and then decreased. The optimum temperature was 800 °C. Similarly, there was some variation from 1 h to 2 h. The optimum residence time and heating rate were 1 h and 2 °C/min, respectively. Therefore, the optimum CA candidate was CA125-800-1-2°. The removal of 1,4-D reached a high percentage of 95.08%, which was 1.1 times higher than that of A560 and 2.8 times higher than that of GAC, respectively. For the sake of convenience for later descriptions, the following discussion uses CA125 referring to CA125-800-1-2°.

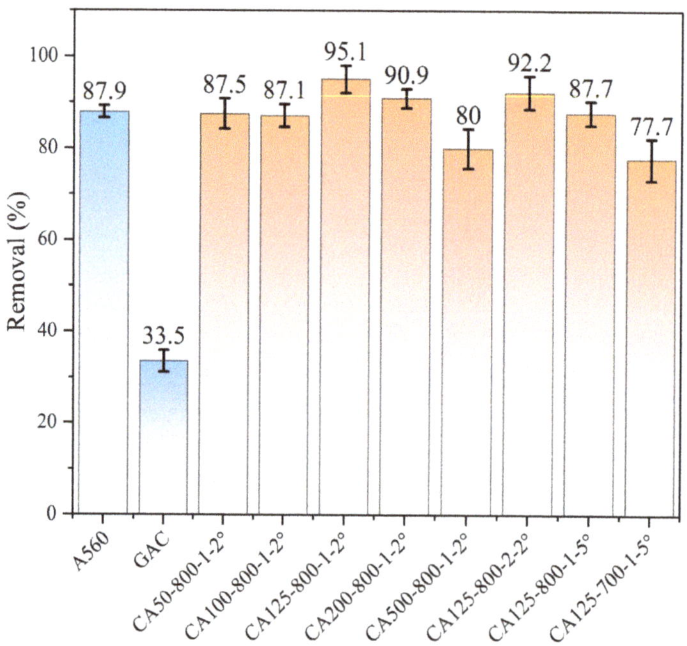


**Figure 2.** Removal effects of CAs, A560, and GAC on 1,4-D.

### 3.2. The Effect of CAs' Porous Structure

The adsorption performance is strongly related to the micro-structural natures of the adsorbent. The crystal structure of CA125 was analyzed via XRD as shown in Figure 3. The diffraction peaks at 23° and 43° correspond to the layered ordered stacked (002) surfaces and the ordered hexagonal carbon structure (100) surfaces. It was shown that CAs are graphite-like microcrystalline carbon materials. The structures resembling graphite exhibit an extended, multi-layered arrangement of unidirectional structural units, and the adsorption process of CAs is enhanced under these conditions [37].

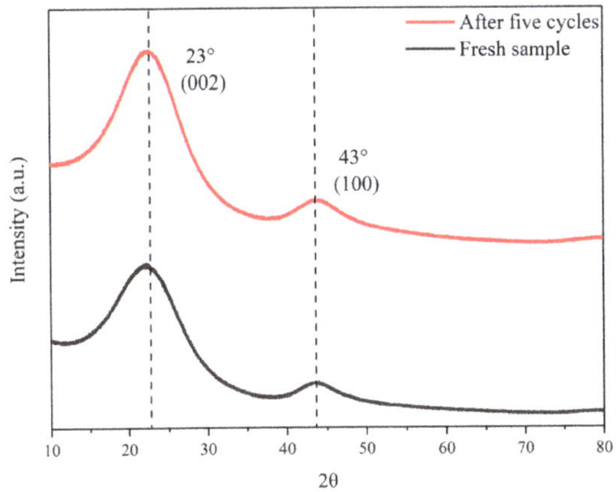


**Figure 3.** XRD spectra of CA125-800-1-2° before and after the reuse experiments.

The specific area (SA) and pore size distribution of the adsorbents showed significance for 1,4-D adsorption. It is believed that not only the surface area but also the pore size distribution determines 1,4-D adsorption. Figure 4a shows a BET surface area isotherm, and the pore structure parameters are listed in detail in Table 3. According to IUPAC, the $N_2$ adsorption–desorption isotherm was the type IV isotherm with the type H3 loop, indicating the formation of developed mesoporous and microporous structures. Among the adsorbents, GAC exhibited the lowest adsorption capacity despite having the largest SA, demonstrating that the surface area is not the key factor for determining the adsorption performance.

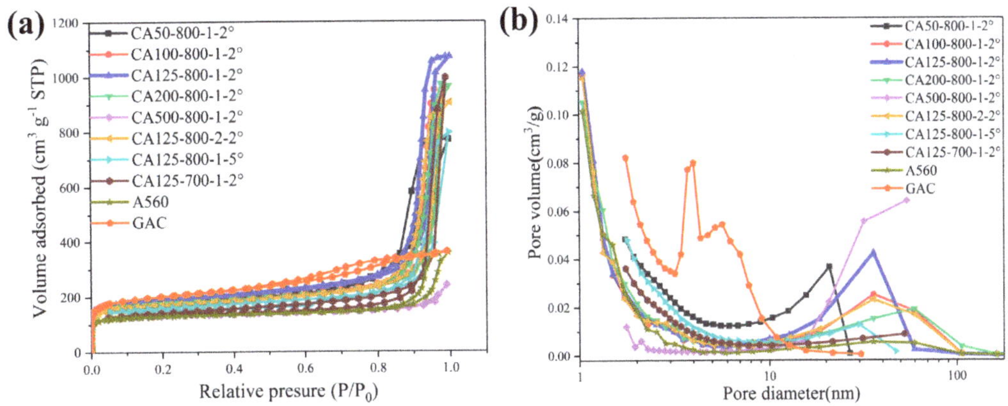


**Figure 4.** Characterizations of CAs, A560 and GAC: nitrogen adsorption–desorption isotherms (**a**) and pore size distribution (**b**).

**Table 3.** Pore structure parameters of absorbents.

| Sample | $S_{BET}$ ($m^2$/g) | $S_{micro}$ ($m^2$/g) | $S_{meso}$ ($m^2$/g) | $V_t$ ($cm^3$/g) | $V_{micro}$ ($cm^3$/g) | $V_{meso}$ ($cm^3$/g) | D (nm) |
|---|---|---|---|---|---|---|---|
| **CA50-800-1-2°** | 572 | 239 | 333 | 1.37 | 0.29 | 1.08 | 12.92 |
| **CA100-800-1-2°** | 638 | 287 | 351 | 1.62 | 0.29 | 1.33 | 9.47 |

**Table 3.** *Cont.*

| Sample | $S_{BET}$ ($m^2/g$) | $S_{micro}$ ($m^2/g$) | $S_{meso}$ ($m^2/g$) | $V_t$ ($cm^3/g$) | $V_{micro}$ ($cm^3/g$) | $V_{meso}$ ($cm^3/g$) | D (nm) |
|---|---|---|---|---|---|---|---|
| **CA125-800-1-2°** | 674 | 275 | 399 | 1.81 | 0.32 | 1.49 | 9.70 |
| **CA200-800-1-2°** | 629 | 288 | 341 | 1.63 | 0.29 | 1.34 | 9.23 |
| **CA500-800-1-2°** | 406 | 311 | 95 | 0.37 | 0.16 | 0.21 | 8.98 |
| **CA125-800-2-2°** | 605 | 258 | 347 | 1.54 | 0.29 | 1.25 | 9.11 |
| **CA125-800-1-5°** | 503 | 304 | 199 | 1.23 | 0.14 | 1.08 | 9.79 |
| **CA125-700-1-2°** | 452 | 236 | 216 | 1.54 | 0.12 | 1.42 | 13.62 |
| **A560** | 480 | 251 | 229 | 0.57 | 0.14 | 0.43 | 4.51 |
| **GAC** | 1086 | 1069 | 17 | 0.42 | 0.39 | 0.03 | 3.83 |

In order to further distinguish the pore size distribution of the carbons, their cumulative pore volume curves are analyzed (Figure 4b). CAs showed a wide pore size distribution which can be attributed to the interval gap between chains of interconnected particles. Furthermore, CA125-800-1-2° showed a sharp mesopore distribution centered at 30 nm.

As can be seen from Table 3, though the total SA of CA is not the largest, it showed the highest $S_{meso}$ and $V_{meso}$. $S_{meso}$ of A560 is also lower than CAs. The order is in accordance with the 1,4-D removal via adsorbents, suggesting the mesopores contribution to 1,4-D adsorption. The 1,4-D molecule possesses two O atoms. It is a polar nonionic compound, according to the literature reported [38], and hydrogen bonds can be formed with $H_2O$ molecules. The diameter of the 1,4-D molecule and $H_2O$ molecule are ca. 0.5 nm and 0.4 nm, respectively, and hydrogen bonds may be formed between few 1,4-D and $H_2O$ molecules. The size of the molecule group is a few nanometers long, so mesopores are fit for 1,4-D molecule transfer. The mesopores are 1,4-D storage and diffusion channels. The existence of the mesopore should be a crucial factor for 11,4-D adsorption.

The observed dependence of 1,4-D uptake on pore size is consistent with previous reports. Huang et al. [39] reported a polystyrene-based hierarchical porous carbon (PS-HPC) for supercapacitors. In their view, one of the important interactions for ion transport is the ion–wall pairs. It depends on the ratio of the pore diameter to the ion diameter. When the ratio is larger than 20, the ions transfer into the pores freely, avoiding collisions between the ion and wall. Therefore, with a larger mesopores proportion, CAs showed a higher 1,4-D uptake amount.

According to the SEM image, as depicted in Figure 5, the CA nanoparticles exhibit a size of approximately 30 nm and display an interconnected three-dimensional network structure. This architecture facilitates the transportation and adsorption of 1,4-D. Importantly, the surface of CAs is characterized by abundant pores and defect structures, leading to an increased number of adsorption sites and an enhanced pollutant adsorption capacity. In contrast, although A560 shares a similar structure with CA, it exhibits significantly fewer surface pores and defects. Therefore, based on the aforementioned characterization analysis, it can be concluded that a high mesoporous porosity and a rich pore structure exert a profound influence on the adsorption behavior towards 1,4-D.

The changes in surface functional groups before and after the carbonization of CAs were investigated using Fourier infrared spectroscopy. The infrared spectra of the aerogel (R/CTAB = 125) before and after carbonization, as shown in Figure 6, reveal a rich variety of surface functional groups in the pre-carbonized aerogel. For instance, the absorption peak at 3342 $cm^{-1}$ corresponds to the stretching vibration of –OH, which is broadened due to hydrogen bonding. The absorption at 2927 $cm^{-1}$ is attributed to the asymmetric stretching vibration of C–H. The peak at 1605 $cm^{-1}$ indicates the characteristic C=C stretching vibration in an asymmetric ring, while peaks at 1473 $cm^{-1}$ represent methylene bonds, and those at 1094 $cm^{-1}$ and 1237 $cm^{-1}$ correspond to methylene ether bonds. Additionally, peaks observed at 1358 $cm^{-1}$ and 1290 $cm^{-1}$ are associated with C–N stretching vibrations. A weak peak detected at 961 $cm^{-1}$ can be assigned to either C–N or C–O stretching vibrations. Furthermore, absorptions within the range of 900~650 $cm^{-1}$ indicate out-of-plane deformation vibrations for both C–H or N–H bonds. However, following the carbonization of the aerogel, most surface functional groups disappear, except for the presence of a tensile vibration peak corresponding to a C–H bond at around 3030 $cm^{-1}$ and a benzene unsaturated bond tensile vibration peak observed near 1559 $cm^{-1}$. These results demonstrate that carbonized CAs primarily consist of benzene rings, with other remaining functional groups being decomposed and removed during high temperature treatment.

**Figure 5.** SEM of CA125 (**a**) and A560 (**b**).

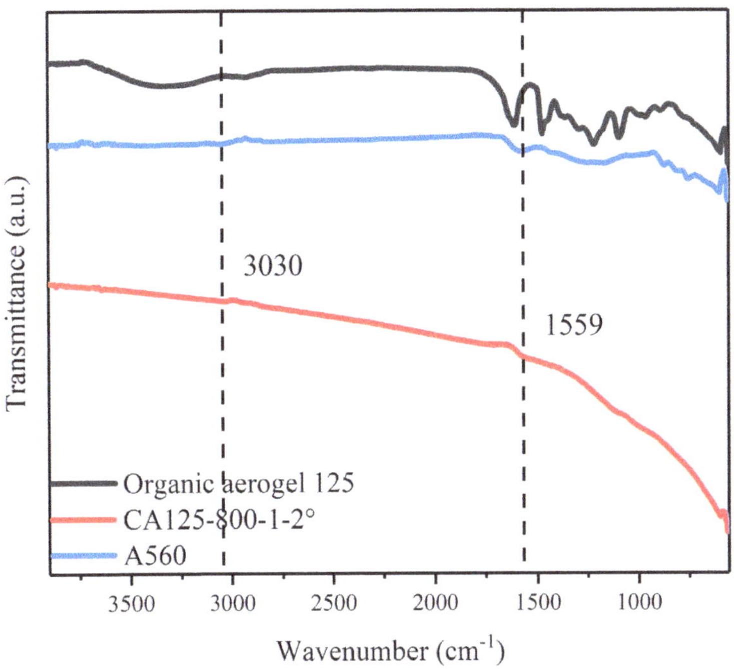


**Figure 6.** FTIR of organic aerogel 125, CA125-800-1-2° and A560.

In summary, CAs belong to a graphite-like amorphous carbon structure, exhibiting a hexagonal carbon framework composed of phenyl rings. They possess a large specific surface area and abundant micro-mesoporous characteristics, rendering them highly effective adsorbents.

### 3.3. 1,4-D Adsorption at Different pH

The effect of the initial solution pH on 1,4-D removal was investigated by varying the solution pH from 2 to 11, and the results are presented in Figure 7. As the pH value increased, the removal did not show obvious differences. The reason is 1,4-D is in the ion form at the entire pH range since the pka of 1,4-D is 2.1. Further, the zero charge point of CAs is 7.1 [40]. Therefore, a pH without adjustment (pH = 6.6) is recommended due to the negligible effects of electrostatic attraction to 1,4-D adsorption by CAs.

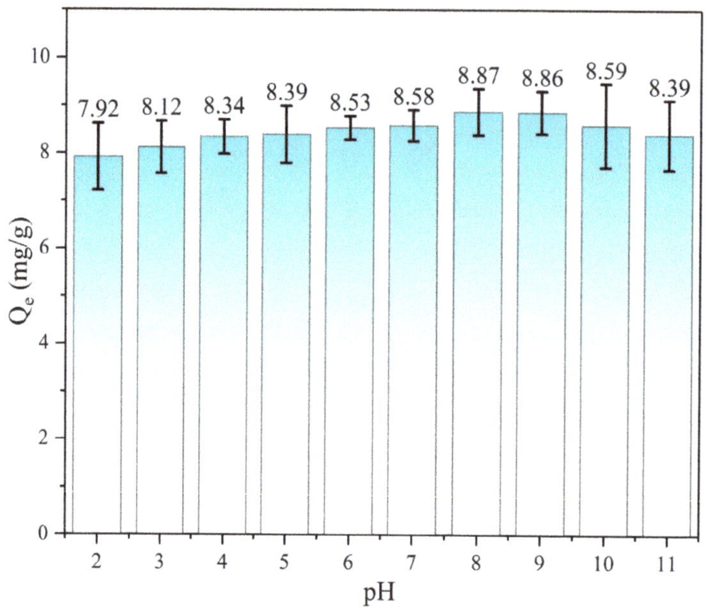


**Figure 7.** Influence of different pH on adsorption capacity.

### 3.4. Adsorption Kinetics and Isotherm

The adsorption of CA125 toward 1,4-D with time is showed in Figure 8. The adsorption performance followed the order of GAC < A560 < CA125. In comparison to A560 and GAC, CA125 exhibited not only a significant increase in adsorption capacity but also a considerably shorter equilibration time. Specifically, the adsorption capacity of CA125 was nearly three times that of GAC, while its equilibration time was only one-sixth of that of A560. Within 30 min, the removal efficiency for 1,4-D reached 66.82%, and complete adsorption equilibrium was achieved within 60 min. Conversely, although A560 demonstrated a higher removal capacity than GAC, it required an extended equilibration time of up to 12 h; on the other hand, GAC showed a shorter equilibration time (60 min) but a relatively lower adsorption capacity.

Figure 9a,b shows the pseudo-first-order and pseudo-second-order kinetic models of adsorption tests on CA125 based on different concentrations of 1,4-D. The related parameters are shown in Table 4. As the results show, the $R^2$ values of pseudo-second-order models were higher than those of the pseudo-first-order model, suggesting that the pseudo-second-order model can preferably describe the adsorption process. In the pseudo second order model, the rate-limiting steps in the surface adsorption involves chemisorption, which means the adsorption process is a physicochemical interaction between 1,4-D and the CA125 surface, where 1,4-D makes contacts with the surface of the CA and binds strongly to the surface.

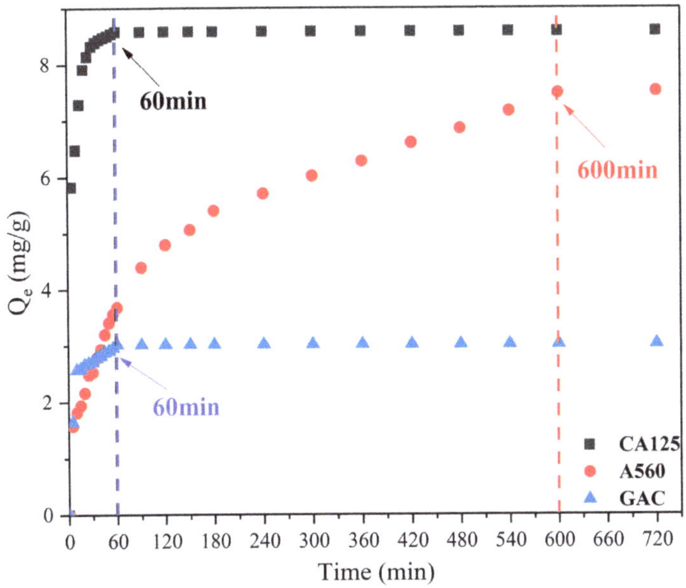


**Figure 8.** The effect of contact time on the adsorption capacity of 1,4-D over CA125-800-1-2°, A560 and GAC.

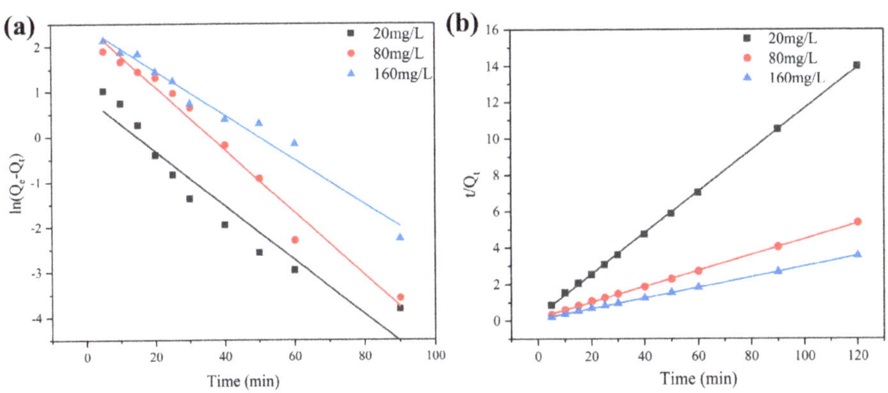


**Figure 9.** Pseudo-first-order (**a**) and pseudo-second-order (**b**) kinetic models for adsorption of 1,4-D on CA125.

**Table 4.** Pseudo-first- and pseudo-second-order kinetic parameters for 1,4-D adsorption on CA.

| $C_0$ (mg/L) | $Q_e$ (exp) (mg/g) | Pseudo-First-Order Models | | | Pseudo-Second-Order Models | | |
|---|---|---|---|---|---|---|---|
| | | $Q_e$ (cal) (mg/g) | $k_1$ ($min^{-1}$) | $R^2$ | $Q_e$ (cal) (mg/g) | $k_2$ (mg/min) | $R^2$ |
| **20** | 8.58 | 5.10 | 0.0987 | 0.9904 | 8.80 | 0.0478 | 0.9997 |
| **80** | 22.40 | 10.93 | 0.0632 | 0.9778 | 23.20 | 0.0122 | 0.9995 |
| **160** | 33.82 | 11.49 | 0.0492 | 0.975 | 34.72 | 0.0096 | 0.9998 |

Langmuir and Freundlich adsorption isotherms (Figure 10) were obtained at different temperatures via batch experiments. As the temperature increased, the adsorption capacity increased, indicating that the adsorption of 1,4-D on CA125 is an endothermic process. Furthermore, as the concentration of the initial aqueous solution of 1,4-Dincreased, the adsorption capacity of CA125 increased. The parameters shown in Table 5. The coefficients ($R^2$) value of the Langmuir model is higher than that of the Freundlich model, indicating that the Langmuir model is more consistent with the experimental data, which showed that the adsorption 1,4-D on CA125 is a monolayer adsorption process, and the maximum adsorption capacity of 1,4-D was 67.28 mg/g under 318 K, which was very close to the experimental value.

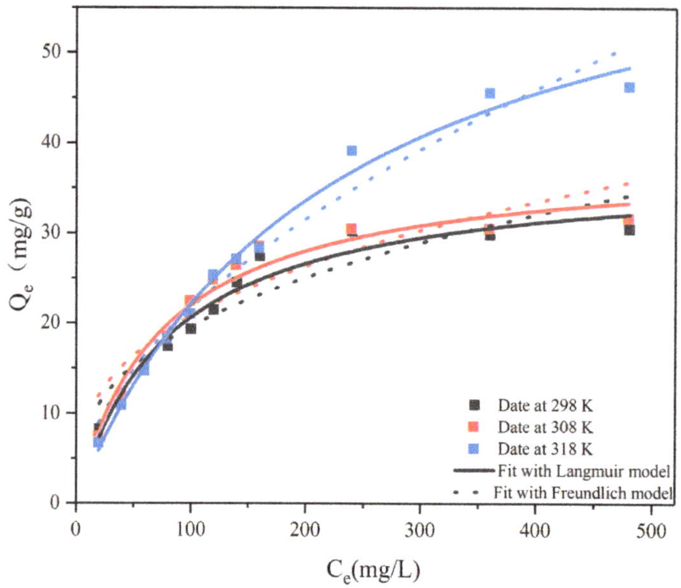


**Figure 10.** The adsorption isotherm model for 1,4-D on CA125-800-1-2°.

**Table 5.** Parameters of Langmuir and Frendlich models for 1,4-D.

| T(K) | Langmuir | | | Freundlich | | |
|---|---|---|---|---|---|---|
| | $R^2$ | $Q_m$ (mg/g) | $K_L$ (L/mg) | $R^2$ | 1/n | $K_F$ |
| **298** | 0.97 | 37.55 | 0.0138 | 0.88 | 0.355 | 3.818 |
| **308** | 0.97 | 38.69 | 0.0113 | 0.86 | 0.410 | 2.360 |
| **318** | 0.99 | 67.28 | 0.0045 | 0.97 | 0.546 | 1.748 |

The fitting equation for LnK, obtained through thermodynamic curve fitting can be expressed as LnK = 6339.6 (1/T) − 10.343. Furthermore, the correlation coefficient is $R^2 = 0.93$. The calculated results of the relevant parameters for thermodynamic analysis are presented in Table 6. All values of Gibbs free energy ΔG are negative, indicating the spontaneous adsorption of 1,4-D by CAs. Furthermore, the magnitude of Gibbs free energy increases with rising temperatures in the adsorption system solution, suggesting a deteriorating adsorption performance of CAs towards 1,4-D [41]. The negative value of ΔS implies that the adsorption process exhibits a decrease in entropy when CAs are used as adsorbents for 1,4-D. This suggests that an enthalpy change drives the adsorption process [42]. In addition, ΔH is negative within the studied adsorption system, signifying an exothermic nature during the adsorption process of 1,4-D. Moreover, it can be concluded that increasing the temperature inhibits this adsorptive behavior based on consistent findings from Gibbs free energy analysis.

**Table 6.** Thermodynamic parameters of adsorption of 1,4-D by CAs.

| T (K) | ΔG (kJ/mol) | ΔH (kJ/mol) | ΔS (J/mol/K) |
|---|---|---|---|
| 298 | −26.83 | −52.71 | −85.99 |
| 308 | −26.75 | | |
| 318 | −25.08 | | |

### *3.5. Competitive Adsorption of 1,4-D and Trichloroethylene*

1,4-D has been predominantly used as a stabilizer for chlorinated solvents such as trichloroethylene (TCE). Hence, the adsorption capacity of CA125 toward 1,4-D was explored with the presence of TCE co-existing in wastewater. As shown in Figure 11, the TCE concentrations were tested between 50 mg/L and 300 mg/L. Compared with the 1,4-D adsorption without TCE, the adsorption capacity was not affected by TCE. Since 1,4-D molecules are neutral amphiphilic, possessing both alkyl and oxygen groups, both hydrophobic and hydrophilic interactions are found to have an important contribution to its adsorption on various adsorbates. The total acidity for the samples of CAs is greater than the total basicity. This is due to the oxidation process undergone by the polymeric gel in the drying process, which induces the formation of more acidic oxygenated functional groups including the carboxylic, lactic, and phenolic groups [43]. The presence of these groups, although in a small proportion, induced the formation of delocalized π electrons and an H bond with 1,4-D molecules. However, TCE is hydrophobic and has a weak interaction with CA, so the adsorption capacity is quite low.

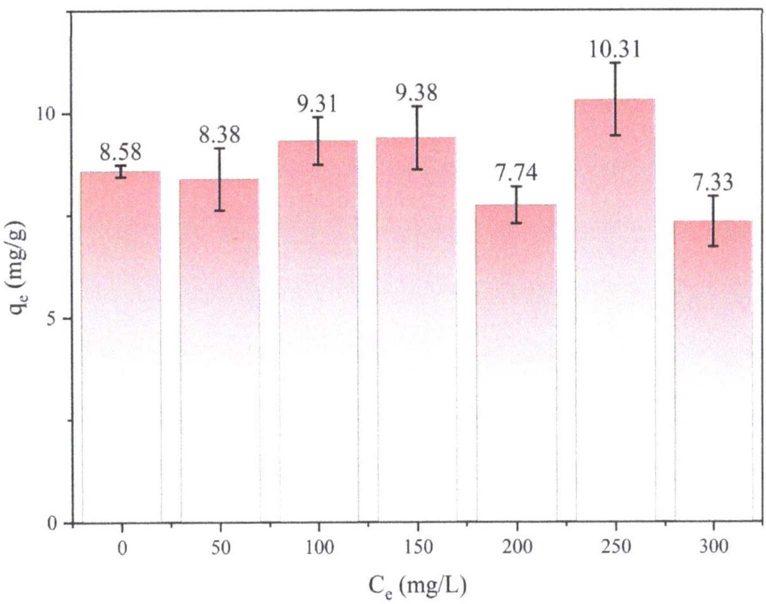


**Figure 11.** The effects of different concentrations of trichloroethylene on 1,4-D adsorption. $C_{1,4\text{-}D}$ = 20 mg/L, T = 25 °C, t = 2 h. Concentrations of TCE are 50, 100, 150, 200, 250, 300 mg/L, respectively.

### *3.6. Regeneration of CAs*

The regeneration performance of adsorbents is one of the important considerations for practical applications. Figure 12 shows the five consecutive adsorption/desorption cycles to remove 1,4-D. The adsorption removal amount was 8.58 mg/g in the first cycle. The adsorption efficacy of the material remains essentially unaltered upon multiple reuses in comparison to its initial adsorption rate. Moreover, when combined with XRD analysis, the configuration of CAs remained essentially unaltered even after repeated utilization. These results showed that the marked retention of the sorption capability is ascribed to the stable structure of CAs. When advantages feature in the CAs' adsorbent capacity in repeated cycles of 1,4-D, loading and elution are taken into consideration. These results illustrate the facile and high regeneration capability of CAs as a promising adsorbent for 1,4-D.

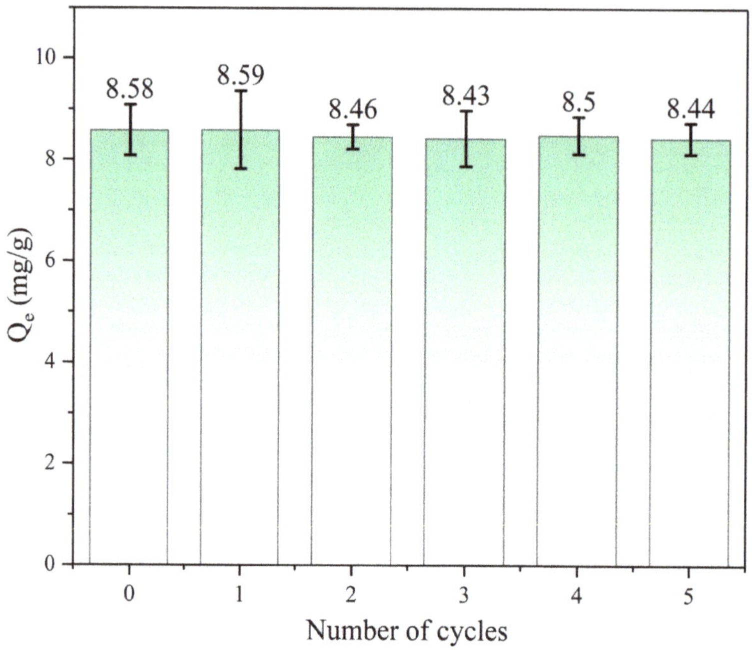


**Figure 12.** Removal quantity of CA125 after regeneration in five cycles. Cycle 1 represents the original CA125. ($C_0$ = 20 mg/L, T = 25 °C, t = 2 h and m = 20 mg).

### 3.7. The Impact of the Adsorbent Quantity

To compare the adsorption effects of CAs, A560 and GAC adsorbents with different weights on 1,4-D, adsorbents weighing 20 mg, 40 mg, and 60 mg were selected to adsorb a solution containing 10 mL of a 20 mg/L concentration of 1,4-D. The obtained results are presented in Table 7. According to the experimental findings, under ambient conditions, the adsorption efficiency of 0.06g CA125-800-1-2° for a 10 mL solution containing 20 mg/L of 1,4-D can attain an impressive rate of 95%.

**Table 7.** Comparison of adsorption rates of CA125-800-1-2°, A560 and GAC for the same concentration of 1,4-D at different mass.

| Mass (mg) | CA125-800-1-2° | A560 | GAC |
|---|---|---|---|
| 20 | 85.81% | 75.71% | 21.54% |
| 40 | 91.13% | 85.99% | 27.45% |
| 60 | 95.08% | 87.89% | 33.47% |

### 3.8. The Investigation of the Mechanism

The interactions between the adsorbent and adsorbate involve various forces, such as van der Waals forces, electrostatic attractions, pore filling, $\pi$-$\pi$ electron donor–acceptor interactions, hydrogen bonding, coordination adsorption, and hydrophobic interactions [44,45].

The adsorption mechanism of 1,4-D caused by CAs is depicted in Figure 13. A specific description for this process is as follows:

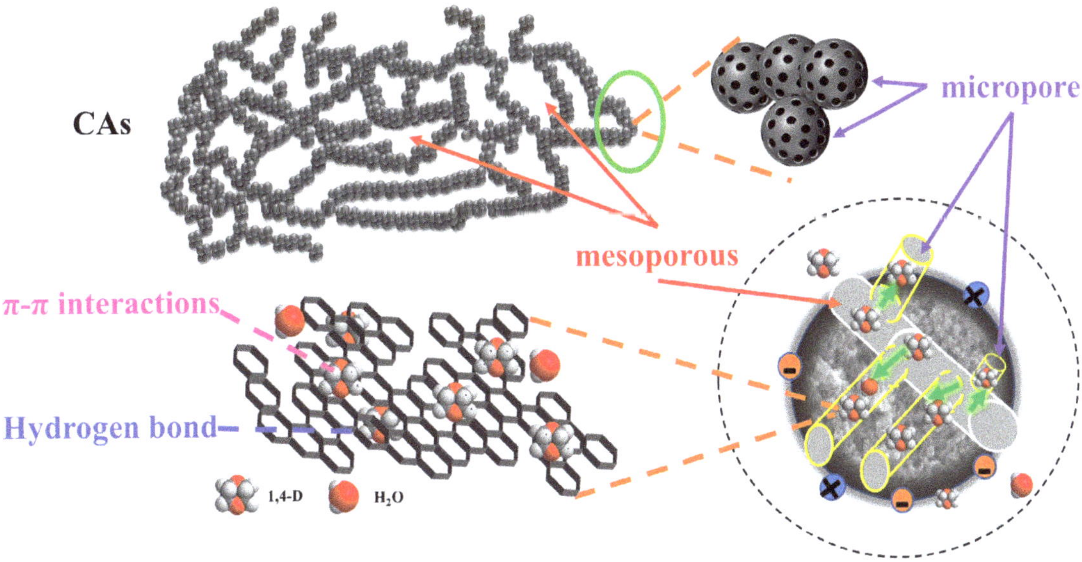


**Figure 13.** Adsorption mechanism diagram of CAs.

(1) Pore filling: according to the conclusions drawn from BET analysis (Table 3), the adsorption rates of 1,4-D by the three adsorbents (Figure 2) and the nitrogen absorption and desorption curve (Figure 4), it can be concluded that the adsorption form of CAs on the refractory organic pollutants is mainly physical adsorption, and the adsorption method is mainly pore filling, and the specific surface area and pore volume of carbon materials are larger. This enables a greater the adsorption capacity of 1,4-D.

(2) Van der Waals force action: according to the adsorption dynamic equilibrium curve of 1,4-D, it can be seen that the adsorption of 1,4-D by CAs is mainly physical adsorption. Since the main force in the physical adsorption process is the van der Waals force, it can be inferred that the adsorption of 1,4-D by CAs has van der Waals forces.

(3) Electrostatic attraction mechanism: the surface of CAs is usually negatively charged, which makes it easier for CAs to electrostatically adsorb positively charged organic compounds to form a stable adsorption interface. From a microscopic point of view, since 1,4-D is a polar solution, and polar molecules hydrogen and oxygen have different abilities to capture electrons, the shared electron pairs of the two will favor oxygen, resulting in a positive charge of oxygen, a negative charge of hydrogen, and a prominent oxygen atom. Therefore, 1,4-D will be absorbed to the surface of CAs due to the electrostatic attraction of the oxygen atom.

(4) Electron donor/acceptor effect: according to infrared spectrum analysis, CAs contain a benzene ring as an electron acceptor and the -O- in the ether bond of 1,4-D as an electron donor. The two can enhance the adsorption capacity between the adsorbent and adsorbent through the electron donor/acceptor complex effect.

## 4. Conclusions

The adsorption behavior of a series of CAs adsorbents and their interactions with 1,4-D were investigated. It was observed that CAs exhibited excellent adsorption capacity for 1,4-D in aqueous solutions. Notably, CA125-800-1-2° possessed a large specific surface area and abundant micro-mesoporous structure, resulting in a remarkable removal efficiency (~95%) for 1,4-D in water. A kinetic simulation revealed the maximum unit adsorption capacities at 298 K and 318 K to be 37.55 mg/g and 67.28 mg/g, respectively. Furthermore, the high adsorption rate coupled with the low desorption tendency indicated that the quasi-second-order kinetics model and the Langmuir isotherm model accurately described the adsorption process. After five adsorption–degradation cycles, CAs still exhibit a good regeneration ability and stability, which provides valuable insights into the potential application of CAs in efficiently removing, controlling, and remediating aqueous solutions contaminated with 1,4-D.

**Author Contributions:** T.L.: methodology, data curation and writing—original draft. H.H.: data curation and writing—original draft. G.L.: writing—review and editing. F.L.: writing—review. R.-j.S.: supervision. Y.C.: data curation. All authors have read and agreed to the published version of the manuscript.

**Funding:** This work was supported by Jiangxi Province Graduate Innovation Special Fund Project (provincial project, YC2021-S664 and YC2022-s716), Jiangxi Province Science and Technology Project (grant numbers 20212AEI91002, 20202ACB203002 and 20212ACB203007). National Natural Science Foundation of China (No. 52270039).

**Institutional Review Board Statement:** Not applicable.

**Informed Consent Statement:** Not applicable.

**Data Availability Statement:** Data are contained within the article.

**Conflicts of Interest:** The authors declare no conflicts of interest.

## References

1. Adamson, D.T.; de Blanc, P.C.; Farhat, S.K.; Newell, C.J. Implications of matrix diffusion on 1,4-dioxane persistence at contaminated groundwater sites. *Sci. Total Environ.* **2016**, *562*, 98–107. [CrossRef]
2. Khan, N.A.; Johnson, M.D.; Carroll, K.C. Spectroscopic methods for aqueous cyclodextrin inclusion complex binding measurement for 1,4-dioxane, chlorinated co-contaminants, and ozone. *J. Contam. Hydrol.* **2018**, *210*, 31–41. [CrossRef]
3. Adamson, D.T.; Pina, E.A.; Cartwright, A.E.; Rauch, S.R.; Hunter Anderson, R.; Mohr, T.; Connor, J.A. 1,4-Dioxane drinking water occurrence data from the third unregulated contaminant monitoring rule. *Sci. Total Environ.* **2017**, *596–597*, 236–245. [CrossRef]
4. Miao, Y.; Johnson, N.W.; Phan, T.; Heck, K.; Gedalanga, P.B.; Zheng, X.; Adamson, D.; Newell, C.; Wong, M.S.; Mahendra, S. Monitoring, assessment, and prediction of microbial shifts in coupled catalysis and biodegradation of 1,4-dioxane and co-contaminants. *Water Res.* **2020**, *173*, 115540. [CrossRef] [PubMed]
5. Xu, X.; Liu, S.; Smith, K.; Wang, Y.; Hu, H. Light-driven breakdown of 1,4-Dioxane for potable reuse: A review. *Chem. Eng. J.* **2019**, *373*, 508–518. [CrossRef]
6. Tang, Y.; Mao, X. Recent Advances in 1,4-Dioxane Removal Technologies for Water and Wastewater Treatment. *Water* **2023**, *15*, 1535. [CrossRef]
7. Bagheri, M.; Mohseni, M. Pilot-scale treatment of 1,4-dioxane contaminated waters using 185 nm radiation: Experimental and CFD modeling. *J. Water Process Eng.* **2017**, *19*, 185–192. [CrossRef]
8. Zhang, S.; Gedalanga, P.B.; Mahendra, S. Advances in bioremediation of 1,4-dioxane-contaminated waters. *J. Environ. Manag.* **2017**, *204 Pt 2*, 765–774. [CrossRef]
9. Yang, I.; Kwon, D.; Kim, M.-S.; Jung, J.C. A comparative study of activated carbon aerogel and commercial activated carbons as electrode materials for organic electric double-layer capacitors. *Carbon* **2018**, *132*, 503–511. [CrossRef]
10. Preethi; Shanmugavel, S.P.; Kumar, G.; Yogalakshmi, K.N.; Gunasekaran, M.; Rajesh, B.J. Recent progress in mineralization of emerging contaminants by advanced oxidation process: A review. *Environ. Pollut.* **2024**, *341*, 122842. [CrossRef] [PubMed]
11. Barndõk, H.; Blanco, L.; Hermosilla, D.; Blanco, Á. Heterogeneous photo-Fenton processes using zero valent iron microspheres for the treatment of wastewaters contaminated with 1,4-dioxane. *Chem. Eng. J.* **2016**, *284*, 112–121. [CrossRef]
12. Barndõk, H.; Merayo, N.; Blanco, L.; Hermosilla, D.; Blanco, Á. Application of on-line FTIR methodology to study the mechanisms of heterogeneous advanced oxidation processes. *Appl. Catal. B Environ.* **2016**, *185*, 344–352. [CrossRef]
13. Ersan, G.; Kaya, Y.; Apul, O.G.; Karanfil, T. Adsorption of organic contaminants by graphene nanosheets, carbon nanotubes and granular activated carbons under natural organic matter preloading conditions. *Sci. Total Environ.* **2016**, *565*, 811–817. [CrossRef]
14. Eder, S.; Müller, K.; Azzari, P.; Arcifa, A.; Peydayesh, M.; Nyström, L. Mass Transfer Mechanism and Equilibrium Modelling of Hydroxytyrosol Adsorption on Olive Pit–Derived Activated Carbon. *Chem. Eng. J.* **2021**, *404*, 126519. [CrossRef]
15. Świderska-Dąbrowska, R.; Piaskowski, K.; Zarzycki, P.K. Preliminary Studies of Synthetic Dye Adsorption on Iron Sludge and Activated Carbons. *J. AOAC Int.* **2018**, *101*, 1429–1436. [CrossRef]
16. Hadi, P.; To, M.H.; Hui, C.W.; Lin, C.S.; McKay, G. Aqueous mercury adsorption by activated carbons. *Water Res.* **2015**, *73*, 37–55. [CrossRef] [PubMed]

17. Cataldo, S.; Iannì, A.; Loddo, V.; Mirenda, E.; Palmisano, L.; Parrino, F.; Piazzese, D. Combination of advanced oxidation processes and active carbons adsorption for the treatment of simulated saline wastewater. *Sep. Purif. Technol.* **2016**, *171*, 101–111. [CrossRef]
18. Fukuhara, T.; Iwasaki, S.; Hasegawa, T.; Ishihara, K.; Fujiwara, M.; Abe, I. Adsorption of 1,4-Dioxane from Aqueous Solutions onto Various Activated Carbons. *J. Water Environ. Technol.* **2011**, *9*, 249–258. [CrossRef]
19. Myers, M.A.; Johnson, N.W.; Marin, E.Z.; Pornwongthong, P.; Liu, Y.; Gedalanga, P.B.; Mahendra, S. Abiotic and bioaugmented granular activated carbon for the treatment of 1,4-dioxane-contaminated water. *Environ. Pollut.* **2018**, *240*, 916–924. [CrossRef] [PubMed]
20. Woodard, S.; Mohr, T.; Nickelsen, M.G. Ambersorb 560 Synthetic Media: A Promising New Treatment Technology for 1,4-Dioxane. *Remediat. J.* **2014**, *24*, 27–40. [CrossRef]
21. Pekala, R.W.; Farmer, J.C.; Alviso, C.T.; Tran, T.D.; Mayer, S.T.; Miller, J.M.; Dunn, B. Carbon aerogels for electrochemical applications. *J. Non-Cryst. Solids* **1998**, *225*, 74–80. [CrossRef]
22. Zhu, X.; Hope-Weeks, L.J.; Yu, Y.; Yuan, J.; Zhang, X.; Yu, H.; Liu, J.; Li, X.; Zeng, X. Effect of concentration of glycidol on the properties of resorcinol-formaldehyde aerogels and carbon aerogels. *RSC Adv.* **2022**, *12*, 20191–20198. [CrossRef] [PubMed]
23. Yu, Z.L.; Li, G.C.; Fechler, N.; Yang, N.; Ma, Z.Y.; Wang, X.; Antonietti, M.; Yu, S.H. Polymerization under Hypersaline Conditions: A Robust Route to Phenolic Polymer-Derived Carbon Aerogels. *Angew. Chem.* **2016**, *55*, 14623–14627. [CrossRef]
24. Molina-Campos, D.F.; Delgadillo, D.P.V.; Giraldo, L.; Moreno-Pirajan, J.C. Removal of metal ions Cd(II), Cr(VI) and Ni(II) from aqueous solution using an organic aerogel and carbon aerogel obtained by acid catalysis. *Mater. Express* **2020**, *10*, 127–139. [CrossRef]
25. Li, Y.C.; Zhou, M.Q.; Waterhouse, G.I.N.; Sun, J.C.; Shi, W.J.; Ai, S.Y. Efficient removal of cadmium ions from water by adsorption on a magnetic carbon aerogel. *Environ. Sci. Pollut. R.* **2021**, *28*, 5149–5157. [CrossRef]
26. Song, Z.; Chen, X.; Gong, X.; Gao, X.; Dai, Q.; Nguyen, T.T.; Guo, M. Luminescent carbon quantum dots/nanofibrillated cellulose composite aerogel for monitoring adsorption of heavy metal ions in water. *Opt. Mater.* **2020**, *100*, 109642. [CrossRef]
27. Gan, G.; Li, X.; Fan, S.; Wang, L.; Qin, M.; Yin, Z.; Chen, G. Carbon Aerogels for Environmental Clean-Up. *Eur. J. Inorg. Chem.* **2019**, *2019*, 3126–3141. [CrossRef]
28. Cotet, L.C.; Maicaneanu, A.; Fort, C.I.; Danciu, V. Alpha-Cypermethrin Pesticide Adsorption on Carbon Aerogel and Xerogel. *Sep. Sci. Technol.* **2013**, *48*, 2649–2658. [CrossRef]
29. Achari, V.S.; Lopez, R.M.; Rajalekshmi, A.S.; Jayasree, S.; Ravindran, B.; Sekkar, V. Scavanging nitrophenol from aquatic effluents with triethyl amine catalyzed ambient pressure dried carbon aerogel. *J. Environ. Chem. Eng.* **2020**, *8*, 103670. [CrossRef]
30. Kalotra, S.; Mehta, R. Carbon aerogel and polyaniline/carbon aerogel adsorbents for Acid Green 25 dye: Synthesis, characterization and an adsorption study. *Chem. Eng. Commun.* **2022**, *209*, 757–773. [CrossRef]
31. Geca, M.; Wisniewska, M.; Nowicki, P. Biochars and activated carbons as adsorbents of inorganic and organic compounds from multicomponent systems—A review. *Adv. Colloid. Interface Sci.* **2022**, *305*, 102687. [CrossRef] [PubMed]
32. DiGuiseppi, W.; Walecka-Hutchison, C.; Hatton, J. 1,4-Dioxane Treatment Technologies. *Remediat. J.* **2016**, *27*, 71–92. [CrossRef]
33. Chen, R.; Liu, C.; Johnson, N.W.; Zhang, L.; Mahendra, S.; Liu, Y.; Dong, Y.; Chen, M. Removal of 1,4-dioxane by titanium silicalite-1: Separation mechanisms and bioregeneration of sorption sites. *Chem. Eng. J.* **2019**, *371*, 193–202. [CrossRef]
34. Liu, Y.; Johnson, N.W.; Liu, C.; Chen, R.; Zhong, M.; Dong, Y.; Mahendra, S. Mechanisms of 1,4-Dioxane Biodegradation and Adsorption by Bio-Zeolite in the Presence of Chlorinated Solvents: Experimental and Molecular Dynamics Simulation Studies. *Environ. Sci. Technol.* **2019**, *53*, 14538–14547. [CrossRef] [PubMed]
35. Yang, X.; Yang, D.; Zhang, G.; Zuo, H. Preparation of mesoporous carbon aerogels via ambient pressure drying using a self-sacrificing melamine-formaldehyde template. *J. Power Sources* **2021**, *482*, 229135. [CrossRef]
36. Wei, H.L.; Han, L.L.; Tang, Y.C.; Ren, J.; Zhao, Z.B.; Jia, L.Y. Highly flexible heparin-modified chitosan/graphene oxide hybrid hydrogel as a super bilirubin adsorbent with excellent hemocompatibility. *J. Mater. Chem. B* **2015**, *3*, 1646–1654. [CrossRef] [PubMed]
37. Chen, H.; Wang, S.; Tang, Y.; Zeng, F.; Schobert, H.H.; Zhang, X. Aromatic cluster and graphite-like structure distinguished by HRTEM in thermally altered coal and their genesis. *Fuel* **2021**, *292*, 120373. [CrossRef]
38. Borowski, P.; Gac, W.; Pulay, P.; Woliński, K. The vibrational spectrum of 1,4-dioxane in aqueous solution–theory and experiment. *New J. Chem.* **2016**, *40*, 7663–7670. [CrossRef]
39. Huang, J.Y.; Liang, Y.R.; Dong, H.W.; Hu, H.; Yu, P.F.; Peng, L.; Zheng, M.T.; Xiao, Y.; Liu, Y.L. Revealing contribution of pore size to high hydrogen storage capacity. *Int. J. Hydrogen Energy* **2018**, *43*, 18077–18082. [CrossRef]
40. Grzyb, B.; Hildenbrand, C.; Berthon-Fabry, S.; Bégin, D.; Job, N.; Rigacci, A.; Achard, P. Functionalisation and chemical characterisation of cellulose-derived carbon aerogels. *Carbon* **2010**, *48*, 2297–2307. [CrossRef]
41. Zhao, X.; Wang, X.; Lou, T. Simultaneous adsorption for cationic and anionic dyes using chitosan/electrospun sodium alginate nanofiber composite sponges. *Carbohydr. Polym.* **2022**, *276*, 118728. [CrossRef]
42. Li, X.; Wang, Z.; Ning, J.; Gao, M.; Jiang, W.; Zhou, Z.; Li, G. Preparation and characterization of a novel polyethyleneimine cation-modified persimmon tannin bioadsorbent for anionic dye adsorption. *J. Environ. Manag.* **2018**, *217*, 305–314. [CrossRef] [PubMed]
43. Fonseca-Correa, R.A.; Giraldo, L.; Moreno-Piraján, J.C. Thermodynamic study of adsorption of nickel ions onto carbon aerogels. *Heliyon* **2019**, *5*, e01789. [CrossRef] [PubMed]

44. Qin, Y.; Chai, B.; Wang, C.; Yan, J.; Fan, G.; Song, G. New insight into remarkable tetracycline removal by enhanced graphitization of hierarchical porous carbon aerogel: Performance and mechanism. *Colloids Surf. A Physicochem. Eng. Asp.* **2022**, *655*, 130197. [CrossRef]
45. Sato, N.; Aoyama, Y.; Yamanaka, J.; Toyotama, A.; Okuzono, T. Particle Adsorption on Hydrogel Surfaces in Aqueous Media due to van der Waals Attraction. *Sci. Rep.* **2017**, *7*, 6099. [CrossRef]

*toxics*

MDPI

*Article*

# $SiO_2$–$TiO_2$ Nanoparticle Aqueous Foam for Volatile Organic Compounds' Suppression

**Jintao Yu and Yuning Xuan ***

Shanghai Institute of Chemical Industry Environmental Engineering Co., Ltd., Guangfu Rd.(W) No.2666, Putuo District, Shanghai 200062, China; yujintao829@163.com
* Correspondence: xuanyninlnsy@gmail.com

**Abstract:** Volatile organic compounds (VOCs) are prevalent soil contaminants. During the ex situ soil remediation process, VOCs may overflow from the soil and cause gas to diffuse into the atmosphere. Moreover, some VOCs, such as trichloromethane, are categorized by the EPA as emerging contaminants, imparting toxicity to organs, and the endocrine and immune systems, and posing a huge threat to human health and the environment. To reduce VOCs' emissions from contaminated soil, aqueous foam suppression is a prospective method that provides a durable mass transfer barrier for VOCs, and it has been widely used in odor control. Based on an aqueous foam substrate, in order to enhance the foam's stability and efficiency of suppression, $SiO_2$–$TiO_2$-modified nanoparticles have been used as stabilizing agents to improve the mechanical strength of liquid film. The nanoparticles are endowed with the ability to photocatalyze after the introduction of titanium dioxide. From SEM imaging, IR, and a series of morphological characterization experiments, the dispersibility of the $SiO_2$–$TiO_2$-modified nanoparticles was significantly improved under the polar solvent, which, in turn, increased the foam duration. The foam dynamic analysis experiments showed that the foam liquid half-life was increased by 4.08 h, and the volume half-life was increased by 4.44 h after adding the novel synthesized nanoparticles to the bulk foam substrate. From the foam VOC suppression test, foam with modified nanoparticles was more efficient in terms of VOCs' suppression, in contrast with its nanoparticle-free counterparts, due to the longer retention time. Moreover, in a bench-scale experiment, the $SiO_2$–$TiO_2$ nanoparticles foam worked against dichloroethane, n–hexane, and toluene for almost 12 h, with a 90% suppression rate, under UV irradiation, which was 2~6 h longer than that of UV-free $SiO_2$–$TiO_2$ nanoparticles, the KH−570-modified nanosilica foam, and the nanoparticle-free bulk foam. XPS and XRD results indicate that in $SiO_2$–$TiO_2$ nanoparticles, the proportion of titanium valence was changed, providing more oxygen vacancies compared to raw titanium dioxides.

**Keywords:** volatile organic compounds; emerging pollutants; aqueous foam suppression; nanoparticle silica; $TiO_2$ photocatalysis

## 1. Introduction

Ex situ schemes account for a large proportion of soil remediation projects and refer to soil transfers that may disturb the soil's ambient surroundings. Some contaminated sites are liable to suffer from pollutant leaking if trenching or digging works are carried out. The volatilizing vapors in VOCs pose an intangible threat to human health, which may cause damage to the human respiratory system and blood circulation [1–5]. Therefore, it is essential to reduce the effusion of such organic contaminants. At present, membranes and adsorbents are considered two reliable materials for VOCs control [6–10], both of which are physical barrier agents. Membrane materials include an inflatable membrane structure; these membranes can create a mass transfer barrier against volatile gases on the soil surface [11]. Some adsorbents, such as porous materials, have abundant adsorption sites that provide the ability to contain gas molecules [12]. Although these two materials are efficient in terms of VOCs control, they still lack applications in contaminated soil plots.

Moreover, their undegradable characteristics make it difficult to dispose of them after the remediation project is completed.

Stabilizing aqueous foam is currently one of the environmentally friendly materials used for volatile gas control and the suppression of pungent odors [13–16]. Compared with membrane materials and adsorbents, aqueous foam has inimitable fluidity and flexibility, and strong degradability. The barrier to VOCs that aqueous foams contain is mainly manifested by the 'circuitous path' generated inside the foam and the sealing of the liquid carried by bubble film [13,17,18]. As the foam carpet covers contaminated soil, liquid draining occurs simultaneously, making the liquid film thinner and thinner, which undermines the VOC suppression performance [19–22]. Modified components should be appended to the substrate to improve the stability of foams, which further prolongs the effective suppressing time [23–25]. In addition to some common additions to the foam formula, such as lamellae stabilizers, viscosities, or bi-surfactants [20], nanoparticles are likewise regarded as effective ingredients, which can curb the possibility of the foam collapsing by forming a coalescence barrier [26]. Moreover, the expanded specific surface area and plentiful hydroxide radicals on nanoparticles can undergo functional modification. In recent reports, a series of hydrophobic groups have been selected to modify the hydroxyl on the surface of nanosilica [27–31]. The strong emerging dispersion of the post-modified silica can improve multifunctionalities, such as stabilizing emulsion, supercritical extraction, or oil removal. Although the method described above has significantly improved the stability of foam film, the process of foam drainage due to gravity is unavoidable. Since the ability of aqueous foam to blcok VOCs mainly comes from the liquid sealing effect of the liquid film, when the foam gradually drains to form an unstable foam skeleton, its ability to block VOCs is significantly weakened. Therefore, with the foam skeleton framing, the time needed for the foam to block VOCs can be effectively extended if a layer of hydrophobic absorbent materials are attached to the film interface.

The vapor recovery agent, which is coupled with the vapor-suppressing scheme, can enhance the suppressing performance of VOC-contaminated soil [32]. As a novel vapor recovery material, nano titanium dioxide(nano-$TiO_2$) possesses the capacity to decompose VOCs because of its photocatalysis performance [33]. In a bench-scale experiment, a $SiO_2$ core–$TiO_2$ shell composite nanoparticle was shown to enhance the photocatalysis of nano-$TiO_2$ [34]. These composite particles are widely used in the environmental or medical fields, have been proven to contain no toxicity, and have no risk of causing secondary pollution [35]. When assembled with the nanoparticle described above, the stabilizing aqueous foam demonstrates a bi-functional property that involves a mass barrier of molecular VOCs, as well as VOC degradation. Moreover, nanoparticles modified by hydrophobic groups show superior dispersibility, meaning they can uniformly distribute on foam lamellae and form an intensive layer to restrain coalescence.

In summary, by adding nanoparticles with functional photocatalytic and improved hydrophobicity to the base of the foam-stabilizing substrate, the particles can spontaneously arrange themselves at the liquid film–air interface, and still retain their original form in a dry foam whose liquid film is completely drained. Ultimately, the particles can create a strong foam skeleton barrier for VOCs' suppression under intermolecular force.

## 2. Materials and Methods

### 2.1. Samples' Preparation

#### 2.1.1. Preparation of Bulk Foaming Solution

To prepare a substrate solution, sodium dodecylbenzene sulfonate (SDBS) and N–lauryl sarcosinate sodium (NLSS) were mixed homogeneously at a molar ratio of 7:3 in a total concentration of 0.1 mol/L. After adding 6 vol% triethanolamine, 7 vol% glycerol, 3 g/L Xanthan gum, and 0.3 mol/L $ZnSO_4$ into the hybrid substrate solution, ultrasonic mixing was carried on for 15 min to obtain a bulk foaming solution. The bulk foaming solution served as a substrate for the $SiO_2$–$TiO_2$ nanoparticle foam. All test samples were sourced from Aladdin.

#### 2.1.2. Preparation of Trimethyl Silane (TMCS)-Modified Nanosilica Particles [36]

Briefly, 25-nanometer silica was added to 15 vol% TMCS n–Hexane or ethanol aqueous solution with a concentration of 50 g/L at ambient temperature. Then, it was activated by sodium hydroxide, and adjusted to ca. pH 9. A further experimental operation was launched after changing the brand-new solution, while the reaction condition was at 60 °C for 2~4 h or more. The post-reaction product was soaked in an n–hexanol solution for 8 h, centrifuged, and dried at 60 °C for 4 h at atmospheric pressure. A gradient heating program was followed to prepare a pure TMCS-modified nanosilica particle with an 80 °C heating program for 3 h and a 100 °C heating program for 2 h.

#### 2.1.3. Sodium Stearate (SS)-Modified Nanosilica Particle Preparation

Briefly, 10 g of 25-nanometer silica was added in 100 mL deionized water and dispersed adequately in a 250 mL beaker. Sodium stearate with an additive amount of 3.5 g was added to the beaker under 80 °C heating for 40 min in a thermostatic water bath, then cooled to ambient temperature after the reaction was in equilibrium. The after-treatment solution was transferred to a centrifuge tube and centrifuged at a rotation speed of 3000 r/min for 5 min, and then washed with alcohol once and deionized water twice. Hydrophobic modified nanosilica was obtained after drying and centrifugation.

#### 2.1.4. Preparation of γ–Methacryloyl Oxy Propyl Trimethoxysilane (KH–570)-Modified Nanosilica Particles

Birefly, 10 g nanosilica and KH–570 (15 wt%) were mixed and dispersed adequately in 100 mL deionized water, followed by heating at 80 °C for 12 h in a thermostatic water bath. The final products were washed three times in ethanol, then centrifuged and dried at 100 °C for 2 h.

#### 2.1.5. Preparation of $SiO_2$–$TiO_2$ Nanoparticles [34]

$SiO_2$–$TiO_2$ nanoparticles were synthesized by adding tetrabutyltitanate dropwise to the hydrophobic nanosilica substrate. The reaction took place in 100 mL ethanol, and 5 g hydrophobic nanosilica was added into the solution. The total dosage of tetrabutyltitanate was 15 vol%. The hydrophobic nanosilica was selected from Sections 2.1.2–2.1.4 by a series of characteristics, as follows. By dropping tetrabutyltitanate constantly, hydrolysis happened on the site of silanol functional group, which is exposed on the hydrophobic nanosilica surface. After an ultrasonic dispersion at ambient temperature for 30 min, the post-reaction product was washed 3 times in deionized water, then centrifuged and dried at 100 °C for 2 h.

#### 2.1.6. Preparation of $SiO_2$–$TiO_2$ Nanoparticle Foam

Briefly, 0.5 g $SiO_2$–$TiO_2$ nanoparticles were added into 100 mL of bulk foam solution, prepared as detailed in Section 2.1.1, in a spotless beaker (200 mL), and dispersed by ultrasonic for 15 min. The solution must be foamed immediately when it is dispersed uniformly.

### 2.2. *Characteristics*

#### 2.2.1. Hydrophobic Performance

The hydrophobic property of nanosilica after surface modification is strongly related to its lipophilicity, and the degree of lipophilicity, which is related to methanol consumption, is a common index used to evaluate the abundance of hydrophobic functional groups on the surface of nanosilica. A higher lipophilic degree means more methanol consumption, reflecting a strong lipophilicity and outstanding hydrophobic performance [35,36]. According to the relative experimental method, 0.2 g of well-prepared hydrophobic nanosilica particles were added into a beaker containing 50 mL of deionized water. As the methanol was slowly dripped into the beaker under magnetic stirring (200 r/min), the modified nanoparticles floated on the surface of the pure water, and infiltrated gradually as the methanol was added dropwise. The insufficient amount of methanol could only cause part of the nanosilica to infiltrate.

When all the particles penetrate totally into the methanol−water phase, the consumption of methanol (V, mL) could be recorded. The calculation method is shown as follows:

$$V_m = \frac{V}{50 + V} \times 100\% \quad (1)$$

where V is the consumption of methanol; $V_m$ is the degree of lipophilicity, which is a dimensionless unit.

2.2.2. Dispersion Performance

To evaluate whether the hydrophobic nanosilica particles have outstanding dispersibility, the dispersion performance of the material was assessed. When the particle size distribution is similar, the sedimentation rate of ultrafine powder can faithfully reflect the dispersion of the particles [37], and the sedimentation half-life is also an effective method of reflecting the sedimentation rate [38]. Briefly, 0.2 g of modified nanosilica particles were placed in a vial containing 25 mL of ethanol solution and dispersed ultrasonically for 15 min; they then settled statically on the horizontal plane. An infrared detector was equipped to monitor the sediments' height in dispersions. The durations were recorded when the sediments' height settled down to half their original height.

2.2.3. Stability Performance

The experiment mainly utilized foam volume half-life ($t_{1/2}{}^V$) and foam drainage half-life ($t_{1/2}{}^L$) to characterize foam stability. Their specific definitions are as follows.

Half-life of foam volume ($t_{1/2}{}^V$): the time it takes for the volume of foam to decay to half of the initial volume of foam.

Foam drainage half-life ($t_{1/2}{}^L$): the time it takes for foam film to drain until the drainage volume reaches half that of the initial foaming solution.

2.2.4. VOC-Suppressing Performance

The suppressing performance was tested using a bench-scale testing system (Figure 1). The testing system contained a switchable UV lamp and a foam VOC-suppressing chamber, and was combined with an Agilent 5977C gas chromatograph–mass spectrometer (GC–MS) which came from the Agilent Technologies Co., Ltd. (Shanghai, China). A soil turbulent stirrer was settled at the bottom of the acrylic jar. Throughout the evaluation process, the stirrer was turned on and mixed the standard soil and contaminants homogenously, with a stirring speed of 50 rpm. After mixing the 1.5 mL volatile organic compounds with 50 g of standard soil in the sealed acrylic jar, the surface was coated with a 4 cm thick aqueous foam, and the foam-covered VOC-contaminated standard soil in the sealed jar was placed into a switchable ultraviolet chamber equipped with a VOC-releasing GC–MS monitor. The chamber was exposed under a 30 W power UV with a wavelength of 254 nm. The gas in the VOC-releasing chamber was extracted every 30 min, and was measured by gas chromatography to obtain corresponding signal values. The relevant software automatically integrated those signals into the peak area ($A_i$) as the experimental points were recorded.

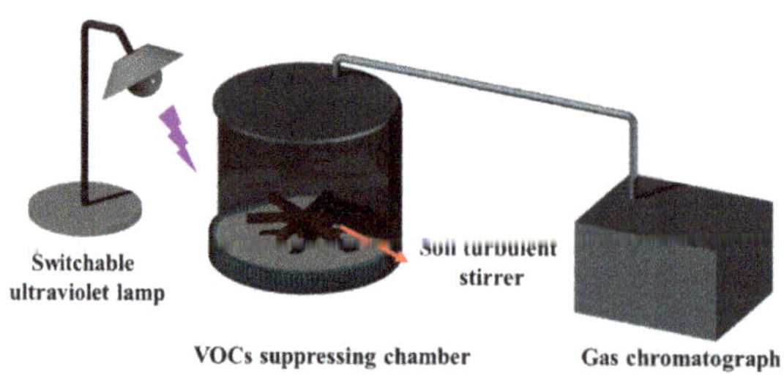


**Figure 1.** System for evaluating the VOC-suppressing foam.

The VOCs' suppression rate at the ith moment is calculated by using the following formula:

$$\theta_i = \left(1 - \frac{\sum_1^i A_i}{\sum_1^n A_i}\right) \times 100\% \quad (2)$$

where $A_i$ represents the response signal integrating the peak area at the ith point. $n$ represents the total number of test points. For evaluation of the VOCs' suppression, this experiment refers to the evaluation criterion of existing studies [14]. We adopt a duration with a suppression rate of 90% (abbreviated to $t_{90}$), and an 80% suppression duration time ($t_{80}$) is similarly used in the suppression assessment.

2.2.5. VOCs Degradation Performance

VOCs contaminant solution with a concentration of 5 mg/L was prepared by diluting the experimental standard agents including dichloroethane, n–hexane, toluene, and trichloromethane. Then, 30 mL of the standard solvents described above were added to a 50 mL vial, and the cap was sealed. Birefly, 0.1 g $SiO_2$–$TiO_2$ nanoparticles were added into each vial containing the contaminants. The test was carried out under the conditions of UV light or in a UV-free environment, and the concentration was tested by headspace gas chromatography–mass spectrometry eventually. The UV exposure time was 4 h, with a wavelength of 254 nm and a radiation intensity of 70 μw/cm$^2$.

2.2.6. Oil Resistivity of Foam

In addition to foam volume and foam stability, the oil resistance of foam also has a significant impact on the ability to suppress VOCs [39,40]. Two main ways of defoaming by oil-phase substances are as follows: (1) A portion of micromolecule oil-phase substances can dissolve in the foaming agent, which reduces the effective concentration of the foaming agent directly. (2) Some oil-phase substances that are insoluble or possess high surface activity may contribute to competitive adsorption with surfactants at the gas–liquid interface. The defoaming ability of the oil-phase material is mainly determined by the entering index ($E$) and spreading index ($S$), and the calculating methods are shown in Formulas (3) and (4). Therein, $\gamma_{AW}$ (mN/m) is the surface tension of the foaming solution, $\gamma_{OW}$ (mN/m) is the interfacial tension of oil–water mixture, and $\gamma_{AO}$ (mN/m) is the surface tension of the oil phase. E indicates the strength of the adsorption capacity of the oil-phase substance at the liquid film, and S indicates the strength of the spreading ability of the oil-phase substance on the liquid film surface. When $E > 0$ and $S < 0$, it is proven that the foam has weak oil resistance to the oil-phase substance. The stronger the defoaming ability of the oil-phase material, the worse the oil resistance of the foam to this material.

$$E = \gamma_{AW} + \gamma_{OW} - \gamma_{AO} \quad (3)$$

$$S = \gamma_{AW} - \gamma_{OW} - \gamma_{AO} \quad (4)$$

## 3. Results and Discussion

### *3.1. Modification of Nanosilica*

Silica with abundant silanol exposed on the surface was synthesized using the vapor phase method, which can provide functionalization sites for surface modification. Some silane coupling agents such as KH–570 can easily react with silanol to form a stable hydrophobic tail. The modified silica has a specific hydrophile–lipophile balance (HLB) [41]. When adjusting the quantity ratio of hydrophobic agents and raw silica, the modified silica showed various dispersion characteristics under different solutions. With the hydrophobic groups partly modified on the surface of the silica, dispersion characteristics were significantly improved, promoting the subsequent synthesis of $TiO_2$–$SiO_2$ nanoparticles.

The modified nanosilica was characterized by Fourier transform infrared spectroscopy with the Thermofisher Nicolet iS5, which is from the Thermo Fisher Scientific (Shanghai, China). The experimental results are shown in Figure 2.

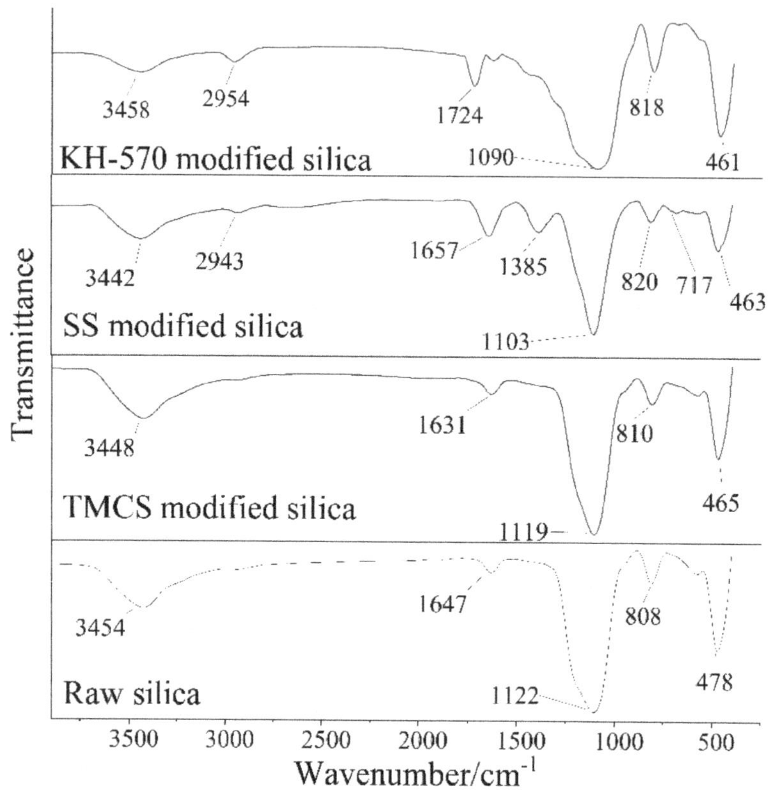


**Figure 2.** Infrared spectra of nanosilica modified with KH−570, SS, and TMCS.

Figure 2 displays the infrared spectra of nanosilica modified with KH–570, SS, and TMCS. Clearly, the characteristic peaks of modified hydrophobic nanosilica were slightly weakened with the shift in the silanol absorption peak (with a wave number of 2800 $cm^{-1}$), except for the TMCS-modified nanoparticles [42]. There are significant shape differences between the modified and unmodified nanosilica at the range of 1750 to 500 $cm^{-1}$. SS has a characteristic peak at 720 $cm^{-1}$, which indicates more than four adjacent –$CH_2$– groups [34,43], and a stretching vibration peak at 1300 $cm^{-1}$ belongs to a carbon skeleton of C–C. The more complex circumstance shown in the KH–570 spectrogram has a miscellaneous peak, resulting in a great contrast with the raw silica.

A surface morphology study was carried out with a Zeiss Merlin Compact scanning electron microscope and an Oxford AztecX–Max80 energy-dispersive spectrometer, both are supplied by Beijing Opton Optical Technology Co., Ltd. (Beijing, China), as depicted in Figure 3. The electronic images of raw silica as well as the modified hydrophobic nanosilica were analyzed. Obviously, the raw nanosilica is liable to crosslinking, forming a large and continuous area. The distribution of modified nanosilica is significantly inconsistent. Hydrophobic modified nanosilica with high porosity shows a relatively longer distance between clusters.

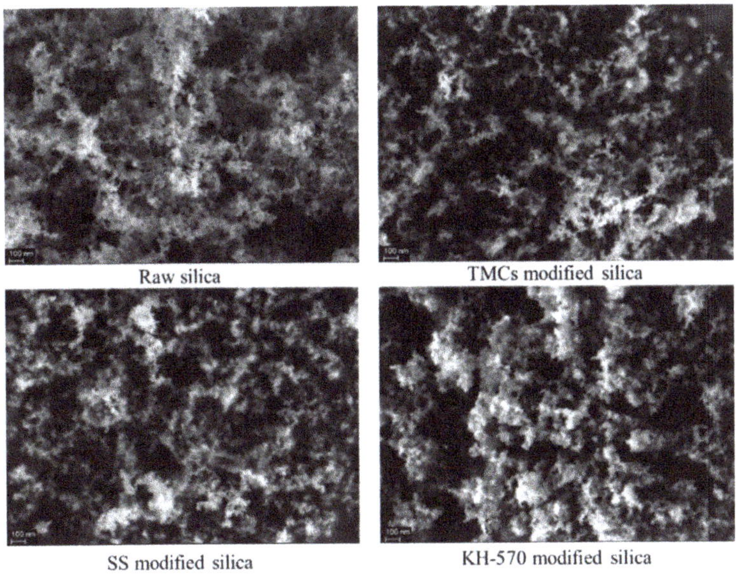


**Figure 3.** SEM images of hydrophobic nanosilica modified by various groups.

The hydrophobicity of a modified material depends on the hydrophobic properties of the functional groups and the degree of functionalization (Figure 4). TMCS, SS, and KH–570 have different chemical structures, which decide the degree of functionalization efficiency during the surface modification process with nanosilica. TMCS mainly undergoes a substitution reaction between Si–Cl and Silanol, generating volatile HCl. This reaction occurs more readily compared to the forward direction, making it easier to functionalize the surface of nanosilica. The long octadecyl hydrophobic chain of SS leads to significant hydrophobic modification. However, during surface modification, the chain tends to curl, wrapping the $COO^-$ at the end of the tail and reducing functionalization efficiency. The $Si(OH)_3$ group on the modified end enables a single KH–570 molecule to simultaneously replace three silanol groups on the nanosilica surface with a higher substitution efficiency.

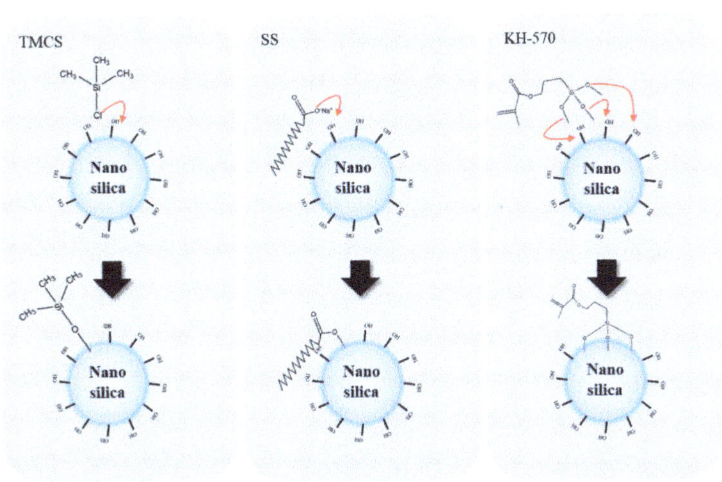


**Figure 4.** Schematic diagram of the surface modification of hydrophobic nanosilica.

The silicon NMR spectra of three modified nanosilica materials, found by the Oxford MQR, are shown in Figure 5. Chemical shifts in various nanosilica materials are decided by the number of bridging oxygen atoms in each material, which reflect the state of surface substituents [11]. The appearance of Q2 and Q3 peaks indicates that the surface of silica has been modified with other functional groups. According to the spectrum results, KH–570 has a richer variety of peaks and has Q2, Q3, and Q4 peaks simultaneously, causing higher functionalization efficiency. Notably, TMCS shows Q3 and Q4 peaks, suggesting that the silica surface has only a single substitution structure, resulting in lower functionalization efficiency than KH–570. No substitution peak was detected in the SS silicon NMR spectrum, demonstrating that SS undergoes pore adsorption of nanosilica.

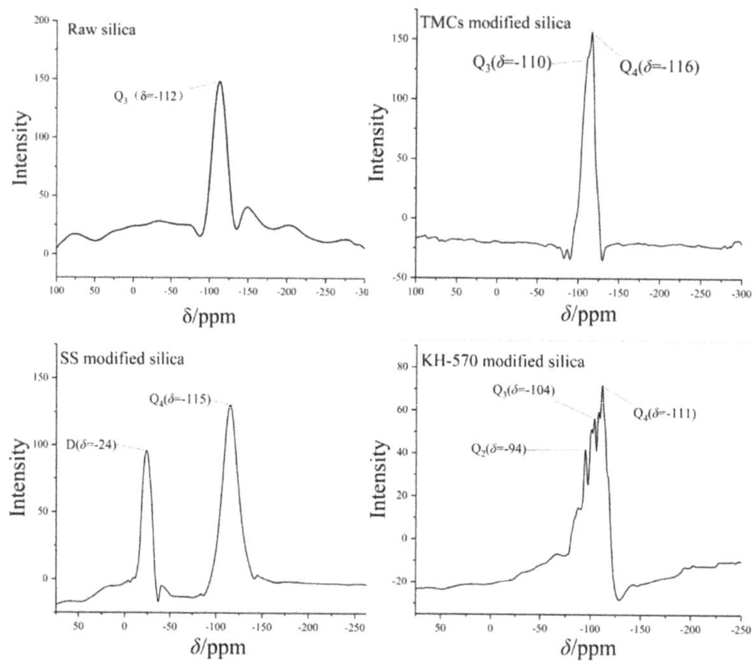


**Figure 5.** Si–NMR images of hydrophobic nanosilica modified by various groups.

Further experiments were carried out to test the characteristic of hydrophobicity alongside the parameter of degree of lipophilicity. Methanol is a typical indicator used to evaluate lipophilicity. By counting different volumes of methanol consumption, the degree of lipophilicity can be calculated using the above formula. According to Table 1, SS has the second-largest degree of lipophilicity at 11.50, while TMCS has the lowest degree of lipophilicity, with a value of 4.94. Thereby, KH–570-modified nanosilica shows the superior hydrophobic performance among the three materials, with a 13.64 degree of lipophilicity, which is consistent with the functionalization efficiency of the Si–NMR reaction.

**Table 1.** Lipophilic degree of modified hydrophobic nanosilica.

| Materials | Methanol Consumption Volume (mL) | Degree of Lipophilicity |
|---|---|---|
| Raw silica | 0.6 | 0.01 |
| TMCS-modified nanosilica | 2.6 | 4.94 |
| SS-modified nanosilica | 6.5 | 11.50 |
| KH–570-modified nanosilica | 7.9 | 13.64 |

In addition to maintaining the high specific surface area of nanoparticles, hydrophobic modification of nanosilica can also enhance the dispersion performance of particles in polar solvents. Since the surface of unmodified nanosilica is rich in silanol, the hydrogen bonding effect and the cross-linking effect between nanoparticles are markedly enhanced in polar solvents (such as water or alcohol solvents), affecting the dispersion of particles. Nanosilica with partial surface hydrophobic modification can use the remaining exposed silanol to form hydrogen bonds with solvent molecules, while the hydrophobic groups repel each other; thus, the particles are evenly dispersed in the solvent.

The same circumstance is shown in the sedimentation experiment. As listed in Table 2, KH–570 has the best dispersion performance, with a sedimentation half-life of 13 min, and SS is the second best, with a sedimentation half-life of 10 min. TMCS has the lowest half-life, at 5 min. The half-lives of the three particles are, respectively, 2.5, 5, and 6.5 times that of the raw nanosilica, dramatically prolonging the duration of sedimentation.

**Table 2.** Sedimentation experiment results of modified hydrophobic nanosilica.

| Materials | Sedimentation Half–Life (min) |
|---|---|
| Raw nanosilica | 2 |
| TMCS-modified nanosilica | 5 |
| SS-modified nanosilica | 10 |
| KH−570-modified nanosilica | 13 |

### 3.2. *$SiO_2$–$TiO_2$ Nanoparticle Characterization*

Based on the above experimental results, KH–570 (15 wt%)-modified nanosilica possesses the best hydrophobic performance, high dispersity in solution, and can prevent aggregation. On the basis of this, a $SiO_2$–$TiO_2$ composite nanoparticle with a core–shell structure was prepared, and the particle's characteristics were studied by IR, SEM, and a bench-scale UV degradation experimental test.

As the experimental results in Figure 6 show, an overlap of peaks can be observed among the three spectrums in the wavelength range of 3500–2500, 1300–1000, and 850 $cm^{-1}$. It can be preliminarily judged that the $TiO_2$–$SiO_2$ nanoparticle contains characteristic groups of both $SiO_2$ and $TiO_2$ components [34], which is consistent with the experimental results of SEM–EDS analysis shown in Figure 7 and Table 3.

In contrast to raw silica, the SEM image in Figure 7 shows that the $SiO_2$–$TiO_2$ nanoparticle has a porous structure, and the surface layer is covered with a continuously polished phase. According to EDS analysis, the surface layer of the particle was coated with silicon and uniformly trace amounts of titanium.

Figure 8b shows that no characteristic peak of $TiO_2$ occurred, which indicates that $SiO_2$ was still the dominant phase in the $SiO_2$–$TiO_2$ nanoparticles. From the XPS data in Figure 8a,c,d, titanium and silicon can be simultaneously detected, but the positions of the binding energy peaks were slightly changed. In comparison with raw $TiO_2$, the electronic orbit of Ti $2p_{3/2}$ was shifted by 0.55 eV. The increase in binding energy led to a higher bond dissociation energy in the Ti–O or Ti–Si covalent bonds, making the $SiO_2$–$TiO_2$ structure more stable; the photocatalytic process could also be carried out under higher-energy UV light instead of visible light, which improved the anti−interference ability of the ambient conditions. The binding energy of $O_{1s}$ showed that most of the oxygen elements form covalent bonds with silicon in $SiO_2$–$TiO_2$ nanoparticles. Figure 8d displays that the energy of $O_{1s}$ ($TiO_2$) in the nanoparticles increased by circa 1.0 eV compared to raw $TiO_2$ (529.1 eV) [44], and the enhanced Ti–O bonds may create more oxygen vacancies for electron transfer, thus improving photocatalytic efficiency [45]. In summary, compared to raw $TiO_2$, $SiO_2$–$TiO_2$ needs UV light instead of visible light or simulated solar light. This change causes the materials to undergo photocatalysis under an external trigger, rather than unstable conditions. This characteristic also means that catalysis of foam constituents under solar irradiation can be avoided. Moreover, $SiO_2$–$TiO_2$ frameworks can provide more abundant active sites on the surface of silica for VOCs' degradation, because of the stronger binding energy of Ti–O [46,47].

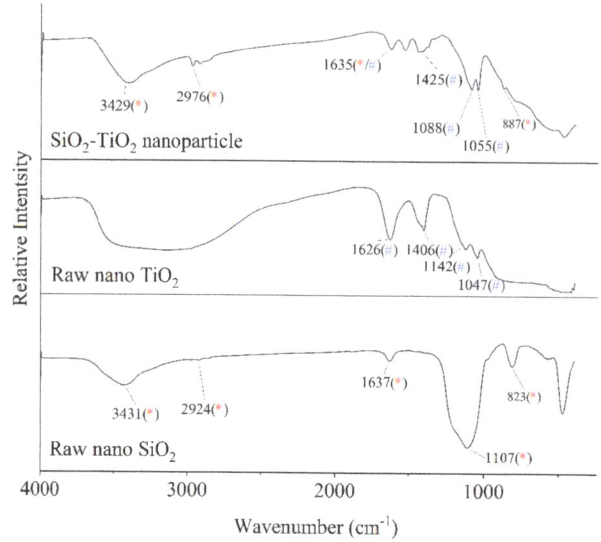


**Figure 6.** Infrared spectrum of the $SiO_2$–$TiO_2$ nanoparticle, raw $TiO_2$, and $SiO_2$ (* marks the characteristic peaks of silica, and # locates the counterparts of $TiO_2$).

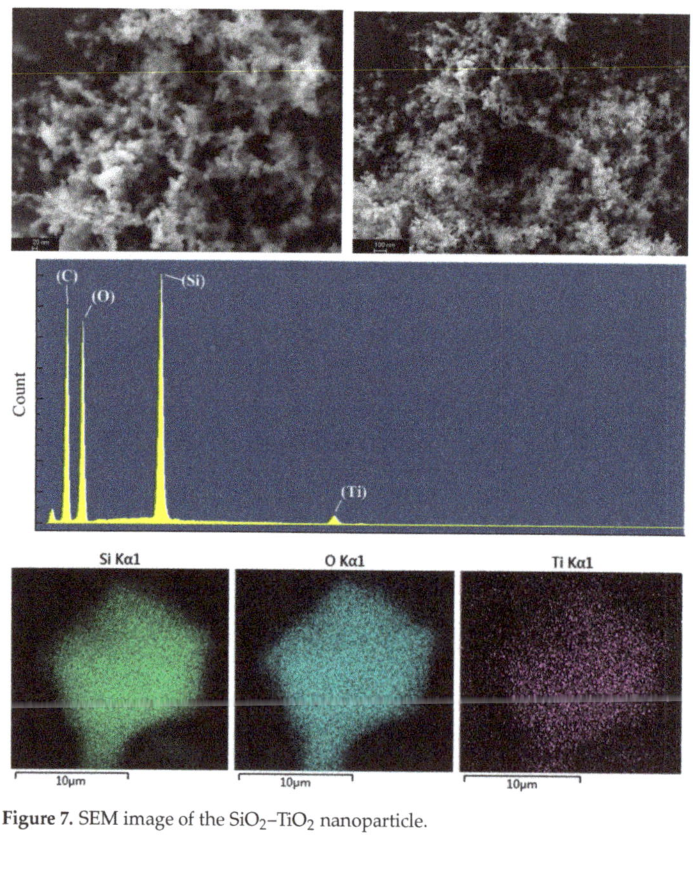


**Figure 7.** SEM image of the $SiO_2$–$TiO_2$ nanoparticle.

**Table 3.** EDS element analysis of the $SiO_2$–$TiO_2$ nanoparticle.

| Chemical Element | wt% |
|---|---|
| O | 49.19 |
| Si | 9.73 |
| Ti | 0.93 |

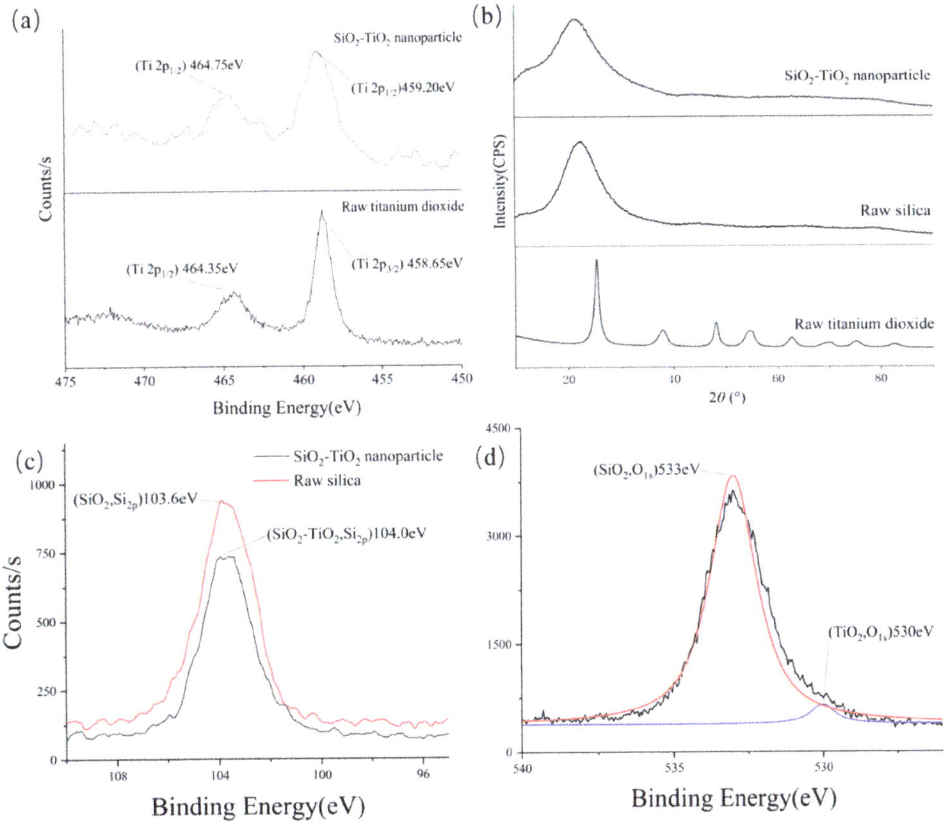


**Figure 8.** XPS/XRD images of the $SiO_2$–$TiO_2$ nanoparticle, raw silica, and raw titanium dioxide. (**a**) XPS images of titanium in the $SiO_2$–$TiO_2$ nanoparticle and raw titanium dioxide; (**b**) XRD images of the $SiO_2$–$TiO_2$ nanoparticle, raw silica, and raw titanium dioxide; (**c**) XPS images of silicon in the $SiO_2$–$TiO_2$ nanoparticle and raw silica; (**d**) XPS images of oxide in the $SiO_2$–$TiO_2$ nanoparticle.

### 3.3. *Aqueous Foam Stability*

When modified nanoparticles are introduced into the stable foam stock solution, the particles will spontaneously move to the air-liquid interface, and the semisphere of particles will be exposed under the air phase due to the hydrophobic groups on the surface of the particles. The exposure area is related to the degree of hydrophobicity. This kind of self-assembled nanoparticle on the surface of the liquid film can form an agglomeration barrier to further prevent the liquid film between foams from coalescing (Figure 9). As a result, the foam continues to maintain a fine size and restrain the Ostwald ripening process.

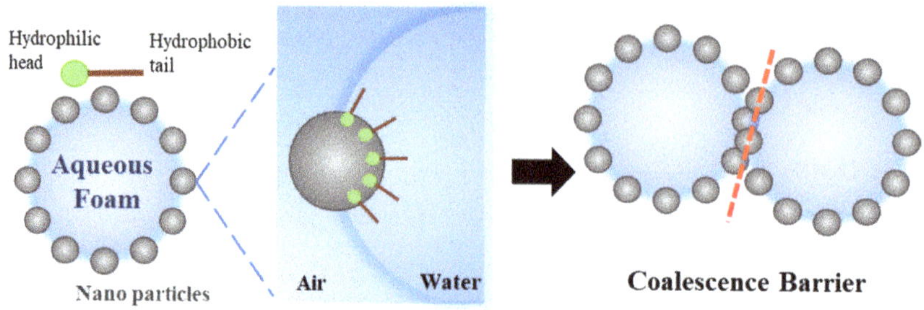


**Figure 9.** Schematic diagram of the coalescence barrier action of nanoparticles.

A Krüss DFA100–FSM dynamic foam analyzer provided by Krüss Scientific Instruments Co., Ltd. (Shanghai, China) was used to observe the morphology of aqueous foam as a function of time. In the results of foam dynamic analysis test, there were great morphological differences between unmodified and hydrophobically modified particle foams. As shown in Figure 10, the liquid volume half-life and foam volume half-life of each column were measured, as shown in Table 4. Looking at the experimental results, after the addition of modified nanosilica particles, the liquid half-life and foam half-life were significantly improved, indicating enhanced stability. Foam structures with different stages were obtained from the dynamic analysis images in Figure 11. When the $SiO_2$–$TiO_2$ nanoparticles were added into the stable bulk foam solution, the average size of the bubbles at the foam volume half-life obviously reduced, thus effectively preventing the foam aggregation process, as well as enhancing the foam's duration. This is consistent with the mechanism of the coalescence barrier.

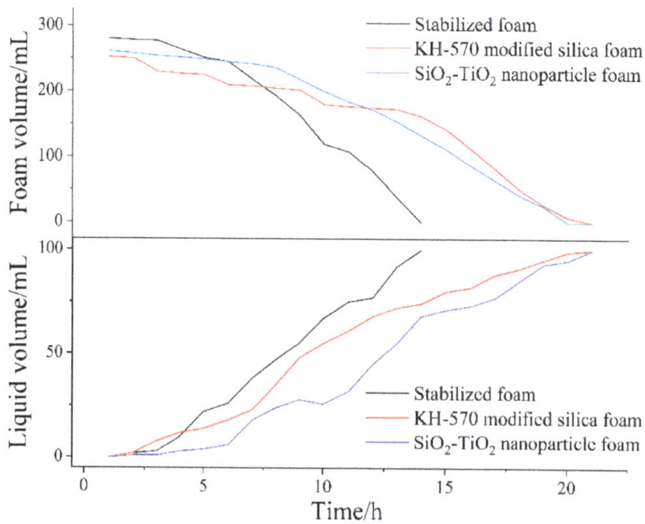


**Figure 10.** Dynamic foam analysis curve of foam/liquid volume as a function of time.

**Table 4.** Foam/liquid half-life of various aqueous foams.

| Foam Stem | Liquid Volume Half-Life/h | Foam Volume Half-Life/h |
|---|---|---|
| Stabilizing foam | 8.46 | 9.62 |
| KH–570-modified silica foam | 10.34 | 14.50 |
| $SiO_2$–$TiO_2$ nanoparticle foam | 12.54 | 14.06 |

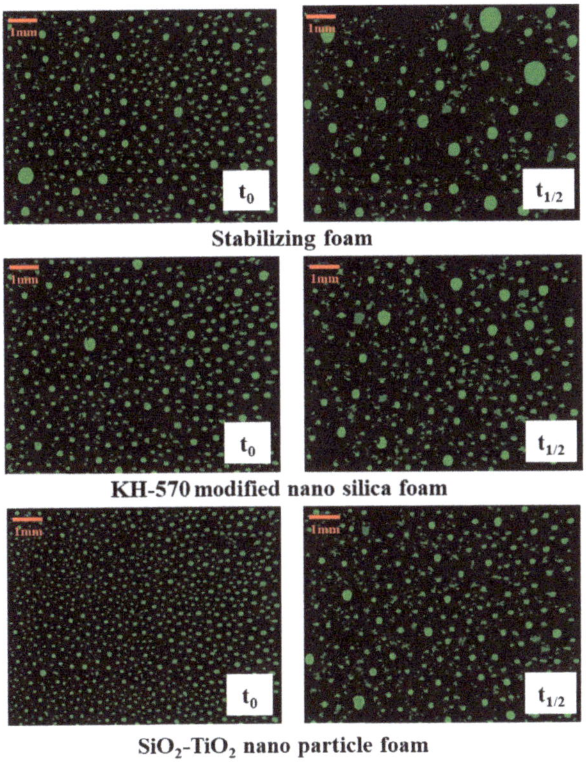


**Figure 11.** Images of dynamic foam analysis.

### 3.4. Aqueous Foam Suppressing VOCs

Aqueous foam can significantly suppress most VOCs by forming a circuitous route for the harmful vapor. The lamellae between the adjacent foam is capable of providing a mass transfer barrier to retard the vapor transition through the vapor solution-volatilization process on the foam membrane surface. However, the suppression ability is completely dependent on the lamellae thickness, which is directly decided by the amount of liquid that the foam membrane can carry. In addition, vapor leakage happens when the adjacent foam coalesces or rearranges, promoting the gas transfer rate and the formation of vacant sites. By adding photocatalytic nanoparticles to the foam solution, the foam membrane was mechanically strengthened and endowed with photocatalytic function [34,48]. Bi-functional foam can make up for the lack of barrier capacity in single liquid film.

Experiments on foam's suppressing performance were carried out using the assessment index of suppressing rate, calculated with Formula (2). Each stabilizing foam solution contains 0.5 wt% nanoparticles or nanosilica. Three stems of aqueous foam (unmodified nanosilica foam, $SiO_2$–$TiO_2$ nanoparticle foam in a UV-free environment, and $SiO_2$–$TiO_2$ nanoparticle foam under UV light) were used to suppress the following typical VOCs: a chlorohydrocarbon (dichloroethane), a petroleum hydrocarbon (n–hexane), a benzene derivative (toluene), and an emerging contaminant (trichloromethane). All the tests of VOCs were carried out by diluting standard laboratory agents.

In order to reflect the photocatalytic effect of the nanoparticles, a bidirectional experiment, including the conservation of the VOC suppressing ratio as a function of time, was carried out with modified nanoparticles and unmodified nanoparticles. The experimental results are as follows. From the dynamic curve, the $t_{80}/t_{90}$ index is presented in Table 4. Compared with unmodified nanosilica, the $t_{80}$ and $t_{90}$ VOC suppressing times of the modified nanoparticle foam were longer than those of the nanoparticles-free stabilizing foam. The $SiO_2$–$TiO_2$ nanoparticle foam under the irradiation of UV light showed an excellent VOC-suppressing ability on four contaminants. Table 5 shows that the suppression times of the four pollutants are obviously different, depending on the degree of oil resistance of foam materials to different pollutants. From Table 6, all the E indexes of foam solution are greater than 0 in the presence of n–hexane, dichloroethane, and toluene, and S indexes are less than 0, indicating that the four pollutants have a defoaming effect. From the indexes of toluene, we observed that this contaminant is more easily adsorbed on the surface of the liquid film, and forms a refractory liquid droplet, which is consistent with the short duration of toluene's blocking rate. The experimental data of n–hexane indicate that n–hexane is more easily adsorbed on the surface of the liquid membrane, affecting the effectiveness of the surface active agent. Therefore, the resistance of the foam to n–hexane is weaker than that of dichloroethane, toluene, and trichloromethane.

**Table 5.** VOC suppression evaluation of various nanoparticle aqueous foams.

| Foam Type | Stem of Contaminants | | | | | | | |
|---|---|---|---|---|---|---|---|---|
| | Dichloroethane | | n–Hexane | | Toluene | | Trichloromethane | |
| | $t_{80}$ | $t_{90}$ | $t_{80}$ | $t_{90}$ | $t_{80}$ | $t_{90}$ | $t_{80}$ | $t_{90}$ |
| Unmodified nanosilica foam | 8.89 | 7.24 | 9.64 | 7.08 | 8.57 | 5.90 | 8.57 | 7.52 |
| Unmodified nanosilica foam (UV) | 9.01 | 7.50 | 9.55 | 6.89 | 8.15 | 6.42 | 9.02 | 7.60 |
| $SiO_2$–$TiO_2$ nanoparticle foam (UV-free) | 10.65 | 10.05 | 11.59 | 10.71 | 12.11 | 11.35 | 10.28 | 8.47 |
| $SiO_2$–$TiO_2$ nanoparticle foam (UV) | 12.45 | 11.94 | 13.32 | 11.19 | 13.15 | 11.84 | 10.71 | 9.16 |

**Table 6.** Entering index and spreading index of various oil–surfactant phases.

| Phase | Surface Tension/(mN · m$^{-2}$) | Entering Index (E) | Spreading Index (S) |
|---|---|---|---|
| n–Hexane | 23.8 | 20.9 | −12.3 |
| n–Hexane/surfactant | 16.6 | | |
| Dichloroethane | 32.2 | 2.2 | −10.4 |
| Dichloroethane/surfactant | 6.3 | | |
| Toluene | 28.5 | 14.6 | −15.4 |
| Toluene/surfactant | 15.0 | | |
| Trichloromethane | 28.9 | 14.7 | −16.3 |
| Trichloromethane/surfactant | 15.5 | | |
| Surfactant solution | 28.10 | – | – |

Convergence barriers were formed during the transfer of nanoparticles to the air-liquid interface of bubble film, which obstructed the gas penetration between foam units. The barriers are also able to strengthen the foam structure, enabling more flexibility to confront outer vapor pressures such as wind. The prolonged retention time increases the suppression time accordingly. At the same time, the titanium in the $SiO_2$–$TiO_2$ composite nanoparticles plays a role in photocatalysis, causing the amounts of VOCs to be further effectively suppressed. The photocatalytic property of the nanoparticles was characterized by the degradation test. Experiments were performed by headspace gas chromatography–mass spectrometry, and results are shown in Figure 12. Apart from the first three typical VOCs, trichloromethane is easily affected by photochemical reactions and can be degraded without the addition of catalysts, producing hydrogen chloride and phosgene [49]. The introduction of nanoparticles may weaken the degradation caused by the photochemical reaction due to the decrease in light transmittance. Furthermore, as the foam lamellae become thinner because of the film liquid drainage, the foam skeleton keeps its initial frame structure, which gradually emerges from top to bottom; it has then lost the function of mass transfer across the barrier. By adding the photocatalytic nanoparticles in the foam solution, the post-forming skeleton of aqueous foam will be endowed with a function of

VOCs' adsorption–degradation, meaning the dry foam is still able to suppress the VOCs. This theory can be testified by the experimental solution shown in Figure 13, in which the subsequent suppression time $t_{80}$ of UV-irradiated foam is slightly delayed by about 1 h compared to the UV-free case.

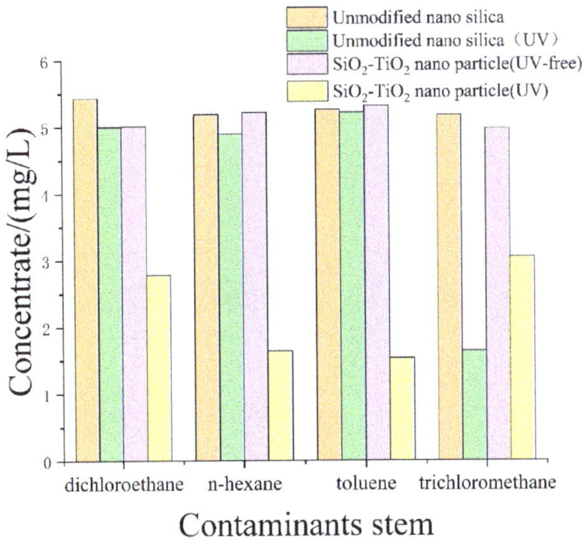


**Figure 12.** Degradation experiment results of unmodified nanosilica, UV-free nanoparticles, and nanoparticles under UV.

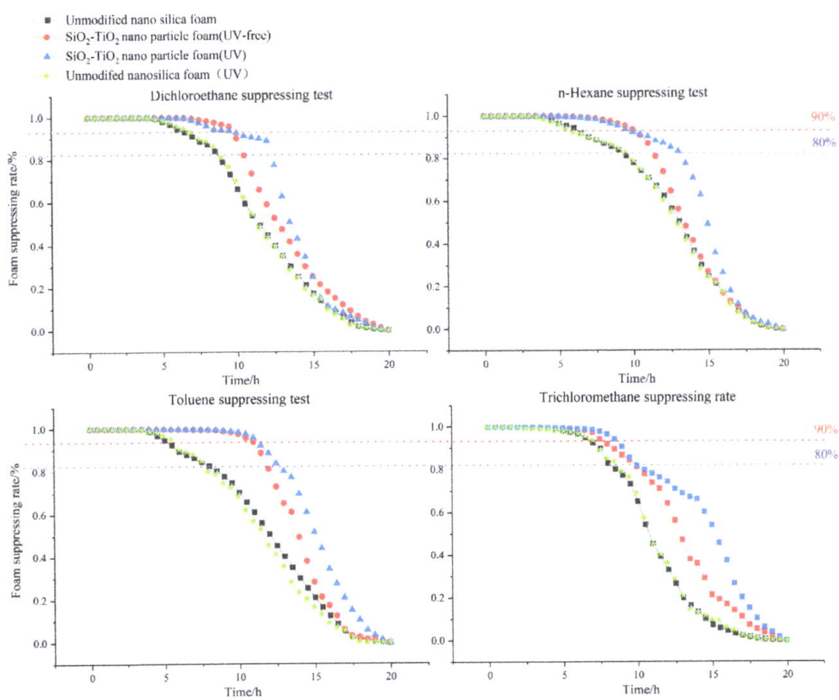


**Figure 13.** Dynamic suppression rates of aqueous foam with four typical contaminants as.

## 4. Conclusions

(1) Three silicon surface modifiers: trimethylchlorosilane, sodium stearate, and KH–570 were used to carry out hydrophobic modification of nanosilica. The dispersion of nanosilica after surface hydrophobic modification was significantly improved, and KH–570-modified nanosilica nanoparticles showed superior hydrophobic and dispersion performance, with a lipophilicity degree of 13.64 and a sedimentation half-life of 13 min in ethanol solvents. From the SEM image, the distance between clusters in the nanosilica aggregate was stretched, meaning a porous structure was formed, and there was a more uniform distribution of clusters. Through a lipophilicity test, a sedimentation half-life test and surface morphology characterization, it can be found that the cross-linking between nano-silica modified by a coupling agent is significantly weakened, and the dispersion in polar solvents is stronger, making it suitable as a synthetic substrate of $SiO_2$–$TiO_2$ nanoparticles.

(2) After the addition of $SiO_2$–$TiO_2$ nanoparticles, the gas permeability between the foams was significantly reduced due to the aggregation barrier, the smaller initial size of the foam, and the delicate and uniform foam. The liquid volume half-life was increased by circa. 2 h, and the foam volume half-life was increased from circa. 9 to 14 h, which indicates greatly improved stability. The liquid volume half-life and the foam volume half-life of the $SiO_2$–$TiO_2$ nanoparticle foam are similar those of the KH–570-modified nanosilica, with a 12.54 h liquid volume half-life and a 14.06 h foam volume half-life.

(3) The effective barrier time was assessed to compare the $SiO_2$–$TiO_2$ nanoparticle foam's VOC suppression ability under different conditions. The VOC suppression properties of the $SiO_2$–$TiO_2$ nanoparticle foam (UV), $SiO_2$–$TiO_2$ nanoparticle foam (UV-free), original nano–$SiO_2$ (UV-free), and original nano–$SiO_2$ (UV) were decreased in order. Compared with unmodified nanosilica foam, the $SiO_2$–$TiO_2$ nanoparticle foam showed stronger suppression performances, especially under UV light irradiation, and the indices of suppression ($t_{80}$, $t_{90}$) of dichloroethane, n–hexane, toluene, and trichloromethane were higher than those of the former foam for 2~6 h.

**Author Contributions:** Conceptualization, J.Y.; methodology, J.Y. and Y.X.; validation, Y.X.; formal analysis, J.Y.; investigation, J.Y. and Y.X.; data curation, Y.X.; writing—original draft preparation, Y.X.; writing—review and editing, J.Y.and Y.X. All authors have read and agreed to the published version of the manuscript.

**Funding:** This research received no external funding.

**Institutional Review Board Statement:** Not applicable.

**Informed Consent Statement:** Not applicable.

**Data Availability Statement:** Our data are unavailable due to privacy.

**Conflicts of Interest:** Jintao Yu and Yuning Xuan were employed by the company Shanghai Institute of Chemical Industry Environmental Engineering Co., Ltd. The authors declare no conflicts of interest.

## References

1. Viegi, G.; Maio, S.; Pistelli, F.; Baldacci, S.; Carrozzi, L. Epidemiology of chronic obstructive pulmonary disease: Health effects of air pollution. *Respirol. Off. J. Asian Pac. Soc. Respirol.* **2006**, *11*, 523–532. [CrossRef] [PubMed]
2. Soni, V.; Singh, P.; Shree, V.; Goel, V. Effects of VOCs on Human Health. In Proceedings of the 1st International Conference on Sustainable Energy and Environmental Challenges (SEEC), Mohali, India, 26–28 February 2017; Springer: Singapore, 2017; pp. 119–142.
3. Li, Y.J.; Mao, T.Y.; Tian, M.J.; Destech Publicat, I. Health Risk Assessment of VOCs Emissions from Petroleum Volatilizes. In Proceedings of the 3rd International Conference on Green Materials and Environmental Engineering (GMEE), Beijing, China, 22–23 October 2017; Destech Publications, Inc.: Beijing, China, 2017; pp. 197–201.
4. Shi, J.W.; Bao, Y.Z.; Xiang, F.; Wang, Z.J.; Ren, L.; Pang, X.C.; Wang, J.; Han, X.Y.; Ning, P. Pollution Characteristics and Health Risk Assessment of VOCs in Jinghong. *Atmosphere* **2022**, *13*, 613. [CrossRef]

5. Ferris Pasquini, V.; Hurtazo, H.; Quintanilla, F.; Cruz-Soto, M. 2,4-Dichlorophenol Shows Estrogenic Endocrine Disruptor Activity by Altering Male Rat Sexual Behavior. *Toxics* **2023**, *11*, 843. [CrossRef] [PubMed]
6. Siegell, J.H. Exploring VOC control options. *Chem. Eng.* **1996**, *103*, 92–96.
7. Zheng, G.Y.; Wei, K.X.; Kang, X.L.; Fan, W.; Ma, N.L.; Verma, M.; Ng, H.S.; Ge, S.B. A new attempt to control volatile organic compounds (VOCs) pollution—Modification technology of biomass for adsorption of VOCs gas. *Environ. Pollut.* **2023**, *336*, 18. [CrossRef]
8. Li, X.; Cui, R.; Yang, B.J.; Xie, S.Y.; Zeng, G.M.; Zheng, H.W.; Zheng, H.L. Research Progress on Indoor VOC Pollution and Control. *Mini-Rev. Org. Chem.* **2023**, *20*, 124–135. [CrossRef]
9. Zhang, W.P.; Li, G.Y.; Yin, H.J.; Zhao, K.; Zhao, H.J.; An, T.C. Adsorption and desorption mechanism of aromatic VOCs onto porous carbon adsorbents for emission control and resource recovery: Recent progress and challenges. *Environ. Sci. Nano* **2022**, *9*, 81–104. [CrossRef]
10. Fan, N.; Liu, C.; Huang, Y.; Li, J.G. Research progress and consideration of VOC pollution control in healthy buildings in China. *Chin. Sci. Bull. Chin.* **2020**, *65*, 263–273. [CrossRef]
11. Sert, O.S.; Fato, K.O.; Gzl, N. Lightweight and highly hydrophobic silica aerogels dried in ambient pressure for an efficient oil/organic solvent adsorption. *J. Hazard. Mater.* **2020**, *408*, 124858–124886. [CrossRef]
12. Sylemez, C.; Türkmen, L.; Zen, L. Achieving Dual-Functionality by Surface Coating of Zeolite with Stearic Acid: Combining Breathability and Odor Control Properties in Polyethylene/Zeolite Composite Films. *Macromol. Res.* **2020**, *28*, 1149–1159. [CrossRef]
13. Gautam, P.S.; Mohanty, K.K. Novel Aqueous Foams for Suppressing VOC Emission. *Environ. Sci. Technol.* **2004**, *38*, 2721. [CrossRef] [PubMed]
14. Gautam, P.S.; Mohanty, K.K. Mass transfer of volatile organic carbons through aqueous foams. *J. Colloid Interface Sci.* **2004**, *273*, 611–625. [CrossRef] [PubMed]
15. Gu, H.; Ma, L.J.; Zhao, T.; Pan, T.; Zhang, P.K.; Liu, B.G.; Chen, X.R. Enhancing protein-based foam stability by xanthan gum and alkyl glycosides for the reduction of odor emissions from polluted soils. *J. Clean Prod.* **2023**, *398*, 7. [CrossRef]
16. Lee, S.R.; Han, J.K.; Choi, Y.J.; Nam, K. Reduction of ammonia and hydrogen sulfide emission from swine manure using aqueous foams amended with microorganisms and chemical additives. *Clean-Soil Air Water* **2007**, *35*, 230–234. [CrossRef]
17. Paria, S. Surfactant-enhanced remediation of organic contaminated soil and water. *Adv. Colloid Interface* **2008**, *138*, 24–58. [CrossRef]
18. Nguyen, P.; Fadaei, H.; Sinton, D. Pore-scale assessment of nanoparticle-stabilized co 2 foam for enhanced oil recovery. *Energy Fuels* **2014**, *28*, 6221–6227. [CrossRef]
19. Farhadi, H.; Riahi, S.; Ayatollahi, S.; Ahmadi, H. Experimental study of nanoparticle-surfactant-stabilized $CO_2$ foam: Stability and mobility control. *Chem. Eng. Res. Des. Trans. Inst. Chem. Eng.* **2016**, *111*, 449–460. [CrossRef]
20. Narsimhan, G. Foam formation and stabilization. *Curr. Opin. Colloid Interface Sci.* **1996**, *1*, 759–763. [CrossRef]
21. Pugh, R.J. Foaming, foam films, antifoaming and defoaming. *Adv. Colloid Interface Sci.* **1996**, *64*, 67–142. [CrossRef]
22. Ozbayoglu, M.E.; Kuru, E.; Miska, S.; Takach, N. A Comparative Study of Hydraulic Models for Foam Drilling. *J. Can. Pet. Technol.* **2000**, *41*, PETSOC-02-06-05.
23. Xu, H.; Xu, P.; Wang, D.; Yang, Y.; Wang, X.; Wang, T.; An, W.; Xu, S.; Wang, Y.Z. A dimensional stable hydrogel-born foam with enhanced mechanical and thermal insulation and fire-retarding properties via fast microwave foaming. *Chem. Eng. J.* **2020**, *399*, 125781. [CrossRef]
24. Yang, L.L.; He, X.B.; Cheng, Y.X.; Jiang, G.C.; Liu, Z.Y.; Wang, S.B.; Qiu, S.X.; Wang, J.H.; Tian, W.G. Eco-friendly aqueous foam stabilized by cellulose microfibers with great salt tolerance and high temperature resistance. *Pet. Sci.* **2023**, *20*, 2499–2511. [CrossRef]
25. Adamova, T.P.; Manakov, A.Y.; Elistratov, D.S.; Pil'nik, A.A.; Chernov, A.A. Experimental study of methane hydrate formation in aqueous foam stabilized by surfactants. *Int. J. Heat Mass Transf.* **2021**, *180*, 8. [CrossRef]
26. Duan, F.Z.; Zhu, Y.F.; Mu, B.; Wang, A.Q. Recent progress and future prospects on aqueous foams stabilized based on clay minerals. *Appl. Clay Sci.* **2023**, *236*, 19. [CrossRef]
27. Mikhienkova, E.I.; Minakov, A.V.; Lysakov, S.V.; Neverov, A.L.; Pryazhnikov, M.I. An Investigation of the Effect of the Addition of Hydrophobic Silica Nanoparticles on the Colloidal Stability of Inverse Emulsions. *Tech. Phys. Lett.* **2021**, *47*, 766–769. [CrossRef]
28. Stöckelhuber, K.W.; Schulze, H.J.; Wenger, A. *Metastable Water Films on Hydrophobic Silica Surfaces*; Springer: Berlin/Heidelberg, Germany, 2001; pp. 11–16.
29. Jung, H.-N.-R.; Lee, Y.K.; Parale, V.G.; Cho, H.H.; Mahadik, D.B.; Park, H.-H. Hydrophobic silica composite aerogels using poly(methyl methacrylate) by rapid supercritical extraction process. *J. Sol-Gel Sci. Technol.* **2017**, *83*, 692–697. [CrossRef]
30. Dargahi-Zaboli, M.; Sahraei, E.; Pourabbas, B. Hydrophobic silica nanoparticle-stabilized invert emulsion as drilling fluid for deep drilling. *Pet. Sci.* **2017**, *14*, 105–115. [CrossRef]
31. Sun, T.; Zhuo, Q.; Liu, X.; Sun, Z.; Wu, Z.; Fan, H. Hydrophobic silica aerogel reinforced with carbon nanotube for oils removal. *J. Porous Mater.* **2014**, *21*, 967–973. [CrossRef]
32. Aguilar-Garrido, A.; Romero-Freire, A.; Paniagua-López, M.; Martínez-Garzón, F.J.; Martín-Peinado, F.J.; Sierra-Aragón, M. Technosols Derived from Mining, Urban, and Agro-Industrial Waste for the Remediation of Metal(loid)-Polluted Soils: A Microcosm Assay. *Toxics* **2023**, *11*, 854. [CrossRef]

33. Ghosh, S.; Patra, R.; Majumdar, D.; Sen, K. Developing scenario of titania-based building materials for environmental remediation. *Int. J. Environ. Sci. Technol.* **2021**, *18*, 2077–2102. [CrossRef]
34. Murashkevich, A.N.; Alisienok, O.A.; Zharskiy, I.M.; Novitskaya, M.S.; Maximovskikh, A.I. Titania sols as precursors in sol-gel technologies of composite materials for photocatalysis, electrorheology, sorption. *J. Sol-Gel Sci. Technol.* **2019**, *92*, 254–263. [CrossRef]
35. Pittol, M.; Tomacheski, D.; Simoes, D.N.; Ribeiro, V.F.; Santana, R.M.C. Evaluation of the Toxicity of Silver/Silica and Titanium Dioxide Particles in Mammalian Cells. *Braz. Arch. Biol. Technol.* **2018**, *61*, e18160667. [CrossRef]
36. Worthen, A.J.; Bagaria, H.G.; Chen, Y.; Bryant, S.L.; Huh, C.; Johnston, K.P. Nanoparticle-stabilized carbon dioxide-in-water foams with fine texture. *J. Colloid Interface Sci.* **2013**, *391*, 142–151. [CrossRef] [PubMed]
37. Il'in, A.P.; Gromov, A.A.; Tikhonov, D.V.; Yablunovskii, G.V.; Il'in, M.A. Properties of ultrafine aluminum powder stabilized by aluminum diboride. *Combust. Explos. Shock Waves* **2002**, *38*, 123–126. [CrossRef]
38. Logan, B.E.; Passow, U.; Alldredge, A.L.; Grossart, H.P.; Simon, M. Rapid formation and sedimentation of large aggregates is predictable from coagulation rates (Half-Lives) of transparent exopolymer particles (TEP). *Deep Sea Res. Part Ii-Top. Stud. Oceanogr.* **1995**, *42*, 203–214. [CrossRef]
39. Sheng, Y.J.; Ma, W.Z.; Yu, X.Y.; Ma, L.; Li, Y. Effect of liquid fuel on foamability and foam stability of mixtures of fluorocarbon and hydrocarbon surfactants. *J. Mol. Liq.* **2023**, *388*, 10. [CrossRef]
40. Guo, J.Z.; Li, Y.; Wang, Y.X.; Lyu, Q.; Zhou, L.F. Identification of an oil-bearing layer by formation water resistivity: A case study of the Jurassic reservoir, Southwest Ordos Basin. *Geophysics* **2023**, *88*, B151–B162. [CrossRef]
41. Phaodee, P.; Weston, J. Review: Implementing the hydrophilic-lipophilic deviation model when formulating detergents and other surfactant-related applications. *J. Surfactants Deterg.* **2023**, *10*, 277–286. [CrossRef]
42. Zhao, F.G.; Wu, X.Q.; Ren, W.; Chen, X.F.; Shi, P.; Yao, X. Preparation and Properties of Ultra Low K Porous Silica Films Modified by Trimethylchlorosilane (TMCS). *Ferroelectrics* **2010**, *406*, 221–227.
43. Teli, M.D.; Annaldewar, B.N. Superhydrophobic and ultraviolet protective nylon fabrics by modified nano silica coating. *J. Text. Inst.* **2017**, *108*, 460–466. [CrossRef]
44. Egerton, T.A.; Parfitt, G.D.; Kang, Y.; Wightman, J.P. XPS analysis of uncoated and silica-coated titanium-dioxide powders. *Colloids Surf.* **1983**, *7*, 311–323. [CrossRef]
45. Cao, T.Q.; Xia, T.Y.; Zhou, L.; Li, G.Q.; Chen, X.; Tian, H.; Zhao, J.L.; Wang, J.O.; Zhang, W.F.; Li, S.F.; et al. Distribution and concentration of surface oxygen vacancy of $TiO_2$ and its photocatalytic activity. *J. Phys. D-Appl. Phys.* **2020**, *53*, 424001. [CrossRef]
46. Li, T.T.; Tao, R.; Wang, Y.X.; Yan, T.; Fan, X.X.; Liu, K.Y. Construction of bismuth oxide iodide (BiOI)/zinc titanium oxide ($Zn_2TiO_4$) p-n heterojunction nanofibers with abundant oxygen vacancies for improving photocatalytic carbon dioxide reduction. *J. Colloid Interface Sci.* **2024**, *655*, 841–851. [CrossRef] [PubMed]
47. Wang, L.; Yu, Y.P.; Hu, H.C.; Xie, B.B.; Du, Y.; Zhu, Q.Y. Quaternized hollow $TiO_2$-enhanced the dye adsorption capacity and photogenerated carrier separation for efficient reactive dye removal. *Appl. Surf. Sci.* **2024**, *644*, 158764. [CrossRef]
48. Babyszko, A.; Wanag, A.; Kusiak-Nejman, E.; Morawski, A.W. Effect of Calcination Temperature of $SiO_2/TiO_2$ Photocatalysts on UV-VIS and VIS Removal Efficiency of Color Contaminants. *Catalysts* **2023**, *13*, 186. [CrossRef]
49. Huber, S.G.; Kotte, K.; Schöler, H.F.; Williams, J. Natural Abiotic Formation of Trihalomethanes in Soil: Results from Laboratory Studies and Field Samples. *Environ. Sci. Technol.* **2009**, *43*, 4934–4939. [CrossRef]

toxics

MDPI

*Article*

# Distribution Characteristics and Risk Assessment of 57 Pesticides in Farmland Soil and the Surrounding Water

**Weiqing Wang [1,2], Donghong Wang [1,2,*], Quanzhen Liu [1], Lihua Lin [1], Yongchang Xie [1,2] and Chuan Du [1,2]**

[1] National Engineering Research Center of Industrial Wastewater Detoxication and Resource Recovery, Research Center for Eco-Environmental Sciences, Chinese Academy of Sciences, Beijing 100085, China; wqwang_st@rcees.ac.cn (W.W.)

[2] University of Chinese Academy of Sciences, Beijing 100049, China

* Correspondence: dhwang@rcees.ac.cn

**Abstract:** To investigate the effect of pesticide use on surface water, the concentration and distribution characteristics of 57 pesticides and 3 degradation products were analyzed in the farmland soil and surface water in the Xingkai Lake area, including water from paddy fields, drainages and the Xingkai Lake, in Heilongjiang Province, China. Forty-three pesticides and three degradation products were detected in farmland soil. In dry field (corn and soybean field) soil, the main detected pesticides were atrazine and acetochlor with mean concentrations of 26.09 $ng \cdot g^{-1}$ and 49.08 $ng \cdot g^{-1}$, respectively. In paddy field soil, oxadiazon, mefenacet and chlorpyrifos were the main detected pesticides with mean concentrations of 14.32 $ng \cdot g^{-1}$, 78.60 $ng \cdot g^{-1}$ and 20.03 $ng \cdot g^{-1}$, respectively. In the surrounding water, including water from paddy fields, drainages and Xingkai Lake, the total concentrations of contaminants detected in the water samples ranged from 71.19 $ng \cdot L^{-1}$ to 10,145.76 $ng \cdot L^{-1}$. Of the three sampling periods, the mean concentration of contaminants in the water exhibited its peak during the vegetative period. In the analysis of the drainage water, the primary pesticides detected were atrazine, acetochlor and buprofezin with mean concentrations of 354.83 $ng \cdot L^{-1}$, 109.09 $ng \cdot L^{-1}$ and 254.56 $ng \cdot L^{-1}$, respectively. Atrazine, simetryn, buprofezin and isoprothiolane were the main pesticides detected in Xingkai Lake water, with the mean concentrations of 222.35 $ng \cdot L^{-1}$, 112.76 $ng \cdot L^{-1}$, 301.87 $ng \cdot L^{-1}$ and 138.02 $ng \cdot L^{-1}$, respectively. The concentrations of contaminants could be correlated with drainage, Da Xingkai Lake and Xiao Xingkai Lake water ($\rho > 0.8$) suggested that the source of these contaminants in drainage and Xingkai Lake water could be the same. The maximum potentially affected fraction (PAF) values of atrazine, chlorpyrifos and prometryn were higher than 5% in Xingkai Lake water, resulting in high ecological risks.

**Keywords:** pesticides; Xingkai Lake; distribution; ecological risk assessment; metabolites

## 1. Introduction

Pesticides are chemical agents used in agriculture to control pests and diseases and regulate plant growth [1]. Pesticides currently in use include organophosphorus, amides, carbamates, azole and so on. As the research on pesticides continues to develop, people have found that organophosphorus insecticides can cause neurotoxic effects by inhibiting acetylcholinesterase. This has adverse effects on the human visual system [2] and even affects the normal development of the fetus [3]. Atrazine in triazine pesticides has endocrine disrupting effects [4] and is classified as an endocrine disruptor by the European Union. Acetochlor in amide herbicides can affect the expression of the thyroid hormone gene in fish [5], while butachlor has toxic and endocrine-disrupting effects on zebrafish embryos [6]. Unfortunately, in China, the annual use of atrazine totals more than 1000 tons [7], and the annual use of alachlor, acetochlor and butachlor, as the three main amide herbicides, totals more than 100,000 tons [8]. Less than 0.1% has an active effect on disease control [9], with the rest forming a residue in the environment.

Such a high concentration of pesticides forms residues in the soil, affects the soil environment and pollutes the water environment through surface runoff, soil leaching [10] and so on, affecting the aquatic organisms. The pollution of surrounding surface water caused by pesticide use is constantly being reported. A total of 83 pesticides have been detected, and imidacloprid was shown to pose a high risk to sensitive ecological species in the northwest of Taihu Lake [11]; in Poyang Lake, neonicotinoid insecticides and organochlorine pesticides were detected [12,13]. The maximum concentration of pesticides detected in Lake Biwa, Japan, was 0.4 $\mu g \cdot L^{-1}$ [14]. The surface water of the northern Indo-Gangetic alluvial plains was affected by agricultural activities, and the concentration of organochlorine pesticides in the surface water was 2.63 $\mu g \cdot L^{-1}$~3.72 $\mu g \cdot L^{-1}$ [15]. They are all the important local surface water resources, but due to the development of agriculture in the surrounding areas, many kinds of pesticides remain in the surface water. However, Xingkai Lake, as the largest freshwater lake in Northeastern Asia [16] with a large amount of farmland surrounding it, has been rarely reported on in relation to pesticide residues.

Xingkai Lake, located in the southeast of Heilongjiang Province, China, is composed of Da Xingkai Lake and Xiao Xingkai Lake and is a located on the boundary between China and Russia. Xingkai Lake is located in a significant local grain production zone, characterized by minimal industrial pollution and predominantly agricultural activities [17,18]. The pollution in the lake is mainly from farmland. According to *The Ecological and Environmental Status Bulletin* of the Chinese government [19], Xingkai Lake is a Class V surface waterbody that is moderately polluted and has a light eutrophic status. In the past, research on the water environment of Xingkai Lake mostly focused on the water quality [20,21]. Although residues of acetochlor, butachlor and organochlorine pesticides have been reported in the farmland soil and surface water of Xingkai Lake [22,23], many other pesticides have not been reported. At the same time, Xingkai Lake is rich in fish and has a developed fishery industry. The effect of the use of agricultural pesticides on the aquatic ecological environment of Xingkai Lake is still unclear.

Therefore, the objective of this study was to explore the current levels of pesticide residues in the Xingkai Lake area and reveal the impact of the use of agricultural pesticides on the water environment of Xingkai Lake. We investigated the concentrations and distribution characteristics of 57 pesticides in the farmland soil and water around Xingkai Lake, including paddy field water, drainage water and Xingkai Lake water, using gas chromatography-mass spectrometry (GC-MS).

## 2. Materials and Methods

### 2.1. Materials and Reagents

The 57 pesticides and 3 degradation products that were used in this study were divided into six categories according to chemical group as shown in Table 1, including amides/anilines, azoles, carbamates, heterocyclic, organophosphates and triazines. The names, CAS numbers and companies from which the 57 pesticides and 3 degradation products were purchased are listed in Table S2. The recovery indicators phenanthrene-d10 and atrazine-d5 were purchased from Accustandard, New Haven, CT, USA. All of the solvents used in this study, including methanol, n-hexane (HEX) and dichloromethane (DCM), were of high-performance liquid chromatography (HPLC) grade (Thermo Fisher Scientific, Waltham, MA, USA). Glass fiber filters with a 0.7 μm aperture were purchased from Millipore, Billerica, MA, USA; anhydrous sodium sulfate was purchased from Sinopharm Chemical Reagent Co., Ltd., Shanghai, China; ultrapure water was prepared using a RephiLe-J24617 system (PURIST PRO, Lefeng Biotechnology Co., Ltd., Shanghai, China); the C18 (6 cc 500 mg) solid phase extraction (SPE) columns and HLB (6 cc 500 mg) SPE columns were purchased from Waters, Milford, MA, USA; celite was purchased from J&K Scientific, Beijing, China; and Florisil (6 mL, 1 g) SPE columns were purchased from Supelco, Merck, Darmstadt, Germany.

**Table 1.** Classification of the 57 pesticides and 3 degradation products.

| Subtype | Pesticide Names | Species Number |
|---|---|---|
| Amides/Anilines | Dimethazone, Oxadiazon, Procymidone, Acetochlor, Alachlor, Bentazon methyl [d], Buprofezin, Butachlor, Cycluron, Dimethachlor, Fenoxanil, Mefenacet, Metalaxyl, Metazachlor, Metolachlor, Pretilachlor, s-Metolachlor | 17 |
| Azoles | Anvil, Epoxiconazole, Paclobutrazol, Propiconazole, Tebuconazole, Tricyclazole, Uniconazole | 7 |
| Carbamates | Baycarb, Isoprocarb, Pirimicarb, Propoxur | 4 |
| Heterocyclic | Fenson, Fludioxonil, Isoprothiolane, Picoxystrobin, Tebuthiuron | 5 |
| Organophosphates | Bolstar, Chlorpyrifos, Demeton, Diazinon, Dichlorvos, Disulfoton, Ethoprop, Fenthion, Malathion, Mevinphos, Phorate, Ronnel, Sulfotep, Tokuthion, Trichloronate | 14 |
| Triazines | Ametryn, Atrazine, Atrazine-desisopropyl [d], Desethylatrazine [d], Gesatamine, Metribuzin, Prebane, Prometon, Prometryn, Propazine, Sebuthylazin, Simazine, Simetryn | 13 |

Note: [d] means pesticide's degradation product.

## 2.2. Study Sites and Sampling

As shown in Figure 1, samples were collected at the 22 sampling sites during May 2022 (sowing period), July 2022 (vegetative period) and September 2022 (maturity period), including surface water samples and farmland soil samples. Detailed information regarding the sampling sites is shown in Table S1.

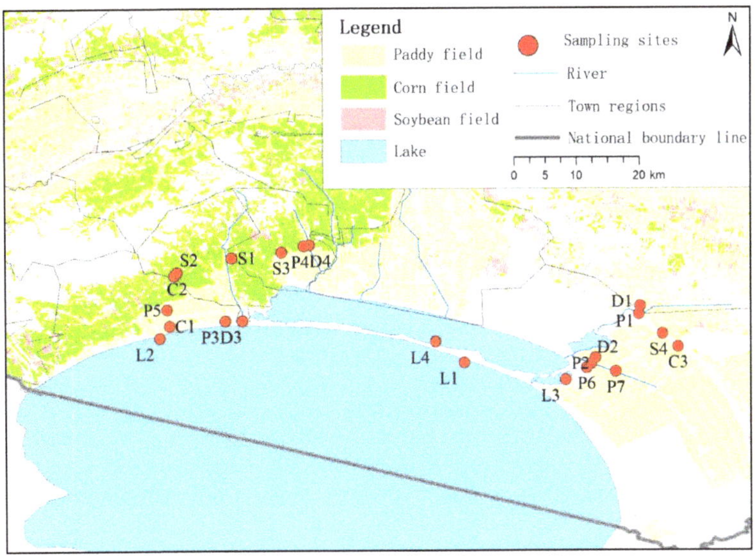


**Figure 1.** Distribution of sampling sites of farmland soil and surface water.

The farmland soil samples were collected at 14 sampling sites, including corn fields (C1~C3), paddy fields (P1~P7) and soybean fields (S1~S4). The soil samples were collected using a five-point sampling method, and the four surrounding sampling points were 20 cm away from the center point. Soil samples were taken 10 cm below the crop root with a stainless steel spoon, and about 100 g was collected at each point. The sampling points were arranged to avoid farmland boundaries, and the collected samples were preserved at a low temperature of −20 °C.

The 26 surface water samples were collected at 12 sampling sites, including paddy field water, drainage water and Xingkai Lake water. The sampling points for paddy field water were designated P1 through to P4, while the drainage water was sampled at D1 through to D4. Additionally, the sampling points for Da Xingkai Lake were identified as L1 and L2, whereas those for Xiao Xingkai Lake were noted as L3 and L4. A total of 4 L of water was collected from each sampling point with a stainless steel vessel, and the sampling points were located 50 cm away from the shore, with a collection depth of 20~30 cm below the surface of the water. Obvious pollution sources were avoided during sampling. The samples were subsequently gathered in brown glass bottles and transported to a storage location at a temperature of −4 °C. Following this, we underwent pretreatment within a 48 h period.

### 2.3. Sample Pretreatment

#### 2.3.1. Soil Sample Pretreatment

The soil was sieved to a particle size of less than 2 mm after being freeze-dried using Freeze Dry Systems (Freezone 4.5, Labconco, TX, USA). Accelerated solvent extraction (ASE) [24,25] was carried out with an ASE 350 extractor (Dionex, Sunnyvale, CA, USA). The extraction program for soil samples was as follows: 5 g of soil was mixed with 2 g of celite in a 34 mL stainless steel vessel and 200 ng each of the recovery indicators phenanthrene-d10 and atrazine-d5 was added; then, acetone: HEX (1:1, $v/v$) was used as the extraction solvent at 100 °C under a pressure of 1500 psi with 5 min of heating and 5 min of static extraction, which was carried out in two cycles; and the extraction cell was flushed with 60% of the cell volume of the solvent and purged with nitrogen for 60 s [26]. The extract was dehydrated using anhydrous sodium sulfate and then concentrated to 2 mL using a rotary evaporator (Heidolph, Schwabach, Bavaria, Germany). After the extraction procedures, the extracts were transferred to a Florisil column that was used to clean up interfering substances. The columns were activated with 10 mL of each of DCM and HEX before use. The target components were eluted with 15 mL of leachate (DCM: HEX = 1:1, $v/v$) [26,27]. The elute was concentrated to dryness with a gentle nitrogen flow, and then the solvent was replaced with HEX, setting the volume to 0.5 mL, and maintained at −20 °C for subsequent instrumental analysis.

#### 2.3.2. Water Sample Pretreatment

Particles were first filtered from the water samples using 0.7 μm glass fiber filters [28]. Next, 2 L of each filtered water sample was taken in a brown glass bottle, and then 100 ng of the recovery indicator phenanthrene-d10 was added for solid phase extraction. The solid phase extraction method was used for the pretreatment of the water sample, and the SPE columns C18 and HLB were selected in tandem. The C18 and HLB columns were activated with 10 mL of each of DCM, methanol and ultrapure water. The C18 and HLB SPE columns were enriched in series using negative pressure. The components were eluted with 10 mL of DCM. The elute was dehydrated using anhydrous sodium sulfate and concentrated to dryness with a gentle nitrogen flow, and then HEX was added to replace the solvent, setting the volume to 0.5 mL. The solution was maintained at −20 °C for subsequent instrumental analysis [29,30].

### 2.4. Instrumental Determination

Gas chromatography-mass spectrometry (GC-MS) was used to analyze the 57 pesticides and 3 degradation products (6890N-5975B, MSD, Agilent Technologies Inc., Santa Clara, CA, USA). The capillary column used was a DB-5MS (30 m × 0.25 mm × 0.25 μm, J&W Scientific, Folsom, CA, USA). Nitrogen was used as the carrier with a constant flow velocity of 1.0 mL·min$^{-1}$. The injection volume was 1.0 μL with a splitless inlet. Qualitative analysis was performed in scan mode, and quantitative analysis was performed in selected ion mode (SIM) [29,30]. Chromatographic quantitative ion ($m/z$) and retention time are shown in Table S2.

### 2.5. Quality Assurance and Quality Control

The methods were verified by measuring pesticide recovery, the limit of detection (LOD), the limit of quantitation (LOQ) and the regression coefficient ($R^2$) before sample analysis. The validation of water and soil sample pretreatment methods was carried out using matrix recovery tests. Ultrapure water was used as the matrix for the water sample method validation experiments, and the actual soil samples were used as the matrix for the soil pretreatment experiments. The mass of the target compound added to the matrix was 150 ng, and the pretreatment process was the same as that of the actual samples. The validity of the method for water and soil samples was confirmed separately. A total of three groups of matrix blank parallel experiments and five groups of matrix addition target object parallel experiments were set up to calculate the target recovery rate. The LOD and LOQ of each pesticide and degradation product were calculated according to the concentrations corresponding to a signal/noise (S/N) of three and ten for pure standard solutions. The recovery rate ranged from 74.8% to 107.8% in water and 60.9% to 109.3% in soil. The recovery rates of the recovery indicators phenanthrene-d10 and atrazine-d5 added during the pretreatment were 86.0%~119.1% and 84.3%~118.8%. The established GC-MS method had a wide linear range and a good correlation coefficient ($R^2 > 0.99$). The detailed information is shown in Table S2.

### 2.6. Ecological Risk Assessment

The species sensitivity distribution (SSD) method was used to assess the ecological risk of the pesticides detected in Xingkai Lake; it combines the toxicity data from multiple single species to predict the concentrations affecting a given number of species in a community [31,32]. It is widely used to evaluate the ecological risk of a pollutant or pollutants to organisms [33]. This study was carried out according to the following steps: (1) toxicity data acquisition; (2) SSD curve fitting; (3) the calculation of the potential affected fraction (PAF); and (4) the ecological risk assessment of single compounds.

The biotoxicity data used in SSD were obtained from the US EPA ECOTOX database (https://cfpub.epa.gov/ecotox/ (accessed on 28 November 2023)), Chinese academic literature (CNKI, http://www.cnki.net/ (accessed on 28 November 2023)) and English academic literature (http://www.sciencedirect.com/ (accessed on 28 November 2023)), and acute toxicity data were used to construct an SSD curve and the lethal concentration ($LC_{50}$) or median effective concentration ($EC_{50}$) as the toxicity endpoint. The exposure medium was fresh water and the exposure time was no more than 4 days.

Although many methods may be used to carry out the cumulative distribution function [34,35], the log-logistic distribution was used since it often fits toxicity data well [31,36].

$$y = \frac{a}{1 + e^{-k(x - x_c)}} \tag{1}$$

where $y$ is the proportion of species affected, which is defined by ranking and numbering toxicity data from smallest to largest and labelling them 1 to $n$, and then calculating $\frac{n}{1+n}$; $x$ is the logarithmic value of the toxicity data; $a$ is the amplitude; $x_c$ is the parameter of location; and $k$ is the parameter of the slope of the curve.

The potentially affected fraction (PAF) of species was calculated for each exposure concentration of the monitored pesticides to evaluate the ecological risk. When the PAF is less than 5%, the ecological risk is low or not significant. However, when the calculated PAF equals or exceeds 5% of the species, the ecological risk escalates significantly [37]. To calculate the $y$, that is, the PAF value, $x$ is taken as the concentration value of the component and brought back to Formula (1).

## 3. Results

### 3.1. Residual Characteristics of Contaminants in Farmland Soil

During the sampling procedure, pesticide bottles containing bentazon were discovered discarded adjacent to the field. Due to the limited response of GC-MS in directly detecting bentazon, an analysis of degraded bentazon methyl was conducted. In this analysis, the detection rate of atrazine was observed to be significantly higher. The degradation products of atrazine, which are atrazine-desisopropyl and desethylatrazine, are highly toxic to aquatic organisms [38]; therefore, we also analyzed atrazine-desisopropyl and desethylatrazine.

A total of 43 pesticides and 3 degradation products were detected in farmland soil, of which amides and anilines were the main component. The total concentrations of contaminants and the types of chemical constituents in the soil samples are shown in Figure 2. In the vegetative period, 44 contaminants were detected with total concentrations of 189.02 ng·g$^{-1}$~845.27 ng·g$^{-1}$, and 41 contaminants were detected with total concentrations of 22.99 ng·g$^{-1}$~281.94 ng·g$^{-1}$ in the maturity period. The concentrations of pesticides and degradation residues in the vegetative period were higher than those in the maturity period, primarily because the vegetative period is the primary application period for pesticides.

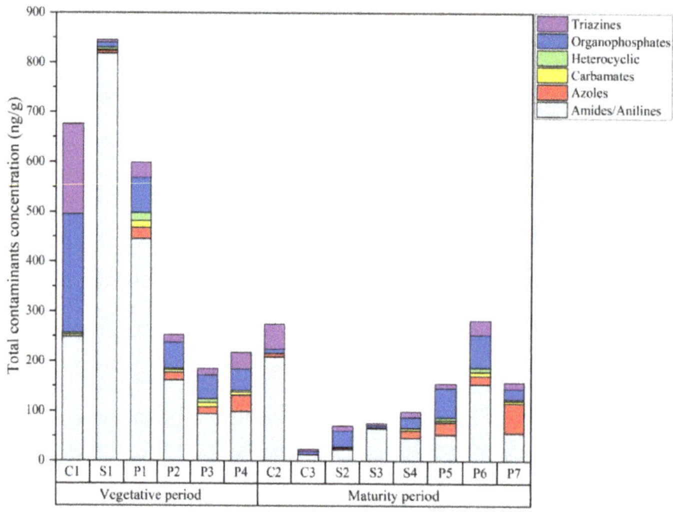


**Figure 2.** Total concentrations of contaminants and the types of chemical constituents in farmland soil at different sampling points.

In dry field (corn and soybean field) soil (Figure 3A), 34 pesticides and 3 degradation products were detected, and the mean concentration of contaminants was 760.62 ng·g$^{-1}$ in the vegetative period and 108.25 ng·g$^{-1}$ in the maturity period. The main detected pesticides were atrazine and acetochlor. Atrazine and acetochlor are applied in the vegetative season. The maximum concentrations of these two herbicides were, respectively, 133.53 ng·g$^{-1}$ and 155.75 ng·g$^{-1}$. Atrazine was regularly reported in dry field soil, and its mean concentration was 26.09 ng·g$^{-1}$ in the dry fields in the Xingkai Lake area. Compared with other studies, this level was notably higher than that found in the agricultural soils of Liaoning (18.38 ng·g$^{-1}$) and in riparian soils in the Songhua River Basin (11.28 ng·g$^{-1}$) [39,40]. Yu's research [22] conducted in the Xingkai Lake area revealed an acetochlor detection rate exceeding 80%, with a maximum recorded concentration of 117.1 ng·g$^{-1}$. This finding aligns with the results of our study.

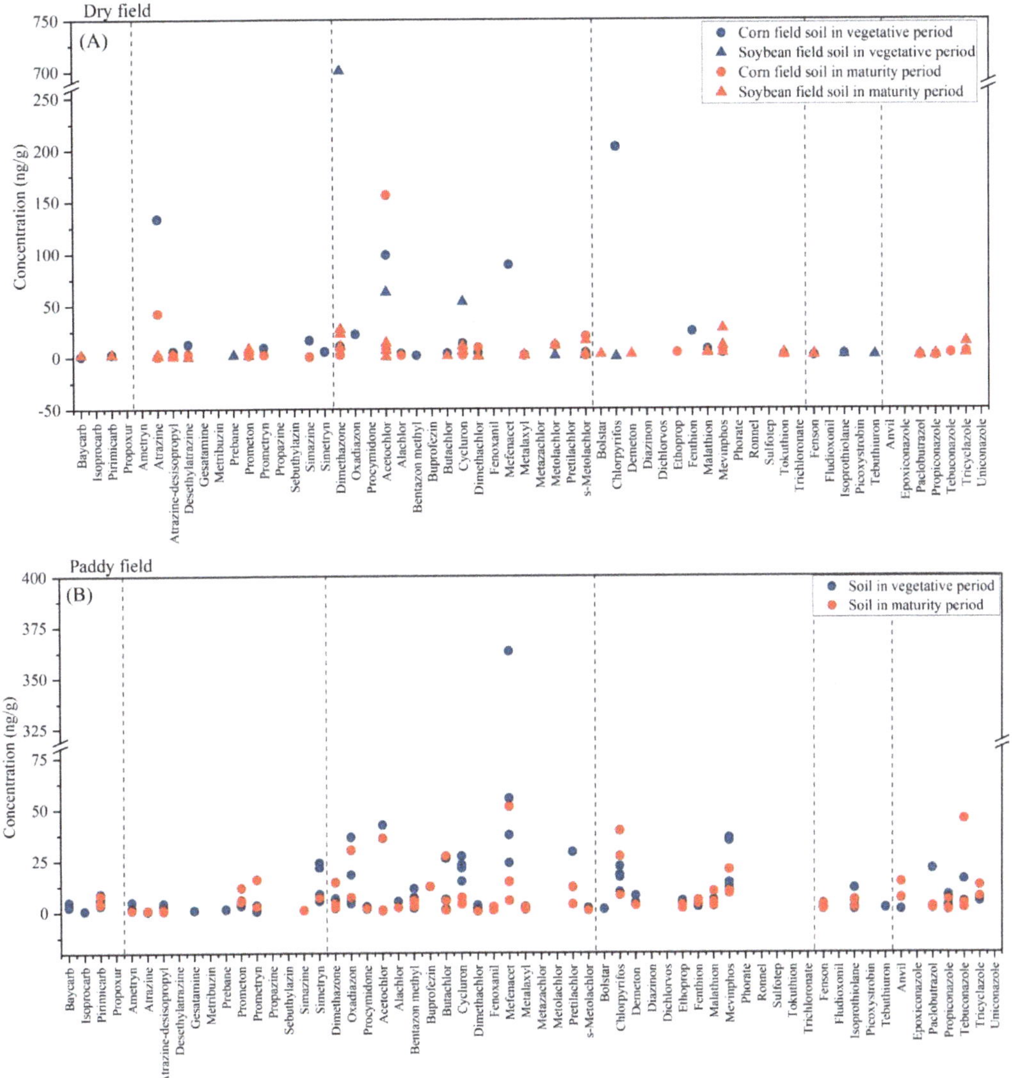


**Figure 3.** Concentrations of the 60 contaminants detected in dry field (**A**) and paddy field (**B**) soil.

In paddy field soil (Figure 3B), 40 pesticides and 2 degradation products were detected. The elevated levels of pesticides detected in paddy fields, compared to dry fields, may be attributed to the distribution of pesticides within the water and soil of these paddy fields. The mean concentration of contaminants was 314.87 ng·g$^{-1}$ in the vegetative period and 198.01 ng·g$^{-1}$ in the maturity period. The main detected pesticides were oxadiazon, mefenacet and chlorpyrifos. The data indicate that mefenacet was the predominant local herbicide in the paddy field, with a peak concentration of 363.00 ng·g$^{-1}$ and an average concentration of 78.60 ng·g$^{-1}$. The maximum observed concentration of butachlor was 27.03 ng·g$^{-1}$, which was lower than the previously reported value of 140.55 ng·g$^{-1}$ [22]. The highest concentration of organophosphates was 26.85 ng·g$^{-1}$, which was lower than the maximum residues in the paddy fields of northern Thailand (58.6 ng·g$^{-1}$) [41]. In paddy field soil, the fungicide detection rate was 50.6%, whereas in dry field soil, it was 22.1%.

In paddy field soil, the fungicide with the maximum concentration was tebuconazole (44.82 ng·g$^{-1}$). In contrast, the most prevalent fungicide in dry field soil was tricyclazole, with a concentration of 13.75 ng·g$^{-1}$. The data suggest that the frequency and concentration of fungicides detected in paddy field soil surpass those found in dry field soil. This discrepancy may be attributed to the humid conditions prevalent in paddy fields, which render them more prone to mildew.

### 3.2. Residual Characteristics of Contaminants in Surrounding Water

In the surrounding water samples, 48 pesticides and 3 degradation products were detected, of which 8 pesticides had a detection rate of 100%, namely atrazine, acetochlor, simetryn, prometryn, isoprothiolane, metolachlor, oxadiazon and fenoxanil. The total concentrations of contaminants and the types of chemical constituents in the water samples are shown in Figure 4. The total concentrations of the contaminants detected in the water samples ranged from 71.19 ng·L$^{-1}$ to 2883.94 ng·L$^{-1}$ in the sowing period, from 585.46 ng·L$^{-1}$ to 10,145.76 ng·L$^{-1}$ in the vegetative period and from 680.38 ng·L$^{-1}$ to 1178.93 ng·L$^{-1}$ in the maturity period. The concentration of residual contaminants in the vegetative period was higher than that in the sowing and maturity periods, which was consistent with increased pesticide use in the vegetative period.

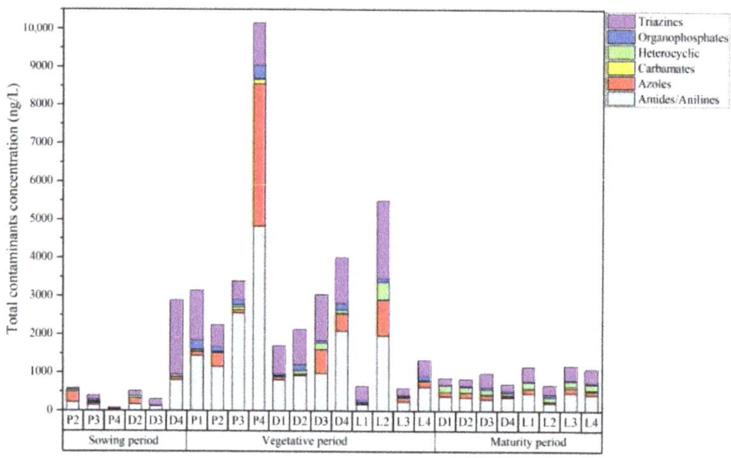


**Figure 4.** Total concentrations of contaminants and the types of chemical constituents in the surrounding water at different sampling points.

In the sowing period (Figure 5A), almost no pesticides were used. The mean total concentration of contaminants was 346.47 ng·L$^{-1}$ in paddy field water and 1233.50 ng·L$^{-1}$ in drainage water. Except for the D4 sampling points, the concentration of contaminants at other sites was minimal, which can be considered to be representative of the environmental background value. The maximum values of atrazine, prometryn, dimethazone and acetochlor were all detected in D4, and they totaled 1594.60 ng·L$^{-1}$, 132.83 ng·L$^{-1}$, 181.88 ng·L$^{-1}$ and 509.60 ng·L$^{-1}$. This may have been due to the fact that the D4 point was close to a corn field that was affected by the use of these substances or other human factors.

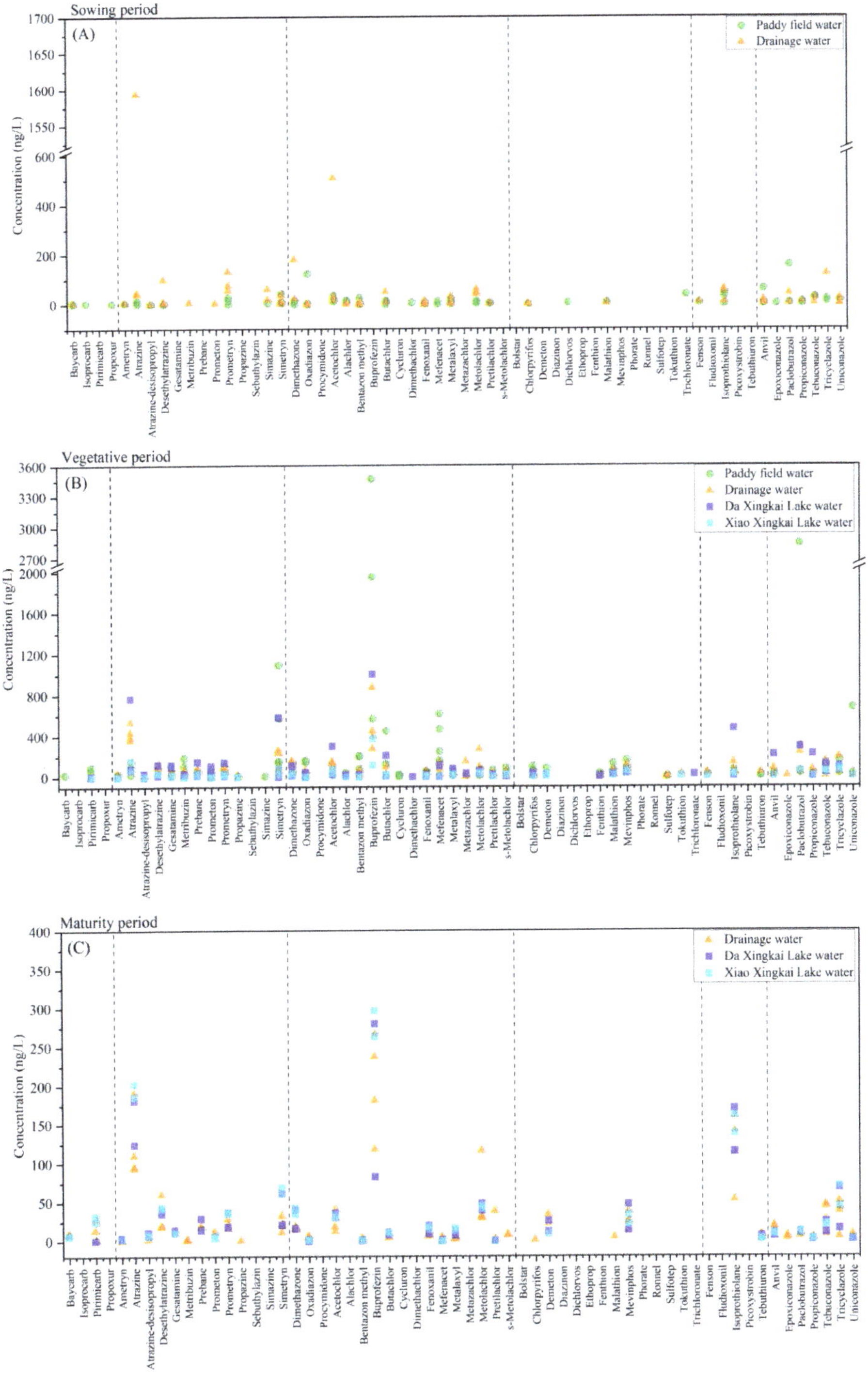


**Figure 5.** Concentrations of the 60 contaminants detected in the sowing (**A**), vegetative (**B**) and maturity (**C**) periods in surrounding water.

In the vegetative period (Figure 5B), the main contaminants were triazines, amides and azole, including seven pesticides: atrazine, simazine, acetochlor, buprofezin, butachlor, mefenacet and paclobutrazol. The mean concentrations across various samples were as follows: 4729.69 ng·L$^{-1}$ in paddy field water, 2717.84 ng·L$^{-1}$ in drainage water, 3067.80 ng·L$^{-1}$ in water from Da Xingkai Lake and 955.91 ng·L$^{-1}$ in water from Xiao Xingkai Lake. The highest concentration of contaminants was observed in paddy field water, a finding that can be attributed to the direct use of pesticides within these fields. According to the results of the study, buprofezin might be the main insecticide used in the Xingkai Lake area, with a mean concentration of 891.18 ng·L$^{-1}$. Buprofezin is typically applied during the summer months, specifically from late July to early August. It is plausible that our sampling occurred during the vegetative period, coinciding with the use of buprofezin. This could have led to elevated residual concentrations in the paddy field water. The primary herbicides detected in the study were simazine, butachlor and mefenacet. The mean concentrations of these herbicides in water were 203.60 ng·L$^{-1}$, 103.10 ng·L$^{-1}$ and 114.43 ng·L$^{-1}$, respectively. These herbicides were also highly prevalent in paddy field soil. During the vegetative period, the mean concentrations of atrazine and paclobutrazol in water were found to be 297.81 ng·L$^{-1}$ and 375.18 ng·L$^{-1}$, respectively.

In the maturity period (Figure 5C), the main contaminants were atrazine, buprofezin and isoprothiolane with mean concentrations of 148.76 ng·L$^{-1}$, 216.19 ng·L$^{-1}$ and 131.56 ng·L$^{-1}$. The mean total concentrations were 841.74 ng·L$^{-1}$ in drainage water, 914.53 ng·L$^{-1}$ in Da Xingkai Lake water and 1133.68 ng·L$^{-1}$ in Xiao Xingkai Lake water. To maintain pesticide residues in crop grains within safe limits, the use of pesticides is progressively halted during the early period prior to harvest. Consequently, the concentration of these residues diminishes during the maturity period compared with the vegetative period.

The use of pesticides on farmland has had a detrimental impact on Xingkai Lake, resulting in the detection of various contaminants in its water. In Da Xingkai Lake and Xiao Xingkai Lake water, 37 pesticides and 3 degradation products were detected. Atrazine, simetryn, buprofezin and isoprothiolane were the main detected pesticides. Atrazine is widespread in surface water in China; the concentration range was 66.80 ng·L$^{-1}$~772.15 ng·L$^{-1}$ in Xingkai Lake water, which was similar to the concentrations in the Songhua River and the Heilongjiang River of 844.9 ng·L$^{-1}$ and 606.5 ng·L$^{-1}$ [42]. The maximum concentration of simetryn was 581.80 ng·L$^{-1}$ and the median concentration was 61.80 ng·L$^{-1}$, while the maximum concentration of simetryn reported in surface water in Wuhan and the Liao-He River was only 10.7 ng·L$^{-1}$ [43,44], which is lower than that in this study. The mean concentration of acetochlor was 70.88 ng·L$^{-1}$, which is lower than that reported in the major growing areas in China (4.3 μg·L$^{-1}$) [8]. Buprofezin was detected with a mean concentration of 301.87 ng·L$^{-1}$, and its maximum concentration reached 1003.44 ng·L$^{-1}$ in the vegetative period. In the Taige Canal basin, buprofezin concentrations ranged from 56 ng·L$^{-1}$ to 390 ng·L$^{-1}$, averaging at 218 ng·L$^{-1}$ [45]. This finding aligns with the results obtained in this study. The contamination of isoprothiolane was notably high during the maturity period, with a mean concentration of 129.66 ng·L$^{-1}$ in the vegetative period and an increase to 146.38 ng·L$^{-1}$ in the maturity period. Wang also reported [11] a higher concentration of isoprothiolane (186 ng·L$^{-1}$) in September, which is consistent with the findings of this research.

Xingkai Lake and Taihu Lake, both freshwater bodies, are significantly impacted by the use of pesticides on farmland. Upon comparing the pesticide levels in Xingkai Lake to those in Taihu Lake, it was observed that the concentration of butachlor was markedly higher than in Taihu Lake. The peak concentration of butachlor in Taihu Lake was recorded at 40 ng·L$^{-1}$, with a mean concentration of 1.5 ng·L$^{-1}$ [11]. In contrast, Xingkai Lake exhibited a maximum concentration of 209.22 ng·L$^{-1}$ for butachlor, accompanied by a mean concentration of 36.58 ng·L$^{-1}$. There were also significant differences between the residues of dichlorvos in Xingkai Lake and in Taihu Lake. Wang [11] reported that the highest average concentration of dichlorvos was measured at 4.2 ng·L$^{-1}$ with a maximum value of 263 ng·L$^{-1}$ in Taihu Lake; Zhou [46] reported that the mean value of dichlorvos in the Meiliangwan Bay of Taihu Lake was 51.6 ng·L$^{-1}$; however, no dichlorvos was detected in Xingkai Lake during this study. These factors could be associated with the cultivated crops and the variations in pesticide usage between the northern and southern regions [8]. Interestingly, a lake in Japan had similar findings to Xingkai Lake. Lake Biwa, as the largest lake in Japan [41], is also heavily impacted by surrounding agricultural activities. In Japan, simetryn and isoprothiolane are also applied in paddy fields. In Lake Biwa, both simetryn and isoprothiolane had high detection rates with the maximum concentrations of 0.41 μg·L$^{-1}$ and 0.15 μg·L$^{-1}$ [41], while dichlorvos was also not detected in the lake. The results of this study were similar to those in Xingkai Lake, possibly because the pesticides used in these two areas are similar.

The proportions of the different types of contaminants in the surrounding water in different periods are illustrated in Figure 6. The concentration of carbamate pesticides was found to be the lowest, potentially due to their high toxicity and thus their being gradually phased out [47]. Triazines and amides were the main detected pesticides in the surrounding water. During both the vegetative and maturity periods, the proportions of various structural contaminants were comparable in both drainage and Xingkai Lake water. The proportions of amide pesticides in drainage, Da Xingkai Lake and Xiao Xingkai Lake water during the maturity period were 40.7%, 39.1% and 36.1%, respectively. The distribution of heterocyclic pesticides exhibited similarity across drainage and Xingkai Lake water. We conducted a Spearman correlation analysis of the contaminants in the drainage and Xingkai Lake water using Origin 2019 as our analytical tool. The results of this correlation, represented by the correlation coefficient (ρ), are shown in Table 2. The results showed that the distribution characteristics of the contaminants could have a correlation in the drainage, Da Xingkai Lake and Xiao Xingkai Lake water ($\rho > 0.8$). Furthermore, the correlation was more pronounced during the maturity period compared to the vegetative period. This suggested that the source of these contaminants in the drainage and Xingkai Lake water could be the same. Similar findings were observed in Taihu Lake, where the cumulative concentration of pesticides detected in its northwest region mirrored that of the upstream river. This suggests that the catchment encompassed by the study area could potentially be the primary source of the pesticides in the northwest region of Taihu Lake [11].

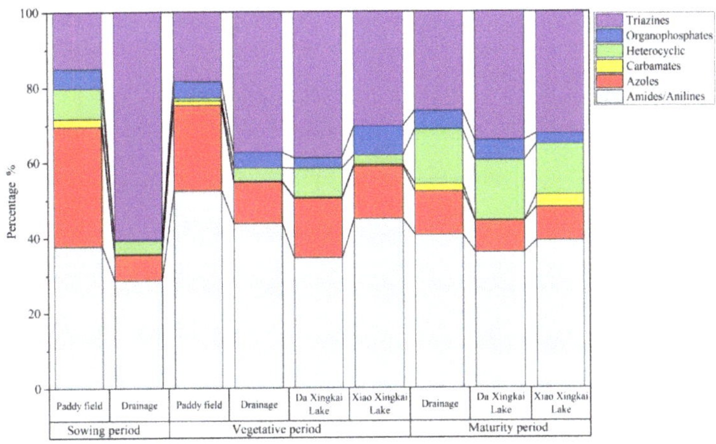


**Figure 6.** Proportion of contaminants with different types of chemical constituents in surrounding water.

**Table 2.** Correlation coefficient of the contaminants in drainage, Da Xingkai Lake and Xiao Xingkai Lake water.

| | Vegetative Period | | | Maturity Period | | |
|---|---|---|---|---|---|---|
| | Drainage Water | Da Xingkai Lake Water | Xiao Xingkai Lake Water | Drainage Water | Da Xingkai Lake Water | Xiao Xingkai Lake Water |
| Drainage water | 1 | | | 1 | | |
| Da Xingkai Lake water | 0.88242 | 1 | | 0.8979 | 1 | |
| Xiao Xingkai Lake water | 0.86558 | 0.80537 | 1 | 0.89893 | 0.91036 | 1 |

*3.3. Ecological Risk Assessment*

To evaluate the possible toxic effects of contaminants on aquatic organisms, an ecological risk assessment was carried out on Da Xingkai Lake and Xiao Xingkai Lake water during different periods. This investigation scrutinized the ecological risks posed by 24 types of highly concentrated and frequently detected pesticides. The findings are presented in Figure 7. The summaries of the regression parameters and the regression coefficient of the SSD curve are shown in Figure S1 and Table S3. The PAF values in the vegetative period were higher than in the maturity period. The maximum PAF values of atrazine, chlorpyrifos and prometryn were higher than 5% in the vegetative period, totaling 5.13%, 7.40% and 5.09%, respectively, so atrazine, chlorpyrifos and prometryn could constitute significant ecological risks. In Taihu Lake, the maximum concentration of atrazine was 614 ng·$L^{-1}$ [46]. When this maximum value was incorporated into our formula, it yielded a PAF value of 4.75%, which is nearly equivalent to 5%. This suggests potential risks to aquatic organisms. C.S. Qu [48] also reported atrazine as one of the greatest hazards to the species in Taihu Lake wetlands. The pesticide metribuzin exhibited a maximum PAF of 4.66%, which was proximate to the threshold of 5%, indicating that it warrants attention. In the maturity period, the PAF values of atrazine and prometryn were still higher than those of other contaminants. Although they did not surpass 5% (3.29% and 3.66%, respectively), these levels still pose a certain risk to the aquatic ecological environment. These results indicated that the harm of atrazine and prometryn to aquatic organisms may exist for a long time Xingkai Lake. To sum up, attention should be paid to the toxic effects of atrazine, chlorpyrifos and prometryn on aquatic organisms.

The aquatic life benchmarks for atrazine, chlorpyrifos and prometryn in freshwater vertebrates and invertebrates were retrieved from USEPA [49]. Detailed information is shown in Table 3. In Xingkai Lake water, the maximum concentrations of atrazine, chlorpyrifos and prometryn were 772.15 ng·$L^{-1}$, 29.71 ng·$L^{-1}$ and 144.00 ng·$L^{-1}$. Both atrazine and prometryn concentrations were below the acute freshwater invertebrate and vertebrate benchmarks, but the chlorpyrifos concentration was above the acute freshwater invertebrate benchmark. This implied that the residual concentration of chlorpyrifos in Xingkai Lake water poses a risk to freshwater invertebrates. Atrazine and prometryn did not present a risk based on the EPA benchmarks, probably due to the fact that the benchmarks are pending an update.

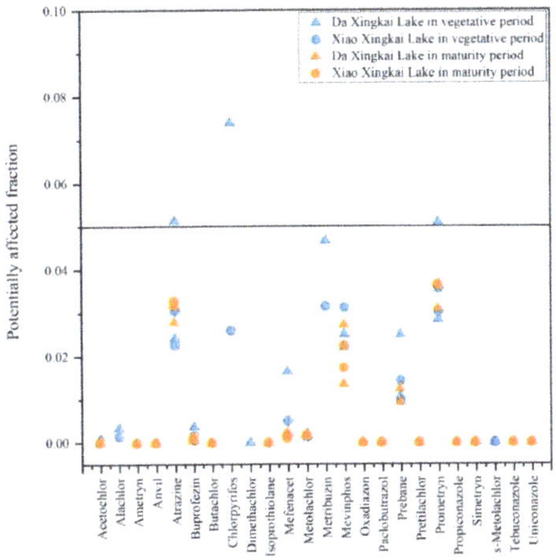


**Figure 7.** Potentially affected fraction of 24 pesticides in Xingkai Lake water.

**Table 3.** Acute aquatic life benchmarks of USEPA.

| Pesticide | Year Updated | Freshwater Vertebrate (μg/L) | Freshwater Invertebrates (μg/L) |
|---|---|---|---|
| Atrazine | 2016 | 2650 | 360 |
| Chlorpyrifos | 2022 | 0.85 | 0.0069 |
| Prometryn | 2014 | 1455 | 4850 |

To better illustrate the characteristics of high-toxicity pesticide (atrazine, chlorpyrifos and prometryn) residues in Xingkai Lake water, we evaluated the proportion of high-toxicity pesticides to total pesticide concentrations. The results are shown in Figure 8. The proportion of high-toxicity pesticides to total pesticide concentrations ranged from 14.71% to 20.43%, with atrazine being the most dominant at 11.97% to 17.18% of the total concentrations. The highest ecological risk was for chlorpyrifos, but the proportion of the maximum concentrations was only 0.48%. Therefore, the residual pesticides in Xingkai Lake water are mainly low-toxicity pesticides.

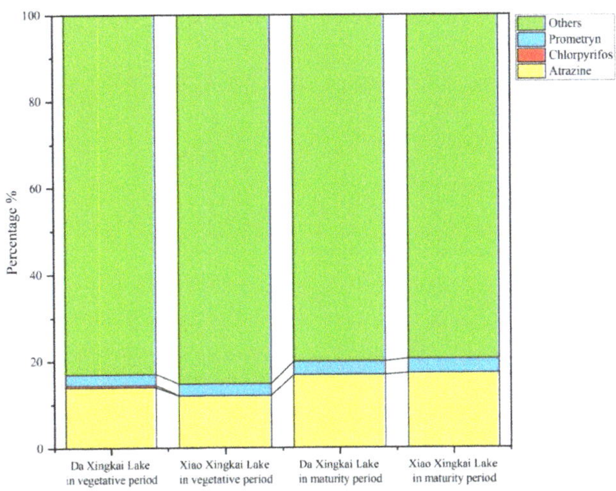


**Figure 8.** The proportion of high-risk pesticide concentrations in Xingkai Lake.

## 4. Conclusions

In the farmland soil of the Xingkai Lake area, 43 pesticides and 3 degradation products were detected. In paddy field soil, 40 pesticides were detected, while in dry field soil, 34 pesticides were identified. The main pesticides were atrazine and acetochlor in dry field soil, and oxadiazon, mefenacet and chlorpyrifos in paddy field soil.

In Xingkai Lake water, atrazine, simetryn, buprofezin and isoprothiolane were the main pesticides among 37 pesticides detected. The pesticides and degradation products detected in Xingkai Lake water were detected in both paddy field water and drainage water. The distribution characteristics of contaminants in the drainage water, Da Xingkai Lake and Xiao Xingkai Lake water could have a correlation, and so the sources could be the same. In the vegetative period, atrazine, chlorpyrifos and prometryn could pose an environmental risk to aquatic organisms in Xingkai Lake water.

The use of pesticides on farmlands may cause pesticide residues in Xingkai Lake water. Over time, the aquatic environment of Xingkai Lake may be seriously affected and even cause species death, which needs to be noticed.

**Supplementary Materials:** The following supporting information can be downloaded at: https://www.mdpi.com/article/10.3390/toxics12010085/s1, Figure S1: Species sensitivity distribution (SSD) of 24 pesticides in Xingkai Lake water; Figure S2: The results of extraction solvent optimization; Figure S3 The results of elution solvent optimization; Table S1: Details of sampling sites; Table S2: Details, quantitative and qualitative ion (m/z), retention time, regression curve parameter, mean recoveries, the limit of detection (LOD), the limit of quantitation (LOQ) and supply company of 57 pesticides and three degradation products; Table S3: Summary of regression parameters and regression coefficient of species sensitivity distribution (SSD); Table S4: The mean concentrations of the 60 contaminants detected in sowing, vegetative and maturity periods in surrounding water ($ng{\cdot}L^{-1}$); Table S5: The mean concentrations of the 60 contaminants detected in dry and paddy field soil ($ng{\cdot}g^{-1}$); Table S6: Octanol-water partition coefficient of 57 pesticides and 3 degradations.

**Author Contributions:** Methodology and Supervision, D.W.; Conceptualization, Methodology, Data Collection and Writing—Original Draft Preparation, W.W.; Data Collection, Q.L.; Data Collection, L.L.; Data Collection, Y.X.; Data Collection, C.D. All authors have read and agreed to the published version of the manuscript.

**Funding:** This research was funded by the Strategic Priority Research Program of the Chinese Academy of Sciences grant number (Grant No. XDA28100503); National Natural Science Foundation of China grant number (42077157) and National Key R&D Program of China grant number (2021YFC3200802-02).

**Institutional Review Board Statement:** Not applicable.

**Informed Consent Statement:** Not applicable.

**Data Availability Statement:** Data are contained within the article and supplementary materials. Other data will be made available on request.

**Conflicts of Interest:** The authors declare that they have no known competing financial interests.

## References

1. Yao, Y.; Tuduri, L.; Harner, T.; Blanchard, P.; Waite, D.; Poissant, L.; Murphy, C.; Belzer, W.; Aulagnier, F.; Li, Y.-F.; et al. Spatial and temporal distribution of pesticide air concentrations in Canadian agricultural regions. *Atmos. Environ.* **2006**, *40*, 4339–4351. [CrossRef]
2. Boyes, W.K.; Padilla, S.; Tandon, P.; Barone, S., Jr. Effects of organophosphates on the visual system of rats. *J. Appl. Toxicol.* **1994**, *14*, 135–143. [CrossRef] [PubMed]
3. Huen, K.; Bradman, A.; Harley, K.; Yousefi, P.; Boyd Barr, D.; Eskenazi, B.; Holland, N. Organophosphate pesticide levels in blood and urine of women and newborns living in an agricultural community. *Environ. Res.* **2012**, *117*, 8–16. [PubMed]
4. Sathiakumar, N.; MacLennan, P.A.; Mandel, J.; Delzell, E. A review of epidemiologic studies of triazine herbicides and cancer. *Crit. Rev. Toxicol.* **2011**, *41* (Suppl. 1), 1–34. [CrossRef]
5. Li, W.; Zha, J.; Li, Z.; Yang, L.; Wang, Z. Effects of exposure to acetochlor on the expression of thyroid hormone related genes in larval and adult rare minnow (*Gobiocypris rarus*). *Aquat. Toxicol.* **2009**, *94*, 87–93. [PubMed]

6. Tu, W.; Niu, L.; Liu, W.; Xu, C. Embryonic exposure to butachlor in zebrafish (*Danio rerio*): Endocrine disruption, developmental toxicity and immunotoxicity. *Ecotoxicol. Environ. Saf.* **2013**, *89*, 189–195.
7. Lin, Z.; Zhen, Z.; Liang, Y.; Li, J.; Yang, J.; Zhong, L.; Zhao, L.; Li, Y.; Luo, C.; Ren, L.; et al. Changes in atrazine speciation and the degradation pathway in red soil during the vermiremediation process. *J. Hazard. Mater.* **2019**, *364*, 710–719.
8. Wang, W.; Man, Y.; Xie, J.; Zhang, Z.; Wang, P.; Liu, X. Occurrence and risk assessment of three chloroamide herbicides in water and soil environment in Northeastern, Eastern and Southern China. *Environ. Res.* **2023**, *219*, 115104. [PubMed]
9. Arias-Estévez, M.; López-Periago, E.; Martínez-Carballo, E.; Simal-Gándara, J.; Mejuto, J.-C.; García-Río, L. The mobility and degradation of pesticides in soils and the pollution of groundwater resources. *Agric. Ecosyst. Environ.* **2008**, *123*, 247–260. [CrossRef]
10. Syafrudin, M.; Kristanti, R.A.; Yuniarto, A.; Hadibarata, T.; Rhee, J.; Al-Onazi, W.A.; Algarni, T.S.; Almarri, A.H.; Al-Mohaimeed, A.M. Pesticides in drinking water—A review. *Int. J. Environ. Res. Public Health* **2021**, *18*, 468. [CrossRef] [PubMed]
11. Wang, T.; Zhong, M.; Lu, M.; Xu, D.; Xue, Y.; Huang, J.; Blaney, L.; Yu, G. Occurrence, spatiotemporal distribution, and risk assessment of current-use pesticides in surface water: A case study near Taihu Lake, China. *Sci. Total Environ.* **2021**, *782*, 146826. [CrossRef]
12. Wang, L.; Deng, Y.; Luo, Y.; Chen, M.; Liu, Y.; Cai, X.; Hou, S.; Guo, C. Occurrence and risk assessment of typical neonicotinoid pesticides in Lake Poyang Basin. *J. Lake Sci.* **2023**, *35*, 909–921.
13. Zhi, H.; Zhao, Z.; Zhang, L. The fate of polycyclic aromatic hydrocarbons (PAHs) and organochlorine pesticides (OCPs) in water from Poyang Lake, the largest freshwater lake in China. *Chemosphere* **2015**, *119*, 1134–1140. [CrossRef] [PubMed]
14. Sudo, M.; Kawachi, T.; Hida, Y.; Kunimatsu, T. Spatial distribution and seasonal changes of pesticides in Lake Biwa, Japan. *Limnology* **2004**, *5*, 77–86. [CrossRef]
15. Singh, K.P.; Malik, A.; Sinha, S. Persistent organochlorine pesticide residues in soil and surface water of northern Indo-Gangetic alluvial plains. *Environ. Monit. Assess.* **2007**, *125*, 147–155.
16. Sun, W.; Zhang, E.; Liu, E.; Chang, J.; Chen, R.; Shen, J. Glacial-interglacial vegetation changes in northeast China inferred from isotopic composition of pyrogenic carbon from Lake Xingkai sediments. *Org. Geochem.* **2018**, *121*, 80–88. [CrossRef]
17. Zhu, L.; Zhou, H.; Xie, X.; Li, X.; Zhang, D.; Jia, L.; Wei, Q.; Zhao, Y.; Wei, Z.; Ma, Y. Effects of floodgates operation on nitrogen transformation in a lake based on structural equation modeling analysis. *Sci. Total Environ.* **2018**, *631–632*, 1311–1320. [CrossRef] [PubMed]
18. Yang, Q.; Huang, X.; Wen, Z.; Shang, Y.; Wang, X.; Fang, C.; Song, K. Evaluating the spatial distribution and source of phthalate esters in the surface water of Xingkai Lake, China during summer. *J. Great Lakes Res.* **2021**, *47*, 437–446. [CrossRef]
19. Department of Ecology and Environment of Heilongjiang Provincial (Ed.) *Ecological and Environmental Status Bulletin of Heilongjiang Province in 2022*; Department of Ecology and Environment of Heilongjiang Provincial: Harbin, China, 2023. Available online: http://sthj.hlj.gov.cn/sthj/c111964/202306/c00_31639046.shtml (accessed on 8 December 2023).
20. Ji, X.; Liu, T.; Liu, J.; Li, J.; Pan, B. Investigation and study on water quality and pollution condition in Lake Xingkai of China. *Environ. Monit. China* **2013**, *29*, 79–84.
21. Yu, S.; Li, X.; Wen, B.; Chen, G.; Hartleyc, A.; Jiang, M.; Li, X. Characterization of water quality in Xiao Xingkai Lake: Implications for trophic status and management. *Chin. Geogr. Sci.* **2021**, *31*, 558–570.
22. Yu, X.; Zheng, S.; Zheng, M.; Ma, X.; Wang, G.; Zou, Y. Herbicide accumulations in the Xingkai lake area and the use of restored wetland for agricultural drainage treatment. *Ecol. Eng.* **2018**, *120*, 260–265. [CrossRef]
23. Piao, D.; Wang, F. Environmental conditions and the protection countermeasures for waters of Lake Xingkai. *J. Lake Sci.* **2011**, *23*, 196–202.
24. Fisher, J.A.; Scarlett, M.J.; Stott, A.D. Accelerated solvent extraction: An evaluation for screening of soils for selected U.S. EPA semivolatile organic priority pollutants. *Environ. Sci. Technol.* **1997**, *31*, 1120–1127.
25. Gan, J.; Papiernik, S.K.; Koskinen, W.C.; Yates, S.R. Evaluation of accelerated solvent extraction (ASE) for analysis of pesticide residues in soil. *Environ. Sci. Technol.* **1999**, *33*, 3249–3253. [CrossRef]
26. Di, S.; Shi, S.; Xu, P.; Diao, J.; Zhou, Z. Comparison of different extraction methods for analysis of 10 organochlorine pesticides: Application of MAE–SPE method in soil from Beijing. *Bull. Environ. Contam. Toxicol.* **2015**, *95*, 67–72.
27. Hubert, A.; Wenzel, K.-D.; Manz, M.; Weissflog, L.; Engewald, W.; Schüürmann, G. High extraction efficiency for POPs in real contaminated soil samples using accelerated solvent extraction. *Anal. Chem.* **2000**, *72*, 1294–1300.
28. Bu, Q.; Wang, D.; Wang, Z.; Gu, J. Identification and ranking of the risky organic contaminants in the source water of the Danjiangkou reservoir. *Front. Environ. Sci. Eng.* **2014**, *8*, 42–53. [CrossRef]
29. Xu, H.; Liu, Y.; Xu, X.; Lan, H.; Qi, W.; Wang, D.; Liu, H.; Qu, J. Spatiotemporal variation and risk assessment of phthalate acid esters (PAEs) in surface water of the Yangtze River Basin, China. *Sci. Total Environ.* **2022**, *836*, 155677. [CrossRef]
30. Li, Q.; Xu, X.; Fang, Y.; Xiao, R.; Wang, D.; Zhong, W. The temporal changes of the concentration level of typical toxic organics in the river sediments around Beijing. *Front. Environ. Sci. Eng.* **2018**, *12*, 8. [CrossRef]
31. Wang, X.; Yan, Z.; Liu, Z.; Zhang, C.; Wang, W.; Li, H. Comparison of species sensitivity distributions for species from China and the USA. *Environ. Sci. Pollut. Res.* **2013**, *21*, 168–176. [CrossRef]
32. Kooijman, S.A.L.M. A safety factor for LC50 values allowing for differences in sensitivity among species. *Water Res.* **1987**, *21*, 269–276. [CrossRef]

33. Nagai, T. Studies on ecological risk assessment of pesticide using species sensitivity distribution. *J. Pestic. Sci.* **2017**, *42*, 124–131. [CrossRef]
34. Chen, L. A conservative, nonparametric estimator for the 5th percentile of the species sensitivity distributions. *J. Stat. Plan. Inference* **2004**, *123*, 243–258. [CrossRef]
35. Hose, G.C.; Van den Brink, P.J. Confirming the species-sensitivity distribution concept for endosulfan using laboratory, mesocosm, and field data. *Arch. Environ. Contam. Toxicol.* **2004**, *47*, 511–520. [PubMed]
36. Feng, C.L.; Wu, F.C.; Dyer, S.D.; Chang, H.; Zhao, X.L. Derivation of freshwater quality criteria for zinc using interspecies correlation estimation models to protect aquatic life in China. *Chemosphere* **2013**, *90*, 1177–1183. [CrossRef]
37. Rico, A.; de Oliveira, R.; Silva de Souza Nunes, G.; Rizzi, C.; Villa, S.; De Caroli Vizioli, B.; Montagner, C.C.; Waichman, A.V. Ecological risk assessment of pesticides in urban streams of the Brazilian Amazon. *Chemosphere* **2022**, *291*, 132821. [CrossRef] [PubMed]
38. Liu, Z.; Wang, Y.; Zhu, Z.; Yang, E.; Feng, X.; Fu, Z.; Jin, Y. Atrazine and its main metabolites alter the locomotor activity of larval zebrafish (*Danio rerio*). *Chemosphere* **2016**, *148*, 163–170.
39. Wang, X.; Liu, Q. Spatial and temporal distribution characteristics of triazine herbicides in typical agricultural regions of Liaoning, China. *Bull. Environ. Contam. Toxicol.* **2020**, *105*, 899–905. [CrossRef]
40. Sun, X.; Zhou, Q.; Ren, W. Herbicide occurrence in riparian soils and its transporting risk in the Songhua River Basin, China. *Agron. Sustain. Dev.* **2013**, *33*, 777–785.
41. Khammanee, N.; Qiu, Y.; Kungskulniti, N.; Bignert, A.; Meng, Y.; Zhu, Z.; Lekew Teffera, Z. Presence and health risks of obsolete and emerging pesticides in paddy rice and soil from Thailand and China. *Int. J. Environ. Res. Public. Health* **2020**, *17*, 3786. [CrossRef]
42. Xu, X.; Li, C.; Sun, J.; Wang, H.; Wang, D.; Song, H.; Wang, Z. Residue characteristics and ecological risk assessment of twenty-nine pesticides in surface water of Major River-Basin in China. *Asian J. Ecotoxicol.* **2016**, *11*, 347–354.
43. Pan, X.; Xu, L.; He, Z.; Wan, Y. Occurrence, fate, seasonal variability, and risk assessment of twelve triazine herbicides and eight related derivatives in source, treated, and tap water of Wuhan, Central China. *Chemosphere* **2023**, *322*, 138158. [CrossRef] [PubMed]
44. Gfrerer, M.; Martens, D.; Gawlik, B.M.; Wenzl, T.; Zhang, A.; Quan, X.; Sun, C.; Chen, J.; Platzer, B.; Lankmayr, E.; et al. Triazines in the aquatic systems of the Eastern Chinese Rivers Liao-He and Yangtse. *Chemosphere* **2002**, *47*, 455–466. [CrossRef] [PubMed]
45. Cheng, X.; Jiang, X.; Liu, L.; Huang, J. Analysis of emerging contaminants in surface water, aquaculture ponds and wastewater treatment facilities in the Taige Canal Basin. *Chem. Res. Chin. Univ.* **2023**, *39*, 516–524.
46. Na, T.; Fang, Z.; Zhanqi, G.; Ming, Z.; Cheng, S. The status of pesticide residues in the drinking water sources in Meiliangwan Bay, Taihu Lake of China. *Environ. Monit. Assess.* **2006**, *123*, 351–370. [CrossRef]
47. Zheng, S.; Chen, B.; Qiu, X.; Chen, M.; Ma, Z.; Yu, X. Distribution and risk assessment of 82 pesticides in Jiulong River and estuary in South China. *Chemosphere* **2016**, *144*, 1177–1192. [CrossRef]
48. Qu, C.S.; Chen, W.; Bi, J.; Huang, L.; Li, F.Y. Ecological risk assessment of pesticide residues in Taihu Lake wetland, China. *Ecol. Model.* **2011**, *222*, 287–292. [CrossRef]
49. USEPA Pesticide Programs' Aquatic Life Benchmarks. Available online: https://www.epa.gov/pesticide-science-and-assessing-pesticide-risks/aquatic-life-benchmarks-and-ecological-risk (accessed on 8 December 2023).

toxics

MDPI

Review

# A Review of the Distribution and Health Effect of Organophosphorus Flame Retardants in Indoor Environments

Xingwei Song [1], Sheng Zhu [2], Ling Hu [1], Xiaojia Chen [3,4], Jiaqi Zhang [5], Yi Liu [6], Qingwei Bu [7,*] and Yuning Ma [8,*]

1 Jiangsu Environmental Monitoring Centre, Nanjing 210019, China; sxw@jshb.gov.cn (X.S.); hul@jshb.gov.cn (L.H.)
2 Quzhou Environmental Monitoring Centre, Quzhou 324000, China; wmaxwell@163.com
3 School of Environment and Architecture, University of Shanghai for Science and Technology, Shanghai 200093, China; chenxiaojia@usst.edu.cn
4 State Environmental Protection Key Laboratory of Formation and Prevention of Urban Air Pollution Complex, Shanghai Academy of Environment Sciences, Shanghai 200233, China
5 School of Environmental Science and Engineering, Shanghai Jiao Tong University, Shanghai 200240, China; zjq560047@163.com
6 Thomas Gosnell School of Life Sciences, Rochester Institution of Technology Rochester, New York, NY 14623, USA; yl5329@rit.edu
7 School of Chemical & Environmental Engineering, China University of Mining & Technology, Beijing 100083, China
8 College of Environmental and Resource Sciences, Zhejiang University, Hangzhou 310058, China
* Correspondence: qingwei.bu@cumtb.edu.cn (Q.B.); julius.yuningma@gmail.com (Y.M.)

**Abstract:** As a replacement for polybrominated diphenyl ethers (PBDEs), organophosphorus flame retardants (OPFRs) have been widely used and detected in different indoor environments all over the world. This paper comprehensively describes the concentration levels and distribution information of 11 kinds of OPFRs from 33 indoor dust and 10 air environments, from which TBOEP, TCIPP, and TDCIPP were observed to have higher concentrations in indoor environments. The ΣOPFRs displayed higher concentrations in indoor dust than in indoor air due to the higher molecular weight and vapor pressure of ΣOPFRs in building decoration materials, specifically for TCIPP and TDCIPP compounds. Considering that it is inevitable that people will be exposed to these chemicals in the indoor environments in which they work and live, we estimated their potential health risks through three human exposure pathways and found that the ingestion exposure to TBOEP for toddlers in Japan may reach up to 1270.80 ng/kg/day, which comprises a significant pathway compared to dermal contact and indoor air inhalation. Specifically, the combined total exposure to OPFRs by air inhalation, dust ingestion, and dermal contact was generally below the RfD values for both adults and toddlers, with a few notable higher exposures of some typical OPFRs.

**Keywords:** organophosphorus flame retardants; indoor environment; human exposure; risk assessment

## 1. Introduction

Flame retardants (FR) are chemicals that are added to various types of consumer goods and materials in order to prevent combustion and the spread of fire [1]. Since the 1970s, polybrominated diphenyl ethers (PBDEs) have been widely applied to consumer goods, as they have a strong flame-retardant property. However, some of them—such as penta-BDE, octa-BDE, and deca-BDE—were mostly banned by the European Union in 2008 due to their persistence, bioaccumulation, and toxicity. Then, as an alternative FR chemical, the emerging halogenated flame retardants (EHFR) were developed. Among the EHFRs, organophosphorus flame retardants (OPFRs), such as tris(2-chloroethyl) phosphate (TCEP) and tris(chloropropyl) phosphate (TCPP), are a group of alternative compounds that have been widely used in the global market to meet the needs of FR products [2].

Many kinds of indoor furniture, electric devices, and plastic products (e.g., sofas, furniture foam, carpets, mattresses, televisions, computers, mobile phones, upholstery, and textiles) may comprise and release a large amount of OPFRs [3,4]. For instance, the production of OPFRs in China reached 70,000 tons in 2007 [5], and the demand for emerging FRs has increased due to the restrictions on legacy PBDEs in the U.K. [6]. OPFRs frequently appear as additives rather than being chemically bonded to the final diverse products, which makes them easy to release [5,7–9]. OPFRs may leach by abrasion, volatilization, and directly contacted migration from the indoor products to their indoor environments [10]. The total concentrations of organic phosphate esters (OPEs) were 2–3 and 1–2 orders of magnitude higher than those of brominated flame retardants (BFRs) in air and dust/window films, respectively, in Canada, the USA, and the Czech Republic [11,12]. From 2012 to 2017, tributoxyethyl phosphate (TBEP) was the major component of the OPFR group in indoor dust environments in China [13–15]. The occurrence of OPFRs in indoor air and dust has been studied deeply in Europe [16,17]. In recent decades, global concern regarding the occurrence of OPFRs has been dramatically increasing, and OPFRs have been detected in more indoor sources and building materials such as beds, car seats, computer screens, TV sets, window film, and upholstery materials [18] Although outdoor pollution can affect indoor air quality, OPFRs are still present at significantly higher concentrations in closed environments than in outdoor areas [19]. Furthermore, some of them have low degradation potential and, as a result, they might be persistent in indoor environments [1]. Meanwhile, OPFRs have been found to be widely distributed in different environmental media (e.g., dust, air, water, soil), as well as in human serum and breast milk [1,20,21]. These studies reported the concentration profiles of OPFRs emitted from different indoor materials. As such, using methods such as principal components analysis (PCA) may help to accurately assess the indoor contamination source and patterns of OPFRs.

At present, many people usually spend over 20 h per day in indoor environments, making exposure to indoor pollution unavoidable in daily life [22]. Ingestion, inhalation, and dermal contact are three important absorptive routes for people to come into contact with these chemicals in indoor environments [23–25]. Regarding exposure to OPFRs, there are two routes: external exposure routes and internal exposure routes. External exposure includes exposures through dietary ingestion, dust ingestion, and air inhalation [26]. Daily intake of seafood and meat led to the exposure of humans to FRs through bioaccumulation, including OPFR metabolites such as DPP and BDCPP [27,28]. In addition, FRs include endocrine-disrupting chemicals (EDCs) that could lead to reproductive disorders and endocrine-related cancers and increase the prevalence of obesity. For instance, Cl-PFRs such as TCIPP, TDCIPP, and TCEP have been found to be carcinogenic and have adverse effects on human health, and TCrP and TNBP pose a potential threat as they may cause thyroid hormone disruption and reproductive toxicity [1,5,9]. Many studies have focused on the acute and long-term toxicity of OPFRs with respect to fish, rats, daphnia, and algae. Du et al. (2015) have found that Aryl-OPFRs, such as TPHP, could cause higher heart toxicity than alkyl-OPFRs by interfering with the expression of transcription regulators in zebrafish [29]. Toxicity data for OPFRs are incomplete, and most of them were obtained through animal studies, making it difficult to connect the toxicity associated with human exposure. Little is known about the combined environmental effect of OPFRs based on indoor concentration profiles and toxicity information. Hence, we plan to use meta-analysis to analyze the connections among their occurrence, concentration, and toxicity data.

This review summarizes the occurrences of and human exposure to OPFRs in indoor microenvironments and establishes an evaluative function that indicates the relationship among these parameters and variables. According to this evaluative function, the pollutant equivalency factors were calculated to conduct an overall assessment of the concentration level and toxicity of OPFRs. Then, several typical OPFRs measured in the indoor environments of different countries are further studied. Finally, the toxicity of PEFs is calculated, providing a basis for understanding which OPFRs should be targeted for control or intervention. To the best of our knowledge, this is the first review reporting an overall evaluation of OPFRs based on their global concentration levels and health effects in indoor environments.

## 2. Materials and Methods

### 2.1. Inclusion Criteria

This review aims to incorporate most of the studies that have been carried out describing the occurrence and distribution of organophosphorus flame retardants in indoor environments and their health effects, including human exposure and toxicity. Based on published literature from the Web of Science Core Collection and the Chinese Science Citation Database, studies were obtained by searching for "organophosphorus flame retardants" or "tris(chloropropyl) phosphate" in the title and "emerging pollutants", "indoor environment", or "phosphorus flame retardant" in the topic. Then, the irrelevant literature was eliminated by reading the titles and abstracts, and we supplemented our literature database by reading the references of the selected studies. The studies had to refer to organophosphorus flame retardants in indoor environments, including living houses and workplaces, through dust and air (as such, some studies referring to OPFRs in cars and building decoration materials were included). Studies reporting risk assessments of OPFRs regarding human exposure and toxicity were also included.

### 2.2. Search Strategy Description

A comprehensive literature search was performed in the bibliographic databases Web of Science Core Collection and SCI Finder, covering studies published using the following keywords (and the combinations of them): (a) regarding organophosphorus flame retardants and indoor environments, "PFRs" (or "phosphorus flame retardant") and "OPFRs" (or "organophosphorus flame retardants") and "indoor" (or "indoor environments" or "indoor air" or "indoor dust" or "building materials"); (b) regarding OPFRs and human exposure to them, "organophosphorus flame retardants" and "human exposure" or "health effect" or "risk assessment"; (c) regarding OPFRs and their toxicity description, "organophosphorus flame retardants" and "toxicity". Attention was paid to the data from different studies, as the conversion of units may be necessary for comparison. Table 1 provides the compound name, CAS number, abbreviation, molar mass, water solubility, vapor pressure, LogKoa, logKow, and instrumental identification ions or $m/z$ of the OPFRs mentioned in this review. Figure 1 presents a schematic presentation of this review on OPFRs in indoor environments.

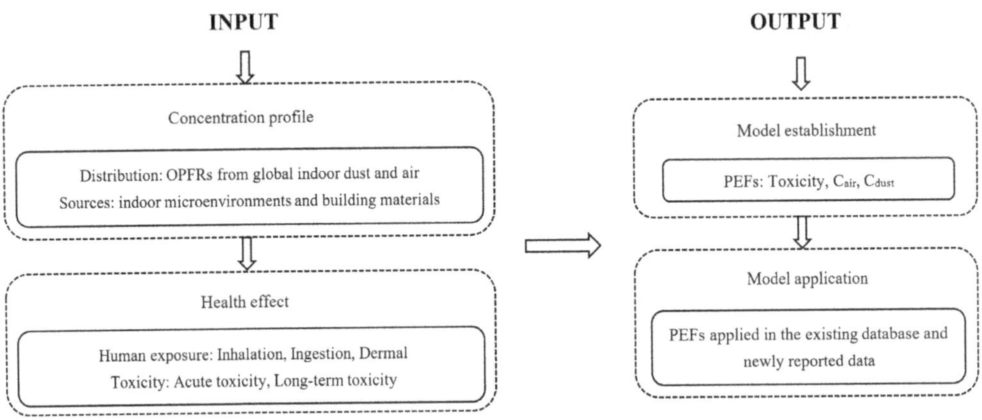


**Figure 1.** A schematic presentation of this review on OPFRs in indoor environments.

**Table 1.** Compound names, abbreviations (alternate names), CAS numbers, and other properties of OPFRs mentioned in this manuscript.

| Name | Abbreviations | Chemical Formula | CAS Number | Molecular Weight | Water Solubility (mg/L) at 25 °C [28,30] | Vapor Pressure (Torr) [16,20] | LogKoa (25 °C) [31] | LogKow (n-Octanol/Water Partition Coefficient) [30] | Identification Ions or *m/z* (EI) |
|---|---|---|---|---|---|---|---|---|---|
| Triethyl phosphate | TEP | $C_6H_{15}O_4P$ | 78-40-0 | 182.15 | $5 \times 10^{-5}$ | $1.77 \times 10^{-1}$ | 2.4 | 0.8 | 99,127,155 |
| Tri-phenyl phosphate | TPHP | $C_{18}H_{15}O_4P$ | 115-86-6 | 326.29 | 1.9 | $6.28 \times 10^{-6}$ | 7.8 | 4.59 | 169.1,325.7 |
| Tripropyl phosphate | TPP | $C_9H_{21}O_4P$ | 513-08-6 | 224.23 | 827 | $2.9 \times 10$ | 3.7 | 2.67 | 170,233,325 |
| Tri-n-butyl phosphate | TNBP (TBP) | $C_{12}H_{27}O_4P$ | 126-73-8 | 266.32 | 280 | $1.13 \times 10^{-3}$ | 5.0 | 4.00 | 211,99,155 |
| Tris(2-chloroethyl) phosphate | TCEP | $C_6H_{12}Cl_3O_4P$ | 115-96-8 | 285.49 | $7 \times 10^{-3}$ | $1.08 \times 10^{-4}$ | 5.2 | 1.44 | 63,143,249 |
| Tris(1-chloro-2-propyl) phosphate | TCIPP (TCPP) | $C_9H_{18}Cl_3O_4P$ | 13674-84-5 | 327.56 | $1.6 \times 10^{-3}$ | $2.02 \times 10^{-5}$ | 5.0 | 2.59 | 279,201 |
| Tris(1,3-dichloro-2-propyl) phosphate | TDCIPP (TDCPP) | $C_9H_{15}Cl_6O_4P$ | 13674-87-8 | 430.90 | 1.5 | $4.07 \times 10^{-8}$ | 7.1 | 3.80 | 381,379,191 |
| Tris(2-butoxyethyl) phosphate | TBOEP (TBEP) | $C_{18}H_{39}O_7P$ | 78-51-3 | 398.48 | $1.2 \times 10^{-3}$ | $2.50 \times 10^{-8}$ | 9.6 | 3.65 | 199,299 |
| 2-ethylhexyl diphenyl phosphate | EHDPP | $C_{20}H_{27}O_4P$ | 1241–94-7 | 362.40 | 1.9 | $2.55 \times 10^{-6}$ | 11.3 | 5.37 | 250.8,77 |
| Tris(2-ethylhexyl) phosphate | TEHP | $C_{24}H_{51}O_4P$ | 78-42-2 | 434.64 | 0.6 | $2.04 \times 10^{-6}$ | 8.5 | 4.22 | 211.2,99 |
| Tris(methylphenyl) phosphate | TMPP (TCP) | $C_{21}H_{21}O_4P$ | 1330-78-5 | 368.37 | 0.36 | $6.00 \times 10^{-7}$ | 8.6 | 5.11 | 368,277,165 |
| Tris(isobutyl) phosphate | TIBP | $C_{12}H_{27}O_4P$ | 126-71-6 | 266.31 | 3.72 | $1.30 \times 10^{-2}$ | 3.6 | 3.60 | 99,155,211 |

## 3. Concentration Profiles

### 3.1. Worldwide Distribution of OPFRs in Indoor Dust and Air

The processes that affect the fate of OPFRs in environments include sorption, volatilization, and biodegradation, which lead to different concentration profiles of organophosphorus flame retardants in indoor air and indoor dust according to the global region and microenvironment (Tables 2 and 3). Vykoukalova et al. (2017) have found that there was no significant difference between the air in bedrooms and living rooms for 13 kinds of OPEs, while in the indoor dust, the concentrations of OPEs and BFRs were correlated [12]. For the dust in daycare centers, TBEP had the highest concentration (1,600,000 ng/g), while in workplaces, chlorinated organophosphate esters in the air (100 ng/$m^3$) had a higher concentration in Stockholm, Sweden [32]. In the Japanese market, TPHP was dominant among 11 kinds of OPFRs, with concentrations ranging from 560 to 14,000,000 ng/g, while TBEP was not detected in an electronic appliance store [33]. However, for floor dust in Japan, TBEP was detected with the highest level in most samples [34], indicating that different indoor sources or locations would have compounds with different kinds and levels. In the Rhine/Main area in Germany, the ΣOPFRs median level in seven indoor microenvironments were as follows: private cars, 180.3 ng/$m^3$; floor/carpet stores, 78.25 ng/$m^3$; offices, 59.32 ng/$m^3$; schools, 36.23 ng/$m^3$; daycare centers, 31.80 ng/$m^3$; building material markets, 31.17 ng/$m^3$; and private homes, 12.51 ng/$m^3$ [35]. The results showed that, in indoor air, the concentration of ΣOPFRs ranged from 3.30 to 751.0 ng/$m^3$, with a median of 40.23 ng/$m^3$, while that in outdoor air was 5.38 ng/$m^3$, suggesting that OPFRs are a pollutant that is dominant in indoor air environments compared to the outdoors [35].

**Table 2.** Concentrations of OPFRs in indoor dust (ng/g) samples from various indoor microenvironments worldwide. Where the median was not available, the geometric mean is given instead.

| | Region | Year | Microenvironments | Sample Number | TEP | TIBP | TNBP | TCEP | TCPP | TDCPP | TBEP | TPHP /TPP | EHDPP | TEHP | TMPP /TCP | ∑OPFRs | References |
|---|---|---|---|---|---|---|---|---|---|---|---|---|---|---|---|---|---|
| Asia | China, Guangzhou | 2015 | AC filter dust | n = 8 | - | - | - | 433 | 272 | 217 | - | 160 | 210 | 143 | - | 1435 | [35] |
| | | | Bed dust | n = 9 | - | - | - | 65.6 | 1005 | 1050 | - | 172 | 86.9 | 368 | - | 2747.5 | |
| | | | Floor dust | n = 9 | - | - | - | 106 | 251 | 327 | - | 281 | 180 | 194 | - | 1339 | |
| | | | Window dust | n = 9 | - | - | - | 167 | 339 | 95.7 | - | 199 | 140 | - | - | 940.7 | |
| | China, Beijing | 2012–2013 | Daycare center room floor dust | n = 9 | 76 | 32 | 124 | 4114 | 435 | 791 | 1010 | 1116 | - | - | 336 | 8034 | [14] |
| | China, Shanghai | 2017 * | Living room | n = 15 | 50 | - | 200 | 1200 | 1500 | 600 | 1600 | 900 | 900 | 1200 | - | 8150 | [15] |
| | | | Bedroom | n = 15 | 30 | - | 300 | 1000 | 1600 | 700 | 2000 | 900 | 900 | 1400 | - | 8830 | |
| | | | Balcony | n = 7 | 70 | - | 200 | 400 | 2200 | 500 | 2000 | 1200 | 300 | 500 | - | 7370 | |
| | China, Nanjing | 2014–2015 | Office dust | n = 5 | - | - | 68 | 166 | 238 | - | - | 30 | - | 32 | - | 534 | [36] |
| | China, Nanjing | 2014–2015 | Office dust | n = 12 | - | - | - | 1530 | 910 | 1330 | - | 900 | - | - | - | 4670 | [37] |
| | | | House dust | n = 6 | - | - | - | 2140 | 720 | 110 | - | 600 | - | - | - | 3570 | |
| | China, Guangzhou and Qingyuan | 2013–2014 | Rural home dust | n = 25 | 60 | - | 140 | 1930 | 1220 | - | 200 | 1090 | 310 | 190 | - | 5140 | [38] |
| | | | Urban home dust | n = 11 | 110 | - | 80 | 3780 | 750 | - | 320 | 150 | 360 | 140 | - | 5690 | |
| | Japan | 2009–2010 | Domestic house dust | n = 10 | - | - | 130 | 2700 | 1700 | 2200 | 82,000 | 820 | 200 | - | 1200 | 90,950 | [39] |
| | Japan, Sapporo | 2009–2010 | Floor dust | n = 48 | - | - | - | - | 740 | - | 30880 | 870 | - | - | - | 32490 | [40] |
| | Japan | 2006 | Floor dust | n = 148 | - | - | 1030 | 5830 | 8690 | 2800 | 508,320 | 4510 | - | 2070 | - | 533,250 | [41] |
| | Philippines | 2008 | House dust (malate) | n = 17 | - | - | 19 | 34 | - | - | - | 89 | 110 | 140 | 18 | 410 | [42] |
| | | | House dust (payatas) | n = 20 | - | - | 20 | 16 | - | - | - | 71 | 34 | 41 | 7.7 | 189.7 | |
| | Pakistan | 2011 | House dust | n = 15 | <5 | 25 | <20 | 15 | <20 | <5 | 16.5 | 175 | 67 | 20 | - | 318.5 | [43] |
| Europe | Netherlands | 2013 | Floor dust | n = 14 | - | - | - | 157 | 815 | 1051 | - | 404 | - | - | 58 | 2485 | [35] |
| | | | Surface dust | n = 14 | - | - | - | 205 | 3641 | 3752 | - | 357 | - | - | 57 | 8012 | |
| | Germany, Rhine | 2015 | Home dust | n = 15 | - | 380 | 250 | - | 4200 | - | 4300 | 1200 | - | - | - | 10,330 | [33] |
| | | | Office dust | n = 11 | 420 | 660 | 280 | - | 8500 | 3200 | 8600 | 2900 | - | - | 340 | 24,900 | |
| | Spain | 2007 * | House dust | n = 8 | - | 210 | 250 | 1700 | 3900 | - | 9900 | 2600 | - | - | - | 18,560 | [44] |

**Table 2.** *Cont.*

| | Region | Year | Microenvironments | Sample Number | TEP | TIBP | TNBP | TCEP | TCPP | TDCPP | TBEP | TPHP/TPP | EHDPP | TEHP | TMPP/TCP | ∑OPFRs | References |
|---|---|---|---|---|---|---|---|---|---|---|---|---|---|---|---|---|---|
| Europe | Romania, Spain, Belgium | 2012 * | Indoor dust | n = 6 | <30 | - | 190 | 680 | 860 | 3180 | 63,000 | 1160 | - | - | 1140 | 70,210 | [1] |
| | Spain, Barcelona | 2016 * | Home dust | n = 5 | - | 143 | 121 | 1790 | 2623 | 706 | - | 1102 | - | 454 | - | 6939 | [45] |
| | Belgium, Flemish | 2011 * | House dust | n = 33 | <50 | 2990 | 130 | 230 | 1380 | 360 | 2030 | 500 | - | - | 240 | 7860 | [1] |
| | Germany | 2011–2012 | Daycare center dust | n = 63 | - | <300 | <300 | 400 | 2680 | - | 225,000 | 500 | - | 500 | - | 229,080 | [46] |
| | Stockholm | 2011 * | Private home dust | n = 10 | - | 1100 | 300 | 2100 | 1600 | 10,000 | 4000 | 1200 | - | - | - | 20,300 | [32] |
| | | | Daycare center dust | n = 10 | 200 | 700 | 1200 | 30,000 | 3100 | 9100 | 1,600,000 | 1900 | - | 100 | - | 1,646,300 | |
| | | | Workplace dust | n = 10 | 100 | 1300 | 200 | 6700 | 19,000 | 17,000 | 87,000 | 5300 | - | - | - | 136,600 | |
| | Norway | 2012 | Residential living room dust | n = 48 | - | - | 55 | 414 | 2680 | - | 13,400 | 981 | 617 | - | 307 | 18,454 | [30] |
| | Norway | 2012 | Household dust | n = 38 | - | - | 54.9 | 403 | 2510 | 501 | 15,000 | 1010 | - | - | 266 | 19,744.9 | [19] |
| | Norway, Oslo | 2016 * | Living room floor dust | n = 61 | - | - | - | 435 | 1997 | 397 | 8146 | 722 | - | 401 | 179 | 12,277 | [47] |
| | | | Living room surface dust | n = 61 | - | - | - | 455 | 5241 | 1130 | 6796 | 1228 | - | 710 | 334 | 15,894 | |
| | Portugal, Aveiro and Coimbra | 2010–2011 | House dust | n = 28 | - | - | 28 | 17 | - | 22 | - | 662.4 | 620 | 1700 | 97 | 3146.4 | [16] |
| America | US, California | 2000–2001 | House dust | n = 125 | - | - | - | 1067 | 410.4 | 2021 | - | 813.5 | - | - | - | 4311.9 | [48] |
| | US, Boston | 2002–2007 | House dust | n = 50 | - | - | - | - | 572 | 1890 | - | 7360 | - | - | - | 9822 | [49] |
| | US, Longview and Vancouver, WA | 2011–2012 | House dust | n = 20 | - | - | - | 1380 | 4820 | 1620 | - | - | - | - | - | 7820 | [23] |
| | Canada, Toronto | 2013 | House dust | n = 23 | - | - | - | 181 | 1470 | 917 | - | 2350 | 754 | 101 | 6.9 | 5779.9 | [12] |
| | Brazil, Sao Paulo State, Araraquara city | 2017 * | Apartment dust | n = 10 | - | 40.1 | 28.1 | 237 | 1870 | 2250 | 22,100 | 3830 | 1750 | 549 | - | 32,654.2 | [31] |
| | | | House dust | n = 10 | - | 30.7 | 12.3 | 230 | 771 | 1370 | 15,900 | 3900 | 1590 | 397 | - | 24,201 | |
| | | | Office dust | n = 5 | - | 51.7 | 40.8 | 237 | 1820 | 4480 | 72,800 | 6420 | 2140 | 500 | - | 88,489.5 | |
| Africa Middle East | KSA, Jeddah | 2016 | House floor dust | n = 15 | - | - | 35 | 410 | 1650 | 500 | 205 | 230 | 220 | 70 | - | 3320 | [50] |
| | | | AC filter dust | n = 15 | - | - | 10 | 820 | 2000 | 7800 | 50 | 600 | 350 | 130 | - | 11,760 | |
| | Egypt, Assiut | 2012–2013 | House dust | n = 20 | - | 23 | 17 | 22 | 28 | 72 | 18 | 67 | 42 | - | - | 289 | [46] |
| | Kuwait | 2011 | House dust | n = 15 | 19 | 54 | 58 | 710 | 1460 | 360 | 855 | 430 | 190 | 65 | - | 4201 | [43] |
| Oceania | New Zealand, Wellington | 2012 * | Indoor dust | n = 34 | - | - | 80 | 110 | 350 | 230 | 4020 | 600 | - | - | 120 | 5510 | [51] |
| | | | | n = 16 | - | - | 70 | 40 | 250 | 110 | 1550 | 240 | - | - | 160 | 2420 | |
| | Australia, Brisbane | 2015 | Indoor dust | n = 3 | - | <30 | 140 | 550 | 1000 | 1500 | 63,000 | 870 | 1300 | <450 | 950 | 69,310 | [51] |

A year accompanied by * is the publication year, and others are sampling years. "-" means no data.

**Table 3.** Concentrations of OPFRs in indoor air (ng/m3) samples from various indoor microenvironments worldwide. Where the median was not available, the geometric mean is given instead.

| | Region | Year | Microenvironments | Sample number | TEP | TIBP | TNBP- | TCEP | TCPP | TDCPP | TBEP | TPHP /TPP | EHDPP | TEHP | TMPP /TCP | ∑OPFRs | References |
|---|---|---|---|---|---|---|---|---|---|---|---|---|---|---|---|---|---|
| Asia | China, Nanjing | 2014–2015 | Office air | n = 5 | - | - | 0.53 | 36 | 10 | - | - | 0.4 | - | 0.3 | - | 47.23 | [36] |
| | | | | n = 9 | - | - | 0.515 | 29.5 | 12.5 | - | - | 0.215 | - | 0.3 | - | 43.03 | |
| | China, Hangzhou | 2013 | Office air | n = 10 | - | - | 0.47 | 3.11 | 7.76 | 0.63 | 0.27 | 1.41 | 0.22 | 0.84 | - | 14.71 | [52] |
| Europe | Norway | 2012 | Residential living room air | n = 47 | - | - | 5.09 | 2.25 | 42.3 | - | 0.598 | 0.258 | 0.119 | - | - | 50.615 | [30] |
| | Norway | 2012 | Household air | n = 38 | - | - | 5.3 | 2.4 | 40.8 | 0.0753 | 0.637 | 0.241 | - | - | 0.0376 | 49.4909 | [18] |
| | Norway, Oslo | 2016 * | Indoor stationary air | n = 58 | - | - | 14 | 3 | 128 | - | - | 1 | - | - | - | 146 | [48] |
| | Germany | 2011–2012 | Daycare center air | n = 63 | - | - | 2.2 | <2 | 2.7 | - | 49 | - | - | - | - | 53.9 | [47] |
| | Stockholm | 2011 * | Private home air | n = 10 | 7.3 | 13 | 9.1 | 4.8 | 5.6 | - | - | - | - | - | - | 39.8 | [31] |
| | | | Daycare center air | n = 10 | 1.7 | 7.2 | 18 | 25 | 8.4 | - | 84 | - | - | - | - | 144.3 | |
| | | | Workplace air | n = 10 | 6.5 | 7.3 | 2.3 | 10 | 100 | 28 | 5.8 | - | - | - | - | 159.9 | |
| | Stockholm | 2010* | Living room air | n = 16 | 7 | 12 | 11 | 3.3 | 8.3 | - | - | - | - | - | - | 41.6 | [31] |
| | Stockholm | 2008 | Apartment air | n = 62 | 5.3 | 8.9 | 13 | 3.9 | 19 | - | - | - | - | - | - | 50.1 | [17] |
| America | Canada Toronto | 2013 | House air | n = 24 | - | - | - | 6.35 | 73.6 | 0.525 | - | 0.723 | 1.71 | 0.042 | 0.007 | 82.957 | [12] |

A year accompanied by * is the publication year, and others are sampling years. "-" means no data.

Summaries of the OPFR contents for indoor dust and indoor air from different countries are reported in Tables 2 and 3, respectively. An overview of the regional levels and trends of the main studied OPFRs is shown in Figure 2, and the compound compositions and patterns of selected OPFRs in indoor dust and indoor air are shown in Figures 3 and 4, respectively. In general, the highest concentration of TBEP appeared in indoor dust, and that of TCPP appeared in indoor air. Dust samples were dominated by OPFRs, including TBEP, TCEP, TPHP, TDCPP, and TEHP. TBEP was found to have the maximum contribution and a global trend of the highest content in indoor dust in this study, which may be explained by its comprehensive range of applicability as a plasticizer; it is often added to floor wax and synthetic rubber[53]. TBEP has a relatively higher molecular weight and a lower vapor pressure, which makes it appear in dust more than in the air. Various factors might impact the OPFR profiles in indoor dust, as their different uses may lead to heterogeneity among countries and regions [54].

In air samples, OPFRs were dominated by TCIPP, TBEP, and TCEP. TCIPP was dominant in Norway (40.8–128 ng/m$^3$); Stockholm, Sweden (100 ng/m$^3$); and Toronto, Canada (73.6 ng/m$^3$). TBEP was dominant in Germany (49 ng/m$^3$). And TCEP was dominant in Nanjing, China (29.5–36 ng/m$^3$). Compared with TCIPP, the content of TCEP was substantially lower than that of TCIPP in European countries, which is likely due to the restrictions from the European Union on TCEP and the associated rising use of TCIPP as an alternative [55]. The abundance of TCIPP in the air was consistent due to its frequent use and relatively high vapor pressure [56].

Figure 2 shows the worldwide concentration distribution of OPFRs in indoor dust, including 25 countries or cities. The highest concentration of ∑OPFRs was found in Stockholm dust (n = 10), with a median level of 1,646,300 ng/g, followed by Japan (533,250 ng/g, n = 148) and Germany (229,080 ng/g, n = 63). After analysis of variance (ANOVA), there was no significant difference between the data from Sweden and Germany ($p > 0.05$), while the differences between the data from Japan and from Sweden and Germany were significant ($p < 0.05$). Figure 2 also shows the concentration distribution of OPFRs in indoor air at six sites, from which we can see that Oslo, Norway (n = 58), has the highest level (146 ng/m$^3$) of ∑OPFRs, followed by Canada (n = 24) with 82.96 ng/m$^3$. The difference between the data from Norway and Canada was significant ($p < 0.05$). This indicates that the distribution characteristics of OPFRs in indoor dust present noticeable regional differences.

Based on the chemical structure of OPFRs, they can be divided into three groups: Chloroalkyl phosphates (ΣChlAlkP) including TCEP, TCIPP, and TDCIPP; alkyl phosphates (ΣAlkP) including TIBP, TNBP, TBOEP, and TEHP; and aryl phosphates (ΣAryP) including TPHP and EHDPP. As depicted in Figure 3, in indoor dust, alkyl phosphates presented the highest level in Japan, Spain, Romania, Germany, Stockholm (Sweden), Norway, Brazil, New Zealand, and Australia. Chloroalkyl phosphates presented significant levels in China, the Netherlands, Barcelona (Spain), California (US), Washington (US), KSA (Jeddah), and Kuwait. Aryl phosphates appeared in higher levels in Pakistan, the Philippines, Boston (US), and Canada. These three groups of OPFRs were mainly present in all of the indoor dust studies but with high variability in their concentration levels.

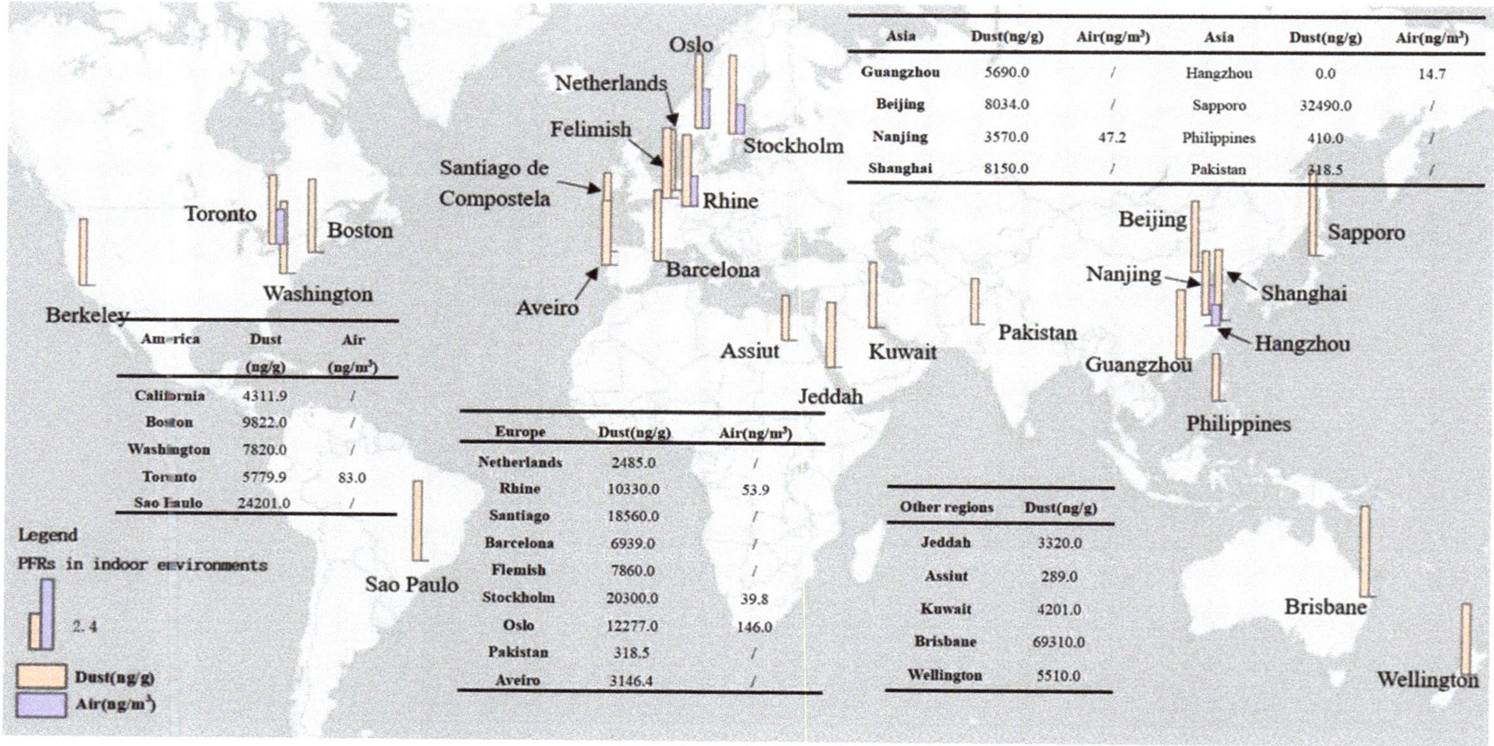


**Figure 2.** The spatial distributions of the concentrations of selected OPFRs in indoor dust (ng/g) and the locations of the six indoor air testing sites. The concentration unit is ng/m$^3$, and the data are all logarithmically processed. "/" represents data not measured in the relevant study.

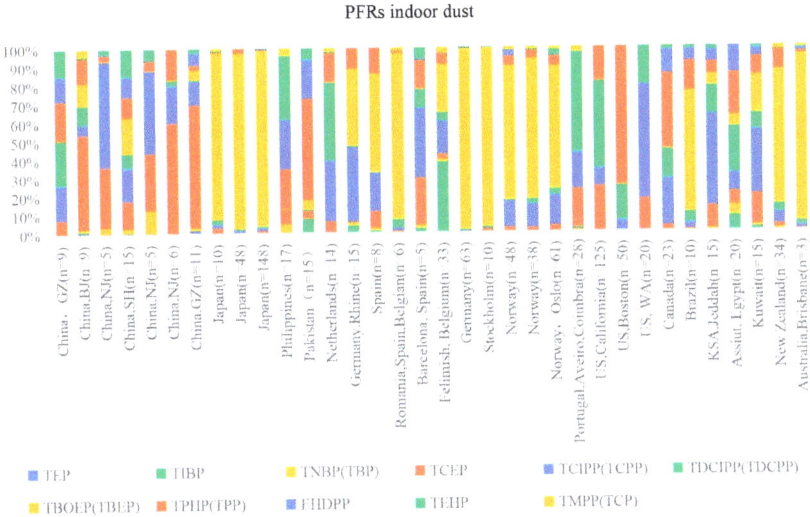


**Figure 3.** Composition profiles of OPFRs in indoor dust collected from different countries with various sampling numbers.

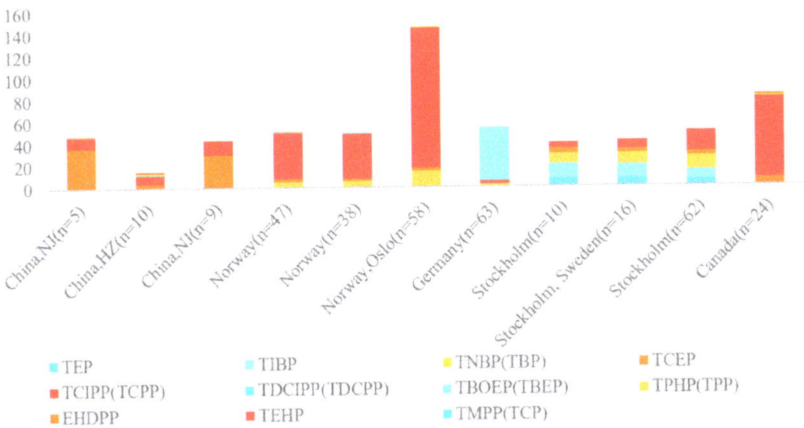


**Figure 4.** Composition profiles of OPFRs in indoor air collected from different countries with various sampling numbers.

Figure 4 presents the compositions of OPFRs in indoor air, and chloroalkyl phosphates present the highest level in most of the places we studied, except for Germany and Stockholm, where alkyl phosphates are expected to be more dominant. These similarities and differences are probably related to the indoor microenvironments, sources, or materials that each study sampled. Correlated relationships among OPEs between air and dust have been discussed and tested using a partitioning model [12]. This Weschler and Nazaroff partitioning model provides a relationship that can be used to calculate gas-phase concentrations using data from dust, which is expressed as

$$C_g = \frac{X_d \rho_d}{K_{OA} f_{OM}}, \quad (1)$$

where $X_d$ is the measured dust concentration (in ng/g), $\rho_d$ is the density of dust ($2.0 \times 10^6$ g/m$^3$), and $f_{OM}$ is the fraction of organic matter in the dust (0.3, as suggested by Bennett and Furtaw in 2004).

Vykoukalova et al. (2017) have shown that an equilibrium is reached for OPEs with log $K_{OA}$ values < 12, and under this condition, the partitioning between dust and air is proportional to a chemical's $K_{OA}$. When the log $K_{OA}$ value is 12, the log dust/air partition coefficient is 3.4 for OPEs, and compared to BFRs, OPEs are more partitioned into the air than in dust [11].

### 3.2. OPFRs from Different Indoor Sources and Materials

Different indoor sources and materials, such as air conditioner (AC) filters, floor coverings, windows, beds, electronic appliances, and building decoration and upholstery materials, were considered for collection and measurement of flame retardants including polybrominated diphenyl ethers (PBDEs), brominated/chlorinated flame retardants (Br/Cl FRs), and organophosphate flame retardants (OPFRs). The concentration and distribution information for major OPFRs in these indoor sources and materials are shown in Table 4. For indoor dust, TBEP was the most abundant OPFR in floors, indicating that its primary source was PVC coverings or floor polishes and waxes. Plastic materials such as computer covers and screens were most likely to be the source of TPP [11]. TCEP was found to be the highest level (94 mg/kg dust) in libraries, indicating that the acoustic ceiling could be a possible source [12]. TCPP is often added to upholstery and is likely emitted from sofas [10]. Consumer products and building materials were collected to measure emissions of OPFRs, and expandable polystyrene (EPS) and extruded polystyrene (XPS) insulating boards were found to be the essential sources of HBCD, while TCPP was observed to be commonly emitted from PU foam products [57]. In Guangzhou, in indoor dust samples from nine bedrooms, ΣPFRs were measured with a level range of 1560–12,600 ng/g in beds; among them, TDCIPP was the predominant component. Moreover, in AC filters, TCEP had the highest concentration, up to 433 ng/g, while in windows, TCIPP had the highest concentration, up to 339 ng/g [12]. In office dust, TPHP was one of the most significant compounds determined in printers and PC tables [25] Car dust collected from car seats was dominated by TDCIPP, with the highest concentration of 1100 μg/g. The primary material of car seats was polyurethane foam, in which TDCIPP and TCIPP were used as additive flame retardants [1]. Regarding building decoration materials in China, OPFRs appeared, in concentrations ordered from high to low, in foam samples, wallpaper, sealing materials, PVC pipes, boards, paints, and wall decoration powder. The Cl-OPFRs had a significantly higher level in foams, while non-Cl-OPFRs were higher in wallpaper (PVC and non-woven) and board samples [5]. TDCIPP, followed by TNBP, had higher median concentrations in the Czech Republic window film samples (Vykoukalova et al., 2017). Electronic equipment also influenced several kinds of organophosphorus flame retardants in houses and cars; for example, Brandsma et al. (2014) found that the concentrations of TPHP and TMPP measured on electronic equipment were much higher than those around electronic equipment in house dust [58]. However, electronics showed a limited contribution to other OPFRs, so the researcher believed that other household materials may have influenced the OPFRs levels in indoor dust. A study in Japan has suggested that the recycling and reuse of electronic products may provide a pathway for OPFRs to go into new products [33].

Principle component analysis (PCA) was applied to obtain a further source apportionment of OPFRs in indoor environments using the indoor dust data collected from previous studies [5,10,12,25,56,57] (see Figure 5). The standardized values used as variables were prefixed with Z. There were three main components in the plot, suggesting a similarity of applications of the three groups of OPFRs in the materials. According to the PCA results and emission sources of OPFRs from Table 4, preliminary source resolution for OPFRs in indoor dust can be performed. The first component was mainly contributed to by TBP, TCEP, TBEP, and TDCPP. TBP and TDCPP had an exceedingly high percentage in foam samples normally used as anti-foaming agents. TDCIPP is commonly used as FR inflexible and rigid PUF [50]. TCEP and TBEP are commonly used in building decoration materials such as acoustic ceilings and non-woven wallpaper, especially for PVC floor coverings and floor waxes, where TBEP is mainly used as a plasticizer and floor polish. The second component was mainly comprised of TEHP, TCPP, and TPP. TEHP and TPP are mainly used as flame retardants, plasticizers, and extraction agents, widely added into plastics and processed fibers, such as furniture upholstery, textile carpet padding, isolation materials, and so on. These were probably the main sources or reservoirs of the second component. EHDPP and TCP had remarkably high percentages in the third component. A significant correlation was observed between TCP and EHDPP (r = 0.794, $p < 0.01$), likely as these two aryl phosphates are commonly used in office PCs or printer tables. TCP is also applied widely as a flame retardant in electrical tables and as a plasticizer for automobile car interiors and furniture upholstery [59].

**Table 4.** Concentrations of major OPFRs in different indoor sources and building materials.

| Different Microenvironments | Major FRs | Concentration Levels | Country | References |
|---|---|---|---|---|
| PVC floor coverings and floor waxes | TBEP | 14–5300 mg/kg dust | Sweden | [10] |
| Computer screens and TV sets | TPP | 3300 mg/m$^2$ | | |
| Acoustic ceilings | TCEP | 0.19–94 mg/kg dust | | |
| Upholstery (sofas) | TCIPP | 50 mg/kg dust | | |
| EPS and XPS insulating boards | HBCD | 0.1–29 ng/m$^2$/h | Europe | [32] |
| PU foam products | TCIPP | 20 ng/m$^2$/h–140 µg/m$^2$/h | | |
| Bedding | TDCIPP | 1050 ng/g dust | South China (Guangzhou) | [25] |
| Bedroom AC filters | TCEP | 433 ng/g | | |
| Bedroom windows | TCIPP | 339 ng/g dust | | |
| Office printer table | TPHP | 5780 ng/g | | |
| Office PC tables | TPHP | 1220 ng/g | | |
| Car seats (PUF) | TDCIPP | 1100 µg/g | Netherlands | [58] |
| | TMPP | 380 µg/g | | |
| Wallpaper (PVC) | TEHP | 9984 ng/g | China | [5] |
| Wallpaper (non-woven) | TNBP | 102,400 ng/g | | |
| Wallpaper (pure paper) | TCIPP | 177.9 ng/g | | |
| Wall decoration powders | TCIPP | 5.84 ng/g | | |
| Decoration paints | TDCIPP | 156.7 ng/g | | |
| Window films | TCIPP | 566 ng/m$^2$ | Czech Republic | [12] |
| | TNBP | 72.6 ng/m$^2$ | | |

Based on the collected concentration and distribution information of OPFRs in worldwide regions and microenvironments, there were more indoor dust studies than indoor air studies. The difference in research purposes was that dust is more often studied as the potential exposure route for ingestion and dermal contact, while air is usually studied with regard to inhalation in human exposure and the degree of enrichment in different media. As for indoor dust samples, Sweden and Japan had the highest concentrations of ΣOPFRs among the countries studied, especially for floor dust-initiated TBOEP. Thus, it could be predicted that TBOEP was mostly emitted from the ground and was a dominant organophosphorus flame retardant in indoor dust. Similarly, in indoor air samples, we observed that ΣOPFRs had a higher level in Norway and Canada, and TCIPP and TDCIPP were the leading components in indoor air in these countries. They were mostly emitted from PU foam products and upholstery, and they are both chlorine flame retardants. It was predicted that they mostly existed in indoor air due to their volatility in building decoration materials. The molecular weight and vapor pressure of different OPFRs also have an impact on the partition equilibrium in dust and air. The above observations show that contamination from OPFRs, and, thus, their impact on human beings, is higher in indoor dust than in indoor air. This review aims to verify this hypothesis by estimating the human exposure to and toxicity information of different OPFRs in indoor dust and air in the next chapter.

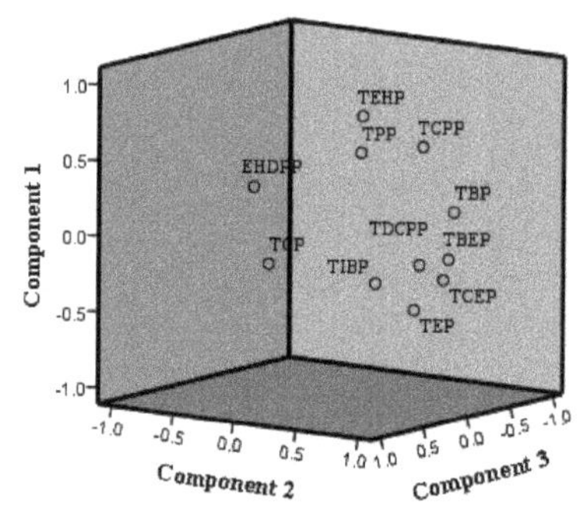


**Figure 5.** The principal component plot of PFRs in indoor dust environments.

## 4. Health Effect

### *4.1. Human Exposure to OPFRs in Indoor Dust and Air*

In indoor environments, there are three main exposure pathways for humans: ingestion, inhalation, and dermal absorption (Thomsen et al., 2001). Some researchers have claimed that FRs are usually associated with human exposure through dust ingestion, while another study suggested that hand-to-mouth contact or dermal absorption might be essential exposure routes for certain Cl-OPFRs such as hydrophilic TCEP and lipophilic TDCIPP [54,59,60]. In Washington State, researchers measured the concentration of ∑Cl-OPFRs with a mean level of 426 ng/m$^3$ and found that inhalation and respiration were important routes for the particulate fraction by estimating the inhalation intake of TCPP, which was up to 4540 ng/day for an adult [23]. Fang et al. (2014) have found that Cl-OPFRs have high bioaccessibility and quickly enter the digestive tract and the tissue in the upper respiratory tract [61]. In comparison to the inhalation of FRs, the exposure routes of dietary intake and dust ingestion were more important, but long-term persistent inhalation would make it a potential pathway for the intake of toxic compounds, especially some OPFRs with high vapor pressures. Non-dietary exposure routes such as inhalation and dermal contact have a potential role in causing unintended consequences [62]. In Japan, an exposure assessment of TBOEP in school and house dust was found to be higher than the reference dose, and the highest hazard quotient value of 1.9 in the dust ingested scenario [39]. For Romanian indoor dust, higher exposures to OPFRs were measured with factors below their corresponding reference dose of 50 (in the case of TBEP) and 165 (in the case of TCP) [54]. In the U.K., human dermal absorption of PFRs was studied by using human ex vivo skin and EPISKIN™ models. TCEP, TCIPP, and TDCIPP presented absorbed fractions of 28%, 25%, and 13% of the applied dose (500 ng/cm$^2$, finite dose), respectively. Furthermore, the estimated dermal contact exposure to PFRs for toddlers was higher than that for adults in indoor dust [60].

Most of the previous studies have estimated exposure pathways in one type of indoor environment. In addition, the data of factors such as recipient body weight, age range, intake rate, and so on differed between each study, leading to uncertainties in human exposure assessment. To simplify the calculation, a settled body weight is considered for toddlers and adults. In this review, a comprehensive assessment method for human exposure to OPFRs through the three pathways in various indoor microenvironments for adults and toddlers is developed.

#### 4.1.1. Inhalation

Inhalation exposure depends on the concentration of OPFRs in the air, recipient inhalation rate, exposure frequency, and body weight. According to the USEPA (The United States Environmental Protection Agency 1998), the inhalation exposure to air pollutants can be estimated using the following equation:

$$\sum Inhalation = \frac{C_{air} \times IR \times EF}{BW \times 365}, \quad (2)$$

where $\sum Inhalation$ is the inhalation exposure to air contaminants (ng/kg/day), $C_{air}$ is the OPFRs concentrations in indoor air (ng/m$^3$), $IR$ is the inhalation rate for an adult (20 m$^3$/day) or toddler (3.8 m$^3$/day), $EF$ is the exposure frequency (350 days/year), and $BW$ is the body weight for an adult (70 kg) or toddler (20 kg) [60,63,64].

The inhalation exposure to ∑OPFRs in indoor air from different regions ranged from 2.53 to 40 ng/kg bw/day. The high exposure values through air inhalation for adults and toddlers were 3.29, 3.84, 8.08, 35.07, 0.17, 13,42, 0.39, 0.47, and 0.01 ng/kg bw/day and 2.19, 2.55, 5.37, 23.32, 0.11, 8.93, 0.26, 0.31, and 0.01 ng/kg bw/day for TIBP, TNBP, TCEP, TCIPP, TDCIPP, TBOEP, TPHP/TPP, EHDPP, and TMPP/TCP, respectively. Among them, TCIPP, TBOEP, TCEP, and TNBP had relatively higher inhalation exposure values. In this study, the inhalation exposure to ∑OPFRs in Oslo (Norway) and Canada was higher than those estimated in China, Germany, and Stockholm. The estimated exposure to TNBP, TCEP, TCIPP, TDCIPP, TBEP, TPP, and TCP was compared with the reference dose (RfD) of the respective compounds given by USEPA. We found that the inhalation exposure was 3–6 orders of magnitude lower than the RfDs, suggesting an insignificant risk is posed to adults and toddlers through the inhalation of indoor air, according to the data used in this study.

In terms of concentration profiles, Cl-OPFRs showed a higher level in most indoor air places. The total intake of Cl-OPFRs through the inhalation exposure route for adults and toddlers ranged from 0.74 to 35.89 ng/kg bw/day (mean 12.62 ng/kg bw/day) and 0.49 to 23.87 ng/kg bw/day (mean 8.40 ng/kg bw/day), respectively, which were higher than other compounds and indicate that the inhalation exposure pathway appears to be of particular importance and should be taken into consideration in the assessment of these chlorinated organophosphate flame retardants. Among them, inhalation exposure to TCIPP in Norway (mean exposure of 23.13 and 15.38 ng/kg bw/day for adults and toddlers, respectively) and Canada (20.16 and 13.41 ng/kg bw/day for adults and toddlers, respectively) was significantly higher than in other regions and was 2–3 orders of magnitude higher than inhalation exposure for other compounds. TBOEP was the most abundant OPFR, and it showed an estimated high intake via inhalation both for adults and toddlers, especially in Germany (13.42 and 8.93 ng/kg bw/day for adults and toddlers, respectively). These exposure values were approximately 0.089% and 0.060% of the mentioned RfD, indicating that daycare centers contributed to higher TBOEP exposure.

Collectively, the estimated human inhalation exposure to OPFRs by indoor air did not pose an immediate and significant health risk to toddlers and adults, as mentioned above. However, some typical OPFRs, such as TCIPP and TBOEP, showed higher levels of air inhalation in the population in these countries [60,63,64]. Further studies are required to fully characterize the overall human exposure to OPFRs through inhalation exposure pathways.

#### 4.1.2. Ingestion

For indoor dust, ingestion is a significant exposure pathway for flame retardants [42]. Dust ingestion exposure occurs as a result of a migration of OPFRs from indoor materials to dust and subsequent ingestion, which depends on the OPFRs concentration in indoor dust, exposure time, daily dust intake, and body weight. Collected concentrations of OPFRs were used to estimate the exposure of adults and toddlers through indoor dust ingestion. The equation that was used to calculate the total daily intake of OPFRs is as follows:

$$\sum \text{Ingestion} = \frac{C_{DI} \times F_I \times I_R}{BW}, \tag{3}$$

where ∑Ingestion is the total daily human exposure to the studied OPFRs through indoor dust ingestion (ng/kg bw/day); $C_{DI}$ is the concentration of OPFRs in indoor dust (ng/g); $F_I$ is the average fraction of daily time spent in indoor environments; $I_R$ is the mean dust ingestion rate (mg/day), set as 20 and 50 for adults and toddlers, respectively [27]; and BW is the body weight (kg), set as 70 and 20 kg for adults and toddlers, respectively. As detailed information on the time spent in different indoor environments was missing, a person was considered to spend 100% of their time in indoor environments, and so, the figures given here represent the highest exposure scenario of dust ingestion for a worst-case assessment.

The high exposure values via dust ingestion for adults and toddlers were 0.31, 0.29, 1.67, 2.48, 2.86, 145.23, 2.10, 0.45, and 0.27 ng/kg bw/day and 2.75, 2.58, 14.58, 21.73, 25.00, 1270.80, 18.40, 3.98, and 2.38 ng/kg bw/day for TIBP, TNBP, TCEP, TCIPP, TDCIPP, TBOEP, TPHP/TPP, EHDPP, and TMPP/TCP, respectively. Among them, TBOEP, TDCIPP, TCIPP, and TPHP had relatively higher ingestion exposure values. For both groups, the estimated exposure levels for most OPFRs were several orders of magnitude lower than their respective reference doses (RfDs) given by the USEPA.

For Cl-OPFRs, the estimated exposure values from dust ingestion for adults and toddlers ranged from 0.01 to 4.95 ng/kg bw/day (mean: 1.10 ng/kg bw/day) and 0.04 to 43.30 ng/kg bw/day (mean: 9.64 ng/kg bw/day), respectively. Contrasting with the inhalation intake of Cl-OPFRs, the total intake by dust ingestion was estimated lower than intake through the inhalation exposure route for adults but was very close and even higher for toddlers, suggesting that Cl-OPFRs have a higher percentage of the amount reaching respiratory tract and a more significant percentage in toddlers, making them very likely to be exposed to these compounds.

Furthermore, the highest ingestion exposure value was found in TBOEP for toddlers in Japan, which was approximately 8.47% of its respective reference dose. The TBOEP exposures in other countries, such as Germany (64.29 and 562.50 ng/kg bw/day for adults and toddlers, respectively) and Australia (18.00 and 157.50 ng/kg bw/day for adults and toddlers, respectively) were higher than those for any other OPFRs. This might be caused by those higher concentrations of TBOEP, which is normally detected in indoor environments. Therefore, ingestion of indoor dust is suggested as a significant exposure pathway to OPFRs, especially for TBOEP.

After estimating inhalation and ingestion exposures through indoor air and indoor dust, we found that the individual OPFRs presented different major pathways. For example, heavier OPFRs such as TBOEP and TPHP were mainly exposed through dust ingestion, while volatile OPFRs such as TCIPP and TCEP had a major inhalation exposure pathway, similar to the results in the study of Xu et al.[48]. The exposure values of toddlers were lower than those of adults, which can be explained by the lower inhalation rate to body weight ratio in toddlers than adults (0.19 $m^3$/day/kg vs. 0.29 $m^3$/day/kg). However, in the ingestion exposure scenario, the exposure values of toddlers were higher as, compared to adults, toddlers ingest more dust due to increased hand-to-mouth contact, their frequent close-to-ground behavior, and their lower personal hygienic standards [60].

#### 4.1.3. Dermal Contact

Dermal contact exposure to dust is a result of OPFRs transferring from products to dust through direct contact, which depends on the concentration of OPFRs in the surface dust, exposed surface area, dust amount adhered to the skin, skin absorption fraction, exposure time, and body weight. Studies have measured concentrations of ORFRs from AC filter dust, beds, windows, balconies, and other surface dust to calculate their dermal exposure values [25]. The equation that was used is as follows:

$$\sum \text{Dermal} = \frac{C \times SA \times DAS \times F \times EF}{BW}, \tag{4}$$

where ∑Dermal is the dermal exposure value (ng/kg bw/day); $C$ is the concentration of OPFRs in indoor dust from AC, beds, or surface dust (ng/g); $SA$ is the skin surface area exposed ($cm^2$), which is 4615 $cm^2$ for an adult and 2564 $cm^2$ for a toddler; $DAS$ is the dust amount adhered to skin ($mg/cm^2$), which is 0.01 $mg/cm^2$ for an adult and 0.04 $mg/cm^2$ for a toddler; $F$ is the fraction absorbed by the skin, which has been reported as 28.3%, 24.7%, and 12.7% for TCEP, TCIPP, and TDCIPP, respectively (as the F values of other OPFRs, including TNBP, TBEP, TPP, TPHP, EHDPP and TMPP, are not available in the literature, the average F of 21.9% was used for TCEP, TCIPP, and TDCIPP); $EF$ is the fraction of time spent in houses, offices, or beds; and BW (kg) is the average body weight, set as 70 kg for an adult and 20 kg for a toddler [12]. Due to a lack of data for the fraction of time spent in each indoor microenvironment, both groups were considered to spend all of their time exposed to surface dust, and so, the reported results are the maximum dermal exposure values.

The estimated dermal exposure levels for most OPFRs were on the same order of magnitude as ingestion exposure values, and both were much lower than their RfDs, suggesting that dermal uptake of OPFRs by surface dust is also a significant pathway of human exposure to indoor contaminants. If people are in contact with surface dust all day, the dermal exposure values of OPFRs for adults and toddlers were estimated to be up to 7.632 ng/kg bw/day for TNBP, TCEP, TCIPP, TDCIPP, TBOEP, TPHP/TPP, EHDPP, and TMPP/TCP, which were still lower than their intake through dust ingestion. Among them, the highest estimated dermal absorption exposure was found for TBOEP for both adults and toddlers, followed by dermal absorption estimates for TCIPP, TDCIPP, and TCEP, indicating that dermal dust contact and absorption is still a major exposure pathway due to the high concentrations of these compounds in dust.

The dermal exposure values for most OPFRs, such as TCIPP, TBOEP, and EHDPP, during sleeping, were very close to their dust ingestion exposure values, which indicates that the exposure risks of OPFRs from beds should be given more attention. In addition, the dermal contact values for beds are higher than those for other surfaces. This is probably because people are in contact with beds longer than other household surfaces, such as air conditioners, windows, or balconies. Moreover, the dermal exposure values of toddlers were higher than those of adults. Compared to adults, more dust adheres to the skin of toddlers, and their exposed skin surface area to body weight ratio is higher (65.93 $cm^2/kg$ vs. 128.20 $cm^2/kg$). Therefore, regular and frequent cleaning of surface dust needs to be conducted in order to dilute the concentration of OPFRs in indoor dust.

The combined total exposure to OPFRs by air inhalation, dust ingestion, and dermal contact was generally below the RfD values for both adults and toddlers, with a few notable higher exposures of some typical OPFRs. It should be noted that these exposure values are not necessarily accurate due to the uncertainties in exposure time fractions and potential physiological behaviors, as well as the lack of some specific data values.

### *4.2. Toxicity of Several Typical OPFRs*

There is limited knowledge related to the toxicity of OPFRs. Animal toxicity tests have been conducted to evaluate the acute toxicity, including $LC_{50}$, $IC_{25/50}$, or $EC_{50}$, of several OPFRs, and it has been found that algae are highly sensitive to OPFRs, as detailed in Table 5. TPHP has been stated to be the most acute toxic triaryl phosphate to water organisms such as fish, shrimps, and daphnia. Results of animal testing have indicated that TPP has low toxicity, and algae are relatively sensitive to TCP [65]. TCP has been shown to be a possible reproductive toxin and was harmful in a salmonella mutagenicity test [1,66,67]. Both the carcinogenic and non-carcinogenic influences of TCPP and TDCPP have been considered, as they exhibit concentration-dependent neurotoxicity, inhibit DNA synthesis, decrease cell number, and alter neurodifferentiation; meanwhile, another Cl-OPFR, TCEP, was considered to be carcinogenic for animals [1,57,68]. Zebrafish embryos are considered a viable and integrative vertebrate model organism for human hazard assessment due to their high throughput (Sipes et al. 2011). Toxicity testing in zebrafish has suggested that several OPFRs have the potential capacity to affect mammalian biology, and their concentrations inducing toxicity in zebrafish could be in the upper range of potential human exposure [68]. For human beings, TCP and TCEP were considered to have negative effects, and TCEP is a human reproductive toxin, while TPHP has a low impact [1]. TDCIPP and TPP in house dust might be associated with altered hormone levels and decreased semen quality in men [69]. The result mainly covered acute and chronic toxicity tests in aquatic organisms, which were limited by data capacity. Fisk et al. have filled most of these data gaps and predicted ecotoxicity values by using quantitative structure–activity relationships (QSARs) and the ECOSAR program [70].

**Table 5.** Toxicity information of several organophosphorus flame retardants.

| | Reported 96 h-$LC_{50}$ to Fish (mg/L) [30] | ECOSAR 96 h-$LC_{50}$ to Fish (mg/L) [30] | Acute Toxicity | Long-Term Toxicity |
|---|---|---|---|---|
| TCIPP | 51–84 | 8.9 | Oral: $LD_{50}$ (rat) 500–4200 mg/kg bw Inhalative: $LD_{50}$ (rat) >4.6 mg/L to >17.8 mg/L Dermal: $LD_{50}$ (rabbit) 1230–5000 mg/kg bw [35] | NOEL = 36 mg/kg bw [35] |
| TDCIPP | 1.1 | 4.7 | 48 h-$EC_{50}$ (daphnia) 3.8–4.6 mg/L Oral:$LC_{50}$ (rat) 2300 mg/kg Dermal:$LC_{50}$ (rat) >2000 mg/kg [45] | NOEL = 15.3 mg/kg bw per day LOEL = 62 mg/kg perday [28] |
| TCEP | 6.3–250 | 35 | | |
| TBOEP | 6.8–24 | 9.5 | 15 min-$IC_{25}$/$IC_{50}$ (bacteria) 15.6–500 mg/L 72 h-$IC_{25}$/$IC_{50}$ (algae) 0.18–91 mg/L 96 h-$LC_{50}$/$EC_{50}$ (invertebrates) 7.8–500 mg/L [7] | |
| TMPP | 0.061–0.75 | 1 | | |
| TPHP | 0.3–0.66 | 1 | $LC_{50}$ (daphnia) 1.0–1.2 mg/L $LC_{50}$ (rats) 3500–10,800 mg/kg [45] | NOEC = 0.1 mg/L (daphnia)/3500–10,800 mg/kg (rat) [28] |
| TEHP | >100 | 0.005 | 15 min-$IC_{25}$/$IC_{50}$ (bacteria) 0.78–100 mg/L 72 h-$IC_{25}$/$IC_{50}$ (algae) 0.36–182 mg/L 96 h-$LC_{50}$/$EC_{50}$ (invertebrates) 3.13–100 mg/L [7] | |

Human exposure values were calculated through the three pathways for various OPFRs, with respect to adults and toddlers in different regions, and toxicity information was collected for some of them. For indoor air, inhalation exposure to eight OPFRs was evaluated, and the results indicated that this kind of exposure was much lower than the RfDs, suggesting an insignificant risk. However, Cl-OPFRs had relatively higher inhalation exposure values and, so, remained a concern. For indoor dust, two exposure pathways were estimated: Ingestion and dermal contact. These two exposure values were found to have similar orders of magnitude, and both were lower than RfDs; however, among them, TBOEP showed a much higher ingestion exposure than any other OPFR, especially in Japan and for toddlers, which indicated that human exposure to dust was a significant pathway that could pose a risk to the exposed population. The estimated exposure results showed that toddlers were exposed to OPFRs more through indoor dust as they interact more closely with dust than adults. Although all of the estimated exposure values were lower than the reference dose-response values, caution should also be given with the increasing application of OPFRs in indoor environments and the contribution of other human exposure pathways, such as food intake. Exposure to OPFR mixtures could lead to dose-additive effects, even if the individual levels of OPFRs are low [71]).

The toxicity of most of OPFRs is not yet completely understood. The toxicity of individual OPFRs and the toxic effects caused by exposure to OPFR mixtures are still unclear and need to be investigated further [35]. It is worth pondering whether a relationship exists between concentration and human exposure or toxicity information of OPFRs for the sake of conducting an overall evaluation of the environmental effects of these pollutants.

## 5. Discussion and Suggestions for Future Research

As flame retardants—especially OPFRs—are widely distributed in our daily lives, it is important to understand their levels of risk from a global perspective. First, we discussed the development and alteration of various types of flame retardants. Then, we found that most studies on OPFRs in indoor environments have been conducted in Europe. Depending on the difference between countries or cities, microenvironments (e.g., dust or air), and sources (e.g., houses or offices and AC filters, floor, windows or beds, household materials, including floor coverings, wallpapers, and electronic equipment, and so on), OPFRs presented a variety of patterns in their distribution and concentration levels. At present, people spend most of their time in indoor environments and are surrounded by various kinds of OPFRs, which may lead to potential diseases and other health effects. We calculated the estimated human exposure to different OPFRs in adults and toddlers by means of inhalation for indoor air and ingestion and dermal contact for indoor dust. The result demonstrated that most of the exposure values were several orders of magnitude lower than the RfDs, from which we can conclude that there is an insignificant risk for both adults and toddlers with respect to air inhalation, dust ingestion, and dermal contact.

There were some limitations to our review. As we only searched studies in the literature published in the English language, we may have missed some studies related to our study published in other languages. We placed emphasis on OPFRs, but not all kinds of flame retardants, according to the needs of our research. Regarding ingestion and dermal exposure, due to a lack of time distribution information, the estimated exposure values were for the worst-case scenario. Finally, we listed the existing toxicity information for several OPFRs, but data for other chemicals in the group were lacking.

**Author Contributions:** Conceptualization, X.S. and S.Z.; methodology, L.H. and J.Z.; software, J.Z.; validation, Y.L.; formal analysis, X.C.; investigation, Y.L.; resources, Y.M.; data curation, Y.M.; writing—original draft preparation, X.S.; writing—review and editing, X.S.; visualization, Q.B.; supervision, Y.M.; project administration, Y.M.; funding acquisition, Y.M. All authors have read and agreed to the published version of the manuscript.

**Funding:** This research was funded by the National Science Foundation of China (Project No. 41807354), State Environmental Protection Key Laboratory of Formation and Prevention of Urban Air Pollution Complex (No. 2021080543), and Shanghai Sailing Program (No. 23YF1428200).

**Institutional Review Board Statement:** Not applicable.

**Informed Consent Statement:** Not applicable.

**Data Availability Statement:** The data associated with this paper are available on request from the corresponding author.

**Conflicts of Interest:** There was no conflict of interest in this research.

## References


1. Van den Eede, N.; Dirtu, A.C.; Ali, N.; Neels, H.; Covaci, A. Multi-residue method for the determination of brominated and organophosphate flame retardants in indoor dust. *Talanta* **2012**, *89*, 292–300. [CrossRef] [PubMed]
2. Bergman, A.; Heindel, J.J.; Kasten, T.; Kidd, K.A.; Jobling, S.; Neira, M.; Zoeller, R.T.; Becher, G.; Bjerregaard, P.; Bornman, R.; et al. The *Impact* of Endocrine Disruption: A Consensus Statement on the State of the Science. *Environ. Health Perspect.* **2013**, *121*, 104–106. [CrossRef] [PubMed]
3. Malliari, E.; Kalantzi, O.-I. Children's exposure to brominated flame retardants in indoor environments—A review. *Environ. Int.* **2017**, *108*, 146–169. [CrossRef] [PubMed]
4. Zhao, H.; Liu, L.; Li, Y.; Zhao, F.; Zhang, S.; Mu, D.; Liu, J.; An, L.; Wan, Y.; Hu, J. Occurrence, Bioaccumulation, and Trophic Transfer of Oligomeric Organophosphorus Flame Retardants in an Aquatic Environment. *Environ. Sci. Technol. Lett.* **2019**, *6*, 323–328. [CrossRef]
5. Wang, Y.; Hou, M.; Zhang, Q.; Wu, X.; Zhao, H.; Xie, Q.; Chen, J. Organophosphorus Flame Retardants and Plasticizers in Building and Decoration Materials and Their Potential Burdens in Newly Decorated Houses in China. *Environ. Sci. Technol.* **2017**, *51*, 10991–10999. [CrossRef] [PubMed]
6. Tao, F.; Abdallah, M.A.-E.; Harrad, S. Emerging and Legacy Flame Retardants in UK Indoor Air and Dust: Evidence for Replacement *of* PBDEs by Emerging Flame Retardants? *Environ. Sci. Technol.* **2016**, *50*, 13052–13061. [CrossRef] [PubMed]
7. Wei, G.L.; Li, D.Q.; Zhuo, M.N.; Liao, Y.S.; Xie, Z.Y.; Guo, T.L.; Li, J.J.; Zhang, S.Y.; Liang, Z.Q. Organophosphorus flame retardants *and* plasticizers: Sources, occurrence, toxicity and human exposure. *Environ. Pollut.* **2015**, *196*, 29–46. [CrossRef]
8. Blum, A.; Behl, M.; Linda, S.; Miriam, L.; Phillips, A.; Singla, V.; Nisha, S.; Heather, M.; Venier, M. Organophosphate Ester Flame Retardants: Are They a Regrettable Substitution for Polybrominated Diphenyl Ethers? *Environ. Sci. Technol. Lett.* **2019**, *6*, 638–649. [CrossRef]
9. Hou, R.; Xu, Y.; Wang, Z. Review of OPFRs in animals and humans: Absorption, bioaccumulation, metabolism, and internal *exposure* research. *Chemosphere* **2016**, *153*, 78–90. [CrossRef]
10. Marklund, A.; Andersson, B.; Haglund, P. Screening of organophosphorus compounds and their distribution in various indoor environments. *Chemosphere* **2003**, *53*, 1137–1146. [CrossRef]
11. Venier, M.; Audy, O.; Vojta, S.; Becanova, J.; Romanak, K.; Melymuk, L.; Kratka, M.; Kukucka, P.; Okeme, J.; Saini, A.; et al. Brominated flame retardants in the indoor environment—Comparative study of indoor contamination from three countries. *Environ. Int.* **2016**, *94*, 150–160. [CrossRef] [PubMed]
12. Vykoukalova, M.; Venier, M.; Vojta, S.; Melymuk, L.; Becanova, J.; Romanak, K.; Prokes, R.; Okeme, J.O.; Saini, A.; Diamond, M.L.; et al. *Organophosphate* esters flame retardants in the indoor environment. *Environ. Int.* **2017**, *106*, 97–104. [CrossRef] [PubMed]
13. Thomsen, C.; Lundanes, E.; Becher, G. Brominated flame retardants in plasma samples from three different occupational groups in Norway. *J. Environ. Monit.* **2001**, *3*, 366–370. [CrossRef] [PubMed]
14. 1u, M.; Yu, G.; Cao, Z.; Wu, D.; Liu, K.; Deng, S.; Huang, J.; Wang, B.; Wang, Y. Characterization and human exposure assessment of organophosphate flame retardants in indoor dust from several microenvironments of Beijing, China. *Chemosphere* **2016**, *150*, 465–471.
15. Peng, C.; Tan, H.; Guo, Y.; Wu, Y.; Chen, D. Emerging and legacy flame retardants in indoor dust from East China. *Chemosphere* **2017**, *186*, 635–643. [CrossRef] [PubMed]
16. Coelho, S.D.; Sousa, A.C.A.; Isobe, T.; Kim, J.-W.; Kunisue, T.; Nogueira, A.J.A.; Tanabe, S. Brominated, chlorinated and phosphate organic contaminants in house dust from Portugal. *Sci. Total Environ.* **2016**, *569*, 442–449. [CrossRef]
17. Luongo, G.; Ostman, C. Organophosphate and phthalate esters in settled dust from apartment buildings in Stockholm. *Indoor Air* **2016**, *26*, 414–425. [CrossRef]
18. Liagkouridis, I.; Cequier, E.; Lazarov, B.; Cousins, A.P.; Thomsen, C.; Stranger, M.; Cousins, I.T. Relationships between estimated flame *retardant* emissions and levels in indoor air and house dust. *Indoor Air* **2017**, *27*, 650–657. [CrossRef]
19. Reemtsma, T.; Benito Quintana, J.; Rodil, R.; Garcia-Lopez, M.; Rodriguez, I. Organophosphorus flame retardants and plasticizers *in* water and air I. Occurrence and fate. *Trac-Trends Anal. Chem.* **2008**, *27*, 727–737. [CrossRef]
20. Kanazawa, A.; Saito, I.; Araki, A.; Takeda, M.; Ma, M.; Saijo, Y.; Kishi, R. Association between indoor exposure to semi-volatile organic *compounds* and building-related symptoms among the occupants of residential dwellings. *Indoor Air* **2010**, *20*, 72–84. [CrossRef]
21. Ma, J.; Zhu, H.; Kannan, K. Organophosphorus Flame Retardants and Plasticizers in Breast Milk from the United States. *Environ. Sci. Technol. Lett.* **2019**, *6*, 525–531. [CrossRef]
22. Bluyssen, P. *The Indoor Environment Handbook: How to Make Buildings Healthy and Comfortable*; Routledge: London, UK, 2009; Volume 95.
23. Schreder, E.D.; Uding, N.; La Guardia, M.J. Inhalation a significant exposure route for chlorinated organophosphate flame retardants. *Chemosphere* **2016**, *150*, 499–504. [CrossRef]
24. Kademoglou, K.; Xu, F.; Padilla-Sanchez, J.A.; Haug, L.S.; Covaci, A.; Collins, C.D. Legacy and alternative flame retardants in Norwegian and UK indoor environment: Implications of human exposure via dust ingestion. *Environ. Int.* **2017**, *102*, 48–56. [CrossRef]

25. Zheng, X.; Qiao, L.; Covaci, A.; Sun, R.; Guo, H.; Zheng, J.; Luo, X.; Xie, Q.; Mai, B. Brominated and phosphate flame retardants (FRs) in indoor dust from different microenvironments: Implications for human exposure via dust ingestion and dermal contact. *Chemosphere* **2017**, *184*, 185–191. [CrossRef] [PubMed]
26. Frederiksen, M.; Vorkamp, K.; Thomsen, M.; Knudsen, L.E. Human internal and external exposure to PBDEs—A review of levels and sources. *Int. J. Hyg. Environ. Health* **2009**, *212*, 109–134. [CrossRef] [PubMed]
27. Lyche, J.L.; Rosseland, C.; Berge, G.; Polder, A. Human health risk associated with brominated flame-retardants (BFRs). *Environ. Int.* **2015**, *74*, 170–180. [CrossRef] [PubMed]
28. Thomas, M.B.; Stapleton, H.M.; Dills, R.L.; Violette, H.D.; Christakis, D.A.; Sathyanarayana, S. Demographic and dietary risk factors in relation to urinary metabolites of organophosphate flame retardants in toddlers. *Chemosphere* **2017**, *185*, 918–925. [CrossRef] [PubMed]
29. Abdallah, M.A.-E.; Covaci, A. Organophosphate Flame Retardants in Indoor Dust from Egypt: Implications for Human Exposure. *Environ. Sci. Technol.* **2014**, *48*, 4782–4789. [CrossRef] [PubMed]
30. Du, Z.K.; Wang, G.W.; Gao, S.X.; Wang, Z.Y. Aryl organophosphate flame retardants induced cardiotoxicity during zebrafish embryogenesis: By the disturbing expression of the transcriptional regulators. *Aquat. Toxicol.* **2015**, *161*, 25–32. [CrossRef] [PubMed]
31. Cristale, J.; Aragao Bele, T.G.; Lacorte, S.; Rodrigues de Marchi, M.R. Occurrence and human exposure to brominated and organophosphorus flame retardants via indoor dust in a Brazilian city. *Environ. Pollut.* **2018**, *237*, 695–703. [CrossRef]
32. Bergh, C.; Torgrip, R.; Emenius, G.; Ostman, C. Organophosphate and phthalate esters in air and settled dust—A multi-location indoor study. *Indoor Air* **2011**, *21*, 67–76. [CrossRef]
33. Zhou, L.; Hiltscher, M.; Puettmann, W. Occurrence and human exposure assessment of organophosphate flame retardants in indoor dust from various microenvironments of the Rhine/Main region, Germany. *Indoor Air* **2017**, *27*, 1113–1127. [CrossRef] [PubMed]
34. Takigami, H.; Suzuki, G.; Hirai, Y.; Ishikawa, Y.; Sunami, M.; Sakai, S.-I. Flame retardants in indoor dust and air of a hotel in Japan. *Environ. Int.* **2009**, *35*, 688–693. [CrossRef] [PubMed]
35. Sugeng, E.J.; Leonards, P.E.G.; van de Bor, M. Brominated and organophosphorus flame retardants in body wipes and house dust, and estimation of house dust hand-loadings in Dutch toddlers. *Environ. Res.* **2017**, *158*, 789–797. [CrossRef] [PubMed]
36. Faiz, Y.; Zhao, W.; Feng, J.; Sun, C.; He, H.; Zhu, J. Occurrence of triphenylphosphine oxide and other organophosphorus compounds in indoor air and settled the dust of an institute building. *Build. Environ.* **2016**, *106*, 196–204. [CrossRef]
37. He, R.; Li, Y.; Xiang, P.; Li, C.; Zhou, C.; Zhang, S.; Cui, X.; Ma, L.Q. Organophosphorus flame retardants and phthalate esters in indoor dust from different microenvironments: Bioaccessibility and risk assessment. *Chemosphere* **2016**, *150*, 528–535. [CrossRef]
38. He, C.; Zheng, J.; Qiao, L.; Chen, S.; Yang, J.; Yuan, J.; Yang, Z.; Mai, B. Occurrence of organophosphorus flame retardants in indoor dust in multiple microenvironments of southern China and implications for human exposure. *Chemosphere* **2015**, *133*, 47–52. [CrossRef]
39. Mizouchi, S.; Ichiba, M.; Takigami, H.; Kajiwara, N.; Takamuku, T.; Miyajima, T.; Kodama, H.; Someya, T.; Ueno, D. Exposure assessment of organophosphorus and organobromine flame retardants via indoor dust from elementary schools and domestic houses. *Chemosphere* **2015**, *123*, 17–25. [CrossRef] [PubMed]
40. Tajima, S.; Araki, A.; Kawai, T.; Tsuboi, T.; Bamai, Y.A.; Yoshioka, E.; Kanazawa, A.; Cong, S.; Kishi, R. Detection and intake assessment of organophosphate flame retardants in house dust in Japanese dwellings. *Sci. Total Environ.* **2014**, *478*, 190–199. [CrossRef]
41. Araki, A.; Saito, I.; Kanazawa, A.; Morimoto, K.; Nakayama, K.; Shibata, E.; Tanaka, M.; Takigawa, T.; Yoshimura, T.; Chikara, H.; et al. Phosphorus flame retardants in indoor dust and their relation to asthma and allergies of inhabitants. *Indoor Air* **2014**, *24*, 3–15. [CrossRef]
42. Kim, J.-W.; Isobe, T.; Sudaryanto, A.; Malarvannan, G.; Chang, K.-H.; Muto, M.; Prudente, M.; Tanabe, S. Organophosphorus flame retardants in house dust from the Philippines: Occurrence and assessment of human exposure. *Environ. Sci. Pollut. Res.* **2013**, *20*, 812–822. [CrossRef] [PubMed]
43. Ali, N.; Ali, L.; Mehdi, T.; Dirtu, A.C.; Al-Shammari, F.; Neels, H.; Covaci, A. Levels and profiles of organochlorines and flame retardants in cars and house dust from Kuwait and Pakistan: Implication for human exposure via dust ingestion. *Environ. Int.* **2013**, *55*, 62–70. [CrossRef]
44. Garcia, M.; Rodriguez, I.; Cela, R. Microwave-assisted extraction of organophosphate flame retardants and plasticizers from indoor dust samples. *J. Chromatogr. A* **2007**, *1152*, 280–286. [CrossRef] [PubMed]
45. Cristale, J.; Hurtado, A.; Gomez-Canela, C.; Lacorte, S. Occurrence and sources of brominated and organophosphorus flame *retardants* in dust from different indoor environments in Barcelona, Spain. *Environ. Res.* **2016**, *149*, 66–76. [CrossRef] [PubMed]
46. Fromme, H.; Lahrz, T.; Kraft, M.; Fembacher, L.; Mach, C.; Dietrich, S.; Burkardt, R.; Voelkel, W.; Goeen, T. Organophosphate flame retardants and plasticizers in the air and dust in German daycare centers and human biomonitoring in visiting children (LUPE 3). *Environ. Int.* **2014**, *71*, 158–163. [CrossRef] [PubMed]
47. Xu, F.; Giovanoulis, G.; van Waes, S.; Padilla-Sanchez, J.A.; Papadopoulou, E.; Magner, J.; Haug, L.S.; Neels, H.; Covaci, A. Comprehensive Study of Human External Exposure to Organophosphate Flame Retardants via Air, Dust, and Hand Wipes: The Importance of *Sampling* and Assessment Strategy. *Environ. Sci. Technol.* **2016**, *50*, 7752–7760. [CrossRef]

48. Castorina, R.; Butt, C.; Stapleton, H.M.; Avery, D.; Harley, K.G.; Holland, N.; Eskenazi, B.; Bradman, A. Flame retardants and their metabolites in the homes and urine of pregnant women residing in California (the CHAMACOS cohort). *Chemosphere* **2017**, *179*, 159–166. [CrossRef]
49. Stapleton, H.M.; Klosterhaus, S.; Eagle, S.; Fuh, J.; Meeker, J.D.; Blum, A.; Webster, T.F. Detection of Organophosphate Flame Retardants in Furniture Foam and US House Dust. *Environ. Sci. Technol.* **2009**, *43*, 7490–7495. [CrossRef]
50. Ali, N.; Eqani, S.A.M.A.S.; Ismail, I.M.I.; Malarvannan, G.; Kadi, M.W.; Albar, H.M.S.; Rehan, M.; Covaci, A. Brominated and organophosphate flame retardants in indoor dust of Jeddah, Kingdom of Saudi Arabia: Implications for human exposure. *Sci. Total Environ.* **2016**, *569*, 269–277. [CrossRef]
51. Ali, N.; Dirtu, A.C.; Van den Eede, N.; Goosey, E.; Harrad, S.; Neels, H.; t Mannetje, A.; Coakley, J.; Douwes, J.; Covaci, A. Occurrence of alternative flame retardants in indoor dust from New Zealand: Indoor sources and human exposure assessment. *Chemosphere* **2012**, *88*, 1276–1282. [CrossRef]
52. Yang, F.; Ding, J.; Huang, W.; Xie, W.; Liu, W. Particle Size-Specific Distributions and Preliminary Exposure Assessments of Organophosphate Flame Retardants in Office Air Particulate Matter. *Environ. Sci. Technol.* **2014**, *48*, 63–70. [CrossRef] [PubMed]
53. Van Esch, G.J.; World Health Organization. *Flame Retardants: Tris(2-butoxyethyl) Phosphate, Tris(2-Ethylhexyl) Phosphate and Tetrakis(hydroxymethyl) Phosphonium Salts*; World Health Organization: Geneva, Switzerland, 2000.
54. Dirtu, A.C.; Ali, N.; Van den Eede, N.; Neels, H.; Covaci, A. Country specific comparison for a profile of chlorinated, brominated and phosphate organic contaminants in indoor dust. Case study for Eastern Romania, 2010. *Environ. Int.* **2012**, *49*, 1–8. [CrossRef]
55. Schramm, E.; Leisewitz, A.; Kruse, H. *Substituting Environmentally Relevant Flame Retardants*; Umweltbundesamt: Berlin, Germany, 2001.
56. Zhang, X.; Suehring, R.; Serodio, D.; Bonnell, M.; Sundin, N.; Diamond, M.L. Novel flame retardants: Estimating the physical-chemical properties and environmental fate of 94 halogenated and organophosphate PBDE replacements. *Chemosphere* **2016**, *144*, 2401–2407. [CrossRef] [PubMed]
57. Kemmlein, S.; Hahn, O.; Jann, O. Emissions of organophosphate and brominated flame retardants from selected consumer products and building materials. *Atmos. Environ.* **2003**, *37*, 5485–5493. [CrossRef]
58. Brandsma, S.H.; de Boer, J.; van Velzen, M.J.M.; Leonards, P.E.G. Organophosphorus flame retardants (PFRs) and plasticizers in house and car dust and the influence of electronic equipment. *Chemosphere* **2014**, *116*, 3–9. [CrossRef] [PubMed]
59. Brommer, S.; Harrad, S.; Van den Eede, N.; Covaci, A. Concentrations of organophosphate esters and brominated flame retardants in German indoor dust samples. *J. Environ. Monit.* **2012**, *14*, 2482–2487. [CrossRef]
60. Abdallah, M.A. Environmental occurrence, analysis and human exposure to the flame retardant tetrabromobisphenol-A (TBBP-A)-A review. *Environ. Int.* **2016**, *94*, 235–250. [CrossRef]
61. Fang, M.; Stapleton, H.M. Evaluating the Bioaccessibility of Flame Retardants in House Dust Using an In Vitro Tenax Bead-*Assisted* Sorptive Physiologically Based Method. *Environ. Sci. Technol.* **2014**, *48*, 13323–13330. [CrossRef]
62. Khan, M.U.; Besis, A.; Li, J.; Zhang, G.; Malik, R.N. New insight into the distribution pattern, levels, and risk diagnosis of FRs in indoor and outdoor air at low- and high-altitude zones of Pakistan: Implications for sources and exposure. *Chemosphere* **2017**, *184*, 1372–1387. [CrossRef] [PubMed]
63. Harrad, S.; Abdallah, M.A.-E. Brominated flame retardants in dust from UK cars—Within-vehicle spatial variability, evidence for degradation and exposure implications. *Chemosphere* **2011**, *82*, 1240–1245. [CrossRef] [PubMed]
64. Yadav, I.C.; Devi, N.L.; Li, J.; Zhang, G. Occurrence and source apportionment of halogenated flame retardants in the indoor air of Nepalese cities: Implication on human health. *Atmos. Environ.* **2017**, *161*, 122–131. [CrossRef]
65. Lassen, C.; Løkke, S.; Andersen, L.I. Brominated flame retardants–substance flow analysis and assessment of alternatives. *Environ. Proj.* **1999**, *494*, 221.
66. World Health Organization. *Flame Retardants: Tris(Chloropropyl) Phosphate and Tris(2-Chloroethyl) Phosphate*; World Health Organization: Geneva, Switzerland, 1998; Volume 71, pp. 687–690.
67. Dishaw, L.V.; Powers, C.M.; Ryde, I.T.; Roberts, S.C.; Seidler, F.J.; Slotkin, T.A.; Stapleton, H.M. Is the PentaBDE Replacement, Tris (1,3-dichloro-2-propyl) Phosphate (TDCPP), a Developmental Neurotoxicant? Studies in PC12 Cells. *Toxicol. Appl. Pharmacol.* **2011**, *256*, 281–289. [CrossRef] [PubMed]
68. Alzualde, A.; Behl, M.; Sipes, N.S.; Hsieh, J.-H.; Alday, A.; Tice, R.R.; Paules, R.S.; Muriana, A.; Quevedo, C. Toxicity profiling of flame retardants in zebrafish embryos using a battery of assays for developmental toxicity, neurotoxicity, cardiotoxicity and hepatotoxicity toward human relevance. *Neurotoxicol. Teratol.* **2018**, *70*, 40–50. [CrossRef] [PubMed]
69. Meeker, J.D.; Stapleton, H.M. House Dust Concentrations of Organophosphate Flame Retardants in Relation to Hormone Levels and Semen Quality Parameters. *Environ. Health Perspect.* **2010**, *118*, 318–323. [CrossRef]
70. Fisk, P.R.; Girling, A.E.; Wildey, R.J. *Prioritisation of Flame Retardants For Environmental Risk Assessment*; Environment Agency: Wallingford, UK, 2003.
71. Kojima, H.; Takeuchi, S.; Itoh, T.; Iida, M.; Kobayashi, S.; Yoshida, T. In vitro endocrine disruption potential of organophos-phate flame retardants via human nuclear receptors. *Toxicology* **2013**, *314*, 76–83. [CrossRef]

MDPI AG
Grosspeteranlage 5
4052 Basel
Switzerland
Tel.: +41 61 683 77 34

*Toxics* Editorial Office
E-mail: toxics@mdpi.com
www.mdpi.com/journal/toxics

www.ingramcontent.com/pod-product-compliance
Lightning Source LLC
Chambersburg PA
CBHW061618210926
51850CB00009B/2365